Android
开发艺术探索

任玉刚 著

电子工业出版社
Publishing House of Electronics Industry
北京·BEIJING

内 容 简 介

本书是一本 Android 进阶类书籍，采用理论、源码和实践相结合的方式来阐述高水准的 Android 应用开发要点。本书从三个方面来组织内容。第一，介绍 Android 开发者不容易掌握的一些知识点；第二，结合 Android 源代码和应用层开发过程，融会贯通，介绍一些比较深入的知识点；第三，介绍一些核心技术和 Android 的性能优化思想。

本书侧重于 Android 知识的体系化和系统工作机制的分析，通过本书的学习可以极大地提高开发者的 Android 技术水平，从而更加高效地成为高级开发者。而对于高级开发者来说，仍然可以从本书的知识体系中获益。

未经许可，不得以任何方式复制或抄袭本书之部分或全部内容。
版权所有，侵权必究。

图书在版编目（CIP）数据

Android 开发艺术探索/任玉刚著. —北京：电子工业出版社，2015.9
ISBN 978-7-121-26939-4

Ⅰ. ①A… Ⅱ. ①任… Ⅲ. ①移动终端－应用程序－程序设计 Ⅳ. ①TN929.53

中国版本图书馆 CIP 数据核字（2015）第 189069 号

责任编辑：陈晓猛
印　　刷：涿州市般润文化传播有限公司
装　　订：涿州市般润文化传播有限公司
出版发行：电子工业出版社
　　　　　北京市海淀区万寿路 173 信箱　邮编：100036
开　　本：787×980　1/16　印张：32.75　字数：733 千字
版　　次：2015 年 9 月第 1 版
印　　次：2022 年 10 月第 24 次印刷
定　　价：79.00 元

凡所购买电子工业出版社图书有缺损问题，请向购买书店调换。若书店售缺，请与本社发行部联系，联系及邮购电话：(010) 88254888，88258888。
质量投诉请发邮件至 zlts@phei.com.cn，盗版侵权举报请发邮件至 dbqq@phei.com.cn。
本书咨询联系方式：010-51260888-819，faq@phei.com.cn。

序 言

与玉刚共事两年，其对技术的热情和执著让人敬佩，其技术进步之快又让人惊叹。如今，他把所掌握的知识与经验成书出版，是一件大幸之事：于作者，此书是他的心血所成，可喜可贺；于读者，可解"工作视野"之困与"百思不得其解"之惑，或许有"啊哈，原来如此"之效，又或许有"技能+1"之得意一笑。

玉刚拥有丰富的 Android 开发经验，对 Android 开发的很多知识点都有深入研究，我相信此书定能为读者带来惊喜。书的内容，大抵有如下几方面：基础知识点之深入理解（例如，Activity 的生命周期和启动模式、Android 的消息机制分析、View 的事件体系、View 的工作原理等章节）；不常见知识点的分析（例如，IPC 机制、理解 Window 和 WindowManager 等章节）；工程实践中的经验（例如，综合技术、Android 性能优化等章节）。因此，此书读者需要有一定的 Android 开发基础和工程经验，否则读起来会比较吃力或者感觉云里雾里。对于想成长为高级或者资深 Android 研发的工程师，书中的知识点都是需要掌握的。

最后，希望读者能够从此书获益，接触到一些工作中未曾了解或者思考的知识点。更进一步，希望读者能够活学活用，并学习此书背后的钻研精神。

涂勇策
百度手机卫士　资深工程师

前言

从目前的形势来看，Android 开发相当火热，但是高级 Android 开发人才却比较少，当然在国内，不仅仅是 Android，其他技术岗位同样面临这个问题。试想下，如果有一本书能够切实有效地提高开发者的技术水平，那该多好啊！纵观市场上的 Android 书籍，很多都是入门类书籍，还有一些 Android 源码分析、系统移植、驱动开发、逆向工程等系统底层类书籍。入门类书籍是目前图书市场中的中坚力量，它们在帮助开发者入门的过程中起到了非常重要的作用，但开发者若想进一步提高技术水平，还需要阅读更深入的书籍。底层书籍包括源码分析、驱动开发、逆向工程等书籍，它们从底层或者某一个特殊的角度来深入地分析 Android，这是很值得称赞和学习的，通过这些书可以极大地提高开发者底层或者相关领域的技术水平。但美中不足的是，系统底层书籍比较偏理论，部分开发者阅读起来可能会有点晦涩难懂。更重要的一点，由于它们往往侧重原理和底层机制，导致它们不能直接为应用层开发服务，毕竟绝大多数 Android 开发岗位都是应用层开发。由于阅读底层类书籍一般只能够加深对底层的认识，而在应用层开发中，还是不能形成直接有效的战斗力，这中间是需要转化过程的。但是，由于部分开发者缺乏相应的技术功底，导致无法完成这个转化过程。

可以发现，目前市场上既能够极大地提高开发者的应用层技术经验，又能够将上层和系统底层的运行机制结合起来的书籍还是比较少的。对企业来说，在业务上有很强的技术能力，同时对 Android 底层也有一定理解的开发人员，是企业比较青睐的技术高手。为了完成这一愿望，笔者写了这本书。通过对本书的深入学习，开发者既能够极大地提高应用层的开发能力，又能够对 Android 系统的运行机制有一定的理解，但如果要深入理解 Android 的底层机制，仍然需要查看相关源码分析的书籍。

本书适合各类开发者阅读，对于初、中级开发者来说，可以通过本书更加高效地达到高级开发者的技术水平。而对于高级开发者，仍然可以从本书的知识体系中获益。本书的书名之所以采用艺术这个词，这是因为在笔者眼中，代码写到极致就是一种艺术。

本文内容

本书共 15 章,所讲述的内容均基于 Android 5.0 系统。

第 1 章介绍 Activity 的生命周期和启动模式以及 IntentFilter 的匹配规则。

第 2 章介绍 Android 中常见的 IPC 机制,多进程的运行模式和一些常见的进程间通信方式,包括 Messenger、AIDL、Binder 以及 ContentProvider 等,同时还介绍 Binder 连接池的概念。

第 3 章介绍 View 的事件体系,并对 View 的基础知识、滑动以及弹性滑动做详细的介绍,同时还深入分析滑动冲突的原因以及解决方法。

第 4 章介绍 View 的工作原理,首先介绍 ViewRoot、DecorView、MeasureSpec 等 View 相关的底层概念,然后详细分析 View 的测量、布局和绘制三大流程,最后介绍自定义 View 的分类以及实现思想。

第 5 章讲述一个不常见的概念 RemoteViews,分别描述 RemoteViews 在通知栏和桌面小部件中的使用场景,同时还详细介绍 PendingIntent,最后深入分析 RemoteViews 的内部机制并探索性地指出 RemoteViews 在 Android 中存在的意义。

第 6 章对 Android 的 Drawable 做一个全面性的介绍,除此之外还讲解自定义 Drawable 的方法。

第 7 章对 Android 中的动画做一个全面深入的分析,包含 View 动画和属性动画。

第 8 章讲述 Window 和 WindowManager,首先分析 Window 的内部工作原理,包括 Window 的添加、更新和删除,其次分析 Activity、Dialog 等类型的 Window 对象的创建过程。

第 9 章深入分析 Android 中四大组件的工作过程,主要包括四大组件的运行状态以及它们主要的工作过程,比如启动、绑定、广播的发送和接收等。

第 10 章深入分析 Android 的消息机制,其中涉及的概念有 Handler、Looper、MessageQueue 以及 ThreadLocal,此外还分析主线程的消息循环模型。

第 11 章讲述 Android 的线程和线程池,首先介绍 AsyncTask、HandlerThread、IntentService 以及 ThreadPoolExecutor 的使用方法,然后分析它们的工作原理。

第 12 章讲述的主题是 Bitmap 的加载和缓存机制,首先讲述高效加载图片的方式,接着介绍 LruCache 和 DiskLruCache 的使用方法,最后通过一个 ImageLoader 的实例来将它们综合起来。

第 13 章是综合技术，讲述一些很重要但是不太常见的技术方案，它们是 CrashHandler、multidex、插件化以及反编译。

第 14 章的主题是 JNI 和 NDK 编程，介绍使用 JNI 和 Android NDK 编程的方法。

第 15 章介绍 Android 的性能优化方法，比如常见的布局优化、绘制优化、内存泄露优化等，除此之外还介绍分析 ANR 和内存泄露的方法，最后探讨如何提高程序的可维护性这一话题。

通过这 15 章的学习，可以让初、中级开发者的技术水平和把控能力提升一个档次，最终成为高级开发者。

本书特色

本书定位为进阶类图书，不会对一些基础知识从头说起，或者说每一章节都不涵盖各种入门知识，但是在向高级知识点过渡的时候，会稍微提及一下基础知识从而做到平滑过渡。开发者在掌握入门知识以后，通过本书可以极大地提高应用层开发的技术水平，同时还可以理解一定的 Android 底层运行机制，并且能够将它们进行升华从而更好地为应用层开发服务。除了这些，开发者还可以掌握一些核心技术和性能优化思想，本书涉及的知识，都是一个合格的高级工程师所必须掌握的。简单地说，本书的目的就是让初、中级开发者更有针对性地掌握高级工程师所应该掌握的技术，能够让初、中级开发者按照正确的道路快速地成长为高级工程师。

致谢

感谢本书的策划编辑陈晓猛，他的高效率是本书得以及时出版的一个重要原因；感谢我的妻子对我写书的支持，接近 1 年的写书时光是她一直陪伴在我身边；感谢百度手机卫士这款产品，它是本书的技术源泉；感谢和我一起奋斗的同事们，和你们在一起工作的时光，我不仅提高了技术水平而且还真正感受到了一种融洽的工作氛围；还要感谢所有关注我的朋友们，你们的鼓励和认可是我前进的动力。

由于技术水平有限，书中难免会有错误，欢迎大家向我反馈：singwhatiwanna@gmail.com，也可以关注我的 CSDN 博客，我会定期在上面发布本书的勘误信息。

本书互动地址

CSDN 博客：http://blog.csdn.net/singwhatiwanna

Github：https://github.com/singwhatiwanna

QQ 交流群：481798332

微信公众号：Android 开发艺术探索

书中源码下载地址：

https://github.com/singwhatiwanna/android-art-res

或者

www.broadview.com.cn/26939

任玉刚
2015 年 6 月于北京

目录

第 1 章　Activity 的生命周期和启动模式 / 1
　1.1　Activity 的生命周期全面分析 / 1
　　　1.1.1　典型情况下的生命周期分析 / 2
　　　1.1.2　异常情况下的生命周期分析 / 8
　1.2　Activity 的启动模式 / 16
　　　1.2.1　Activity 的 LaunchMode / 16
　　　1.2.2　Activity 的 Flags / 27
　1.3　IntentFilter 的匹配规则 / 28

第 2 章　IPC 机制 / 35
　2.1　Android IPC 简介 / 35
　2.2　Android 中的多进程模式 / 36
　　　2.2.1　开启多进程模式 / 36
　　　2.2.2　多进程模式的运行机制 / 39
　2.3　IPC 基础概念介绍 / 42
　　　2.3.1　Serializable 接口 / 42
　　　2.3.2　Parcelable 接口 / 45
　　　2.3.3　Binder / 47
　2.4　Android 中的 IPC 方式 / 61
　　　2.4.1　使用 Bundle / 61
　　　2.4.2　使用文件共享 / 62

2.4.3 使用 Messenger / 65
2.4.4 使用 AIDL / 71
2.4.5 使用 ContentProvider / 91
2.4.6 使用 Socket / 103
2.5 Binder 连接池 / 112
2.6 选用合适的 IPC 方式 / 121

第 3 章 View 的事件体系 / 122

3.1 View 基础知识 / 122
  3.1.1 什么是 View / 123
  3.1.2 View 的位置参数 / 123
  3.1.3 MotionEvent 和 TouchSlop / 125
  3.1.4 VelocityTracker、GestureDetector 和 Scroller / 126

3.2 View 的滑动 / 129
  3.2.1 使用 scrollTo/scrollBy / 129
  3.2.2 使用动画 / 131
  3.2.3 改变布局参数 / 133
  3.2.4 各种滑动方式的对比 / 133

3.3 弹性滑动 / 135
  3.3.1 使用 Scroller / 136
  3.3.2 通过动画 / 138
  3.3.3 使用延时策略 / 139

3.4 View 的事件分发机制 / 140
  3.4.1 点击事件的传递规则 / 140
  3.4.2 事件分发的源码解析 / 144

3.5 View 的滑动冲突 / 154
  3.5.1 常见的滑动冲突场景 / 155
  3.5.2 滑动冲突的处理规则 / 156
  3.5.3 滑动冲突的解决方式 / 157

第 4 章 View 的工作原理 / 174

4.1 初识 ViewRoot 和 DecorView / 174

4.2 理解 MeasureSpec / 177

    4.2.1 MeasureSpec / 177

    4.2.2 MeasureSpec 和 LayoutParams 的对应关系 / 178

4.3 View 的工作流程 / 183

    4.3.1 measure 过程 / 183

    4.3.2 layout 过程 / 193

    4.3.3 draw 过程 / 197

4.4 自定义 View / 199

    4.4.1 自定义 View 的分类 / 200

    4.4.2 自定义 View 须知 / 201

    4.4.3 自定义 View 示例 / 202

    4.4.4 自定义 View 的思想 / 217

第 5 章 理解 RemoteViews / 218

5.1 RemoteViews 的应用 / 218

    5.1.1 RemoteViews 在通知栏上的应用 / 219

    5.1.2 RemoteViews 在桌面小部件上的应用 / 221

    5.1.3 PendingIntent 概述 / 228

5.2 RemoteViews 的内部机制 / 230

5.3 RemoteViews 的意义 / 239

第 6 章 Android 的 Drawable / 243

6.1 Drawable 简介 / 243

6.2 Drawable 的分类 / 244

    6.2.1 BitmapDrawable / 244

    6.2.2 ShapeDrawable / 247

    6.2.3 LayerDrawable / 251

    6.2.4 StateListDrawable / 253

    6.2.5 LevelListDrawable / 255

    6.2.6 TransitionDrawable / 256

    6.2.7 InsetDrawable / 257

    6.2.8 ScaleDrawable / 258

        6.2.9　ClipDrawable / 260
    6.3　自定义 Drawable / 262

第 7 章　Android 动画深入分析 / 265
    7.1　View 动画 / 265
        7.1.1　View 动画的种类 / 265
        7.1.2　自定义 View 动画 / 270
        7.1.3　帧动画 / 272
    7.2　View 动画的特殊使用场景 / 273
        7.2.1　LayoutAnimation / 273
        7.2.2　Activity 的切换效果 / 275
    7.3　属性动画 / 276
        7.3.1　使用属性动画 / 276
        7.3.2　理解插值器和估值器 / 280
        7.3.3　属性动画的监听器 / 282
        7.3.4　对任意属性做动画 / 282
        7.3.5　属性动画的工作原理 / 288
    7.4　使用动画的注意事项 / 292

第 8 章　理解 Window 和 WindowManager / 294
    8.1　Window 和 WindowManager / 294
    8.2　Window 的内部机制 / 297
        8.2.1　Window 的添加过程 / 298
        8.2.2　Window 的删除过程 / 301
        8.2.3　Window 的更新过程 / 303
    8.3　Window 的创建过程 / 304
        8.3.1　Activity 的 Window 创建过程 / 304
        8.3.2　Dialog 的 Window 创建过程 / 308
        8.3.3　Toast 的 Window 创建过程 / 311

第 9 章　四大组件的工作过程 / 316
    9.1　四大组件的运行状态 / 316

9.2 Activity 的工作过程 / 318

9.3 Service 的工作过程 / 336

 9.3.1 Service 的启动过程 / 336

 9.3.2 Service 的绑定过程 / 344

9.4 BroadcastReceiver 的工作过程 / 352

 9.4.1 广播的注册过程 / 353

 9.4.2 广播的发送和接收过程 / 356

9.5 ContentProvider 的工作过程 / 362

第 10 章 Android 的消息机制 / 372

10.1 Android 的消息机制概述 / 373

10.2 Android 的消息机制分析 / 375

 10.2.1 ThreadLocal 的工作原理 / 375

 10.2.2 消息队列的工作原理 / 380

 10.2.3 Looper 的工作原理 / 383

 10.2.4 Handler 的工作原理 / 385

10.3 主线程的消息循环 / 389

第 11 章 Android 的线程和线程池 / 391

11.1 主线程和子线程 / 392

11.2 Android 中的线程形态 / 392

 11.2.1 AsyncTask / 392

 11.2.2 AsyncTask 的工作原理 / 395

 11.2.3 HandlerThread / 402

 11.2.4 IntentService / 403

11.3 Android 中的线程池 / 406

 11.3.1 ThreadPoolExecutor / 407

 11.3.2 线程池的分类 / 410

第 12 章 Bitmap 的加载和 Cache / 413

12.1 Bitmap 的高效加载 / 414

12.2 Android 中的缓存策略 / 417

12.2.1　LruCache / 418
12.2.2　DiskLruCache / 419
12.2.3　ImageLoader 的实现 / 424
12.3　ImageLoader 的使用 / 441
12.3.1　照片墙效果 / 441
12.3.2　优化列表的卡顿现象 / 446

第 13 章　综合技术 / 448

13.1　使用 CrashHandler 来获取应用的 crash 信息 / 449
13.2　使用 multidex 来解决方法数越界 / 455
13.3　Android 的动态加载技术 / 463
13.4　反编译初步 / 469
13.4.1　使用 dex2jar 和 jd-gui 反编译 apk / 470
13.4.2　使用 apktool 对 apk 进行二次打包 / 470

第 14 章　JNI 和 NDK 编程 / 473

14.1　JNI 的开发流程 / 474
14.2　NDK 的开发流程 / 478
14.3　JNI 的数据类型和类型签名 / 484
14.4　JNI 调用 Java 方法的流程 / 486

第 15 章　Android 性能优化 / 489

15.1　Android 的性能优化方法 / 490
15.1.1　布局优化 / 490
15.1.2　绘制优化 / 493
15.1.3　内存泄露优化 / 493
15.1.4　响应速度优化和 ANR 日志分析 / 496
15.1.5　ListView 和 Bitmap 优化 / 501
15.1.6　线程优化 / 501
15.1.7　一些性能优化建议 / 501
15.2　内存泄露分析之 MAT 工具 / 502
15.3　提高程序的可维护性 / 506

第 1 章　Activity 的生命周期和启动模式

作为本书的第 1 章，本章主要介绍 Activity 相关的一些内容。Activity 作为四大组件之首，是使用最为频繁的一种组件，中文直接翻译为"活动"，但是笔者认为这种翻译有些生硬，如果翻译成界面就会更好理解。正常情况下，除了 Window、Dialog 和 Toast，我们能见到的界面的确只有 Activity。Activity 是如此重要，以至于本书开篇就不得不讲到它。当然，由于本书的定位为进阶书，所以不会介绍如何启动 Activity 这类入门知识，本章的侧重点是 Activity 在使用过程中的一些不容易搞清楚的概念，主要包括生命周期和启动模式以及 IntentFilter 的匹配规则分析。其中 Activity 在异常情况下的生命周期是十分微妙的，至于 Activity 的启动模式和形形色色的 Flags 更是让初学者摸不到头脑，就连隐式启动 Activity 中也有着复杂的 Intent 匹配过程，不过不用担心，本章接下来将一一解开这些疑难问题的神秘面纱。

1.1　Activity 的生命周期全面分析

本节将 Activity 的生命周期分为两部分内容，一部分是典型情况下的生命周期，另一部分是异常情况下的生命周期。所谓典型情况下的生命周期，是指在有用户参与的情况下，Activity 所经过的生命周期的改变；而异常情况下的生命周期是指 Activity 被系统回收或者由于当前设备的 Configuration 发生改变从而导致 Activity 被销毁重建，异常情况下的生命周期的关注点和典型情况下略有不同。

1.1.1 典型情况下的生命周期分析

在正常情况下，Activity 会经历如下生命周期。

（1）onCreate：表示 Activity 正在被创建，这是生命周期的第一个方法。在这个方法中，我们可以做一些初始化工作，比如调用 setContentView 去加载界面布局资源、初始化 Activity 所需数据等。

（2）onRestart：表示 Activity 正在重新启动。一般情况下，当当前 Activity 从不可见重新变为可见状态时，onRestart 就会被调用。这种情形一般是用户行为所导致的，比如用户按 Home 键切换到桌面或者用户打开了一个新的 Activity，这时当前的 Activity 就会暂停，也就是 onPause 和 onStop 被执行了，接着用户又回到了这个 Activity，就会出现这种情况。

（3）onStart：表示 Activity 正在被启动，即将开始，这时 Activity 已经可见了，但是还没有出现在前台，还无法和用户交互。这个时候其实可以理解为 Activity 已经显示出来了，但是我们还看不到。

（4）onResume：表示 Activity 已经可见了，并且出现在前台并开始活动。要注意这个和 onStart 的对比，onStart 和 onResume 都表示 Activity 已经可见，但是 onStart 的时候 Activity 还在后台，onResume 的时候 Activity 才显示到前台。

（5）onPause：表示 Activity 正在停止，正常情况下，紧接着 onStop 就会被调用。在特殊情况下，如果这个时候快速地再回到当前 Activity，那么 onResume 会被调用。笔者的理解是，这种情况属于极端情况，用户操作很难重现这一场景。此时可以做一些存储数据、停止动画等工作，但是注意不能太耗时，因为这会影响到新 Activity 的显示，onPause 必须先执行完，新 Activity 的 onResume 才会执行。

（6）onStop：表示 Activity 即将停止，可以做一些稍微重量级的回收工作，同样不能太耗时。

（7）onDestroy：表示 Activity 即将被销毁，这是 Activity 生命周期中的最后一个回调，在这里，我们可以做一些回收工作和最终的资源释放。

正常情况下，Activity 的常用生命周期就只有上面 7 个，图 1-1 更详细地描述了 Activity 各种生命周期的切换过程。

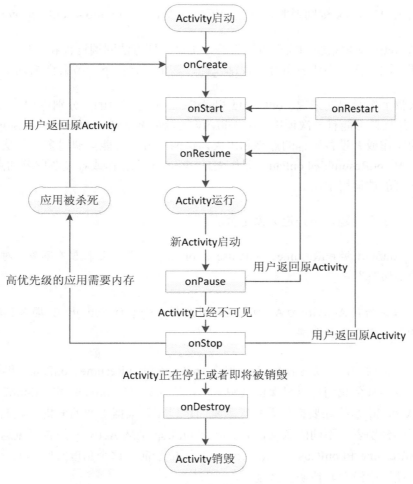

图 1-1　Activity 生命周期的切换过程

针对图 1-1，这里再附加一下具体说明，分如下几种情况。

（1）针对一个特定的 Activity，第一次启动，回调如下：onCreate -> onStart -> onResume。

（2）当用户打开新的 Activity 或者切换到桌面的时候，回调如下：onPause -> onStop。这里有一种特殊情况，如果新 Activity 采用了透明主题，那么当前 Activity 不会回调 onStop。

（3）当用户再次回到原 Activity 时，回调如下：onRestart -> onStart -> onResume。

（4）当用户按 back 键回退时，回调如下：onPause -> onStop -> onDestroy。

（5）当 Activity 被系统回收后再次打开，生命周期方法回调过程和（1）一样，注意只是生命周期方法一样，不代表所有过程都一样，这个问题在下一节会详细说明。

（6）从整个生命周期来说，onCreate 和 onDestroy 是配对的，分别标识着 Activity 的创建和销毁，并且只可能有一次调用。从 Activity 是否可见来说，onStart 和 onStop 是配对的，随着用户的操作或者设备屏幕的点亮和熄灭，这两个方法可能被调用多次；从 Activity 是否在前台来说，onResume 和 onPause 是配对的，随着用户操作或者设备屏幕的点亮和熄灭，这两个方法可能被调用多次。

这里提出 2 个问题，不知道大家是否清楚。

问题 1：onStart 和 onResume、onPause 和 onStop 从描述上来看差不多，对我们来说有什么实质的不同呢？

问题 2：假设当前 Activity 为 A，如果这时用户打开一个新 Activity B，那么 B 的 onResume 和 A 的 onPause 哪个先执行呢？

先说第一个问题，从实际使用过程来说，onStart 和 onResume、onPause 和 onStop 看起来的确差不多，甚至我们可以只保留其中一对，比如只保留 onStart 和 onStop。既然如此，那为什么 Android 系统还要提供看起来重复的接口呢？根据上面的分析，我们知道，这两个配对的回调分别表示不同的意义，onStart 和 onStop 是从 Activity 是否可见这个角度来回调的，而 onResume 和 onPause 是从 Activity 是否位于前台这个角度来回调的，除了这种区别，在实际使用中没有其他明显区别。

第二个问题可以从 Android 源码里得到解释。关于 Activity 的工作原理在本书后续章节会进行介绍，这里我们先大概了解即可。从 Activity 的启动过程来看，我们来看一下系统源码。Activity 的启动过程的源码相当复杂，涉及 Instrumentation、ActivityThread 和 ActivityManagerService（下面简称 AMS）。这里不详细分析这一过程，简单理解，启动 Activity 的请求会由 Instrumentation 来处理，然后它通过 Binder 向 AMS 发请求，AMS 内部维护着一个 ActivityStack 并负责栈内的 Activity 的状态同步，AMS 通过 ActivityThread 去同步 Activity 的状态从而完成生命周期方法的调用。在 ActivityStack 中的 resumeTopActivity-InnerLocked 方法中，有这么一段代码：

```
// We need to start pausing the current activity so the top one
// can be resumed...
boolean dontWaitForPause = (next.info.flags&ActivityInfo.FLAG_RESUME_
WHILE_PAUSING) != 0;
boolean pausing = mStackSupervisor.pauseBackStacks(userLeaving, true,
dontWaitForPause);
if (mResumedActivity != null) {
    pausing |= startPausingLocked(userLeaving, false, true, dontWait-
    ForPause);
    if (DEBUG_STATES) Slog.d(TAG, "resumeTopActivityLocked: Pausing " +
    mResumedActivity);
}
```

从上述代码可以看出，在新 Activity 启动之前，栈顶的 Activity 需要先 onPause 后，新 Activity 才能启动。最终，在 ActivityStackSupervisor 中的 realStartActivityLocked 方法会调用如下代码。

```
app.thread.scheduleLaunchActivity(new Intent(r.intent), r.appToken,
        System.identityHashCode(r), r.info, new Configuration(mService.
        mConfiguration),
        r.compat, r.task.voiceInteractor, app.repProcState, r.icicle,
        r.persistentState,
        results, newIntents, !andResume, mService.isNextTransition-
        Forward(),
        profilerInfo);
```

我们知道，这个 app.thread 的类型是 IApplicationThread，而 IApplicationThread 的具体实现是 ActivityThread 中的 ApplicationThread。所以，这段代码实际上调到了 ActivityThread 的中，即 ApplicationThread 的 scheduleLaunchActivity 方法，而 scheduleLaunchActivity 方法最终会完成新 Activity 的 onCreate、onStart、onResume 的调用过程。因此，可以得出结论，是旧 Activity 先 onPause，然后新 Activity 再启动。

至于 ApplicationThread 的 scheduleLaunchActivity 方法为什么会完成新 Activity 的 onCreate、onStart、onResume 的调用过程，请看下面的代码。scheduleLaunchActivity 最终会调用如下方法，而如下方法的确会完成 onCreate、onStart、onResume 的调用过程。

源码：ActivityThread# handleLaunchActivity

```
private void handleLaunchActivity(ActivityClientRecord r, Intent custom-
Intent) {
    // If we are getting ready to gc after going to the background, well
    // we are back active so skip it.
    unscheduleGcIdler();
    mSomeActivitiesChanged = true;

    if (r.profilerInfo != null) {
        mProfiler.setProfiler(r.profilerInfo);
        mProfiler.startProfiling();
    }

    // Make sure we are running with the most recent config.
    handleConfigurationChanged(null, null);

    if (localLOGV) Slog.v(
        TAG, "Handling launch of " + r);

    //这里新Activity被创建出来,其onCreate和onStart会被调用
    Activity a = performLaunchActivity(r, customIntent);

    if (a != null) {
        r.createdConfig = new Configuration(mConfiguration);
        Bundle oldState = r.state;
        //这里新Activity的onResume会被调用
        handleResumeActivity(r.token, false, r.isForward,
            !r.activity.mFinished && !r.startsNotResumed);
    //省略
}
```

从上面的分析可以看出,当新启动一个Activity的时候,旧Activity的onPause会先执行,然后才会启动新的Activity。到底是不是这样呢?我们写个例子验证一下,如下是2个Activity的代码,在MainActivity中单击按钮可以跳转到SecondActivity,同时为了分析我们的问题,在生命周期方法中打印出了日志,通过日志我们就能看出它们的调用顺序。

代码: MainActivity.java

```
public class MainActivity extends Activity {

    private static final String TAG = "MainActivity";
```

```
//省略
@Override
protected void onPause() {
    super.onPause();
    Log.d(TAG, "onPause");
}

@Override
protected void onStop() {
    super.onStop();
    Log.d(TAG, "onStop");
}
}
```

代码：SecondActivity.java

```
public class SecondActivity extends Activity {
    private static final String TAG = "SecondActivity";

    @Override
    protected void onCreate(Bundle savedInstanceState) {
        super.onCreate(savedInstanceState);
        setContentView(R.layout.activity_second);
        Log.d(TAG, "onCreate");
    }

    @Override
    protected void onStart() {
        super.onStart();
        Log.d(TAG, "onStart");
    }

    @Override
    protected void onResume() {
        super.onResume();
        Log.d(TAG, "onResume");
    }
}
```

我们来看一下 log，是不是和我们上面分析的一样，如图 1-2 所示。

Level	Time	PID	TID	Application	Tag	Text
D	02-01 01:37:33.051	724	724	com.ryg.chapter_1	MainActivity	onPause
D	02-01 01:37:33.111	724	724	com.ryg.chapter_1	SecondActivity	onCreate
D	02-01 01:37:33.111	724	724	com.ryg.chapter_1	SecondActivity	onStart
D	02-01 01:37:33.111	724	724	com.ryg.chapter_1	SecondActivity	onResume
D	02-01 01:37:33.431	724	724	com.ryg.chapter_1	MainActivity	onStop

图 1-2　Activity 生命周期方法的回调顺序

通过图 1-2 可以发现，旧 Activity 的 onPause 先调用，然后新 Activity 才启动，这也证实了我们上面的分析过程。也许有人会问，你只是分析了 Android5.0 的源码，你怎么知道所有版本的源码都是相同逻辑呢？关于这个问题，我们的确不大可能把所有版本的源码都分析一遍，但是作为 Android 运行过程的基本机制，随着版本的更新并不会有大的调整，因为 Android 系统也需要兼容性，不能说在不同版本上同一个运行机制有着截然不同的表现。关于这一点我们需要把握一个度，就是对于 Android 运行的基本机制在不同 Android 版本上具有延续性。从另一个角度来说，Android 官方文档对 onPause 的解释有这么一句：不能在 onPause 中做重量级的操作，因为必须 onPause 执行完成以后新 Activity 才能 Resume，从这一点也能间接证明我们的结论。通过分析这个问题，我们知道 onPause 和 onStop 都不能执行耗时的操作，尤其是 onPause，这也意味着，我们应当尽量在 onStop 中做操作，从而使得新 Activity 尽快显示出来并切换到前台。

1.1.2　异常情况下的生命周期分析

上一节我们分析了典型情况下 Activity 的生命周期，本节我们接着分析 Activity 在异常情况下的生命周期。我们知道，Activity 除了受用户操作所导致的正常的生命周期方法调度，还有一些异常情况，比如当资源相关的系统配置发生改变以及系统内存不足时，Activity 就可能被杀死。下面我们具体分析这两种情况。

1. 情况 1：资源相关的系统配置发生改变导致 Activity 被杀死并重新创建

理解这个问题，我们首先要对系统的资源加载机制有一定了解，这里不详细分析系统的资源加载机制，只是简单说明一下。拿最简单的图片来说，当我们把一张图片放在 drawable 目录后，就可以通过 Resources 去获取这张图片。同时为了兼容不同的设备，我们可能还需要在其他一些目录放置不同的图片，比如 drawable-mdpi、drawable-hdpi、drawable-land 等。这样，当应用程序启动时，系统就会根据当前设备的情况去加载合适的

Resources 资源，比如说横屏手机和竖屏手机会拿到两张不同的图片（设定了 landscape 或者 portrait 状态下的图片）。比如说当前 Activity 处于竖屏状态，如果突然旋转屏幕，由于系统配置发生了改变，在默认情况下，Activity 就会被销毁并且重新创建，当然我们也可以阻止系统重新创建我们的 Activity。

在默认情况下，如果我们的 Activity 不做特殊处理，那么当系统配置发生改变后，Activity 就会被销毁并重新创建，其生命周期如图 1-3 所示。

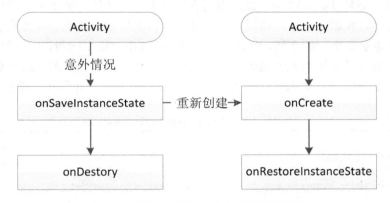

图 1-3　异常情况下 Activity 的重建过程

当系统配置发生改变后，Activity 会被销毁，其 onPause、onStop、onDestroy 均会被调用，同时由于 Activity 是在异常情况下终止的，系统会调用 onSaveInstanceState 来保存当前 Activity 的状态。这个方法的调用时机是在 onStop 之前，它和 onPause 没有既定的时序关系，它既可能在 onPause 之前调用，也可能在 onPause 之后调用。需要强调的一点是，这个方法只会出现在 Activity 被异常终止的情况下，正常情况下系统不会回调这个方法。当 Activity 被重新创建后，系统会调用 onRestoreInstanceState，并且把 Activity 销毁时 onSaveInstanceState 方法所保存的 Bundle 对象作为参数同时传递给 onRestoreInstanceState 和 onCreate 方法。因此，我们可以通过 onRestoreInstanceState 和 onCreate 方法来判断 Activity 是否被重建了，如果被重建了，那么我们就可以取出之前保存的数据并恢复，从时序上来说，onRestoreInstanceState 的调用时机在 onStart 之后。

同时，我们要知道，在 onSaveInstanceState 和 onRestoreInstanceState 方法中，系统自动为我们做了一定的恢复工作。当 Activity 在异常情况下需要重新创建时，系统会默认为我们保存当前 Activity 的视图结构，并且在 Activity 重启后为我们恢复这些数据，比如文本框中用户输入的数据、ListView 滚动的位置等，这些 View 相关的状态系统都能够默认为我们恢复。具体针对某一个特定的 View 系统能为我们恢复哪些数据，我们可以查看 View 的源码。和 Activity 一样，每个 View 都有 onSaveInstanceState 和 onRestoreInstanceState 这两个方法，看一下它们的具体实现，就能知道系统能够自动为每个 View 恢复哪些数据。

关于保存和恢复 View 层次结构，系统的工作流程是这样的：首先 Activity 被意外终止时，Activity 会调用 onSaveInstanceState 去保存数据，然后 Activity 会委托 Window 去保存数据，接着 Window 再委托它上面的顶级容器去保存数据。顶层容器是一个 ViewGroup，一般来说它很可能是 DecorView。最后顶层容器再去一一通知它的子元素来保存数据，这样整个数据保存过程就完成了。可以发现，这是一种典型的委托思想，上层委托下层、父容器委托子元素去处理一件事情，这种思想在 Android 中有很多应用，比如 View 的绘制过程、事件分发等都是采用类似的思想。至于数据恢复过程也是类似的，这里就不再重复介绍了。接下来举个例子，拿 TextView 来说，我们分析一下它到底保存了哪些数据。

源码：TextView# onSaveInstanceState

```java
@Override
public Parcelable onSaveInstanceState() {
    Parcelable superState = super.onSaveInstanceState();

    // Save state if we are forced to
    boolean save = mFreezesText;
    int start = 0;
    int end = 0;

    if (mText != null) {
        start = getSelectionStart();
        end = getSelectionEnd();
        if (start >= 0 || end >= 0) {
            // Or save state if there is a selection
            save = true;
        }
    }

    if (save) {
        SavedState ss = new SavedState(superState);
        // XXX Should also save the current scroll position!
        ss.selStart = start;
        ss.selEnd = end;

        if (mText instanceof Spanned) {
```

```
            Spannable sp = new SpannableStringBuilder(mText);

            if (mEditor != null) {
                removeMisspelledSpans(sp);
                sp.removeSpan(mEditor.mSuggestionRangeSpan);
            }

            ss.text = sp;
        } else {
            ss.text = mText.toString();
        }

        if (isFocused() && start >= 0 && end >= 0) {
            ss.frozenWithFocus = true;
        }

        ss.error = getError();

        return ss;
    }

    return superState;
}
```

从上述源码可以很容易看出，TextView 保存了自己的文本选中状态和文本内容，并且通过查看其 onRestoreInstanceState 方法的源码，可以发现它的确恢复了这些数据，具体源码就不再贴出了，读者可以去看看源码。下面我们看一个实际的例子，来对比一下 Activity 正常终止和异常终止的不同，同时验证系统的数据恢复能力。为了方便，我们选择旋转屏幕来异常终止 Activity，如图 1-4 所示。

图 1-4　Activity 旋转屏幕后数据的保存和恢复

通过图 1-4 可以看出，当我们旋转屏幕以后，Activity 被销毁后重新创建，我们输入的文本"这是测试文本"被正确地还原，这说明系统的确能够自动地做一些 View 层次结构方面的数据存储和恢复。下面再用一个例子，来验证我们自己做数据存储和恢复的情况，代码如下：

```java
@Override
protected void onCreate(Bundle savedInstanceState) {
    super.onCreate(savedInstanceState);
    setContentView(R.layout.activity_main);
    if (savedInstanceState != null) {
        String test = savedInstanceState.getString("extra_test");
        Log.d(TAG, "[onCreate]restore extra_test:" + test);
    }
}

@Override
protected void onSaveInstanceState(Bundle outState) {
    super.onSaveInstanceState(outState);
    Log.d(TAG, "onSaveInstanceState");
    outState.putString("extra_test", "test");
}

@Override
protected void onRestoreInstanceState(Bundle savedInstanceState) {
    super.onRestoreInstanceState(savedInstanceState);
    String test = savedInstanceState.getString("extra_test");
    Log.d(TAG, "[onRestoreInstanceState]restore extra_test:" + test);
}
```

上面的代码很简单，首先我们在 onSaveInstanceState 中存储一个字符串，然后当 Activity 被销毁并重新创建后，我们再去获取之前存储的字符串。接收的位置可以选择 onRestoreInstanceState 或者 onCreate，二者的区别是：onRestoreInstanceState 一旦被调用，其参数 Bundle savedInstanceState 一定是有值的，我们不用额外地判断是否为空；但是 onCreate 不行，onCreate 如果是正常启动的话，其参数 Bundle savedInstanceState 为 null，所以必须要额外判断。这两个方法我们选择任意一个都可以进行数据恢复，但是官方文档的建议是采用 onRestoreInstanceState 去恢复数据。下面我们看一下运行的日志，如图 1-5 所示。

Level	PID	TID	Application	Tag	Text
D	8534	8534	com.ryg.chapter_1	MainActivity	onPause
D	8534	8534	com.ryg.chapter_1	MainActivity	onSaveInstanceState
D	8534	8534	com.ryg.chapter_1	MainActivity	onStop
D	8534	8534	com.ryg.chapter_1	MainActivity	onDestroy
D	8534	8534	com.ryg.chapter_1	MainActivity	[onCreate]restore extra_test:test
D	8534	8534	com.ryg.chapter_1	MainActivity	[onRestoreInstanceState]restore extra_test:test

图 1-5 系统日志

如图 1-5 所示，Activity 被销毁了以后调用了 onSaveInstanceState 来保存数据，重新创建以后在 onCreate 和 onRestoreInstanceState 中都能够正确地恢复我们之前存储的字符串。这个例子很好地证明了上面我们的分析结论。针对 onSaveInstanceState 方法还有一点需要说明，那就是系统只会在 Activity 即将被销毁并且有机会重新显示的情况下才会去调用它。考虑这么一种情况，当 Activity 正常销毁的时候，系统不会调用 onSaveInstanceState，因为被销毁的 Activity 不可能再次被显示。这句话不好理解，但是我们可以对比一下旋转屏幕所造成的 Activity 异常销毁，这个过程和正常停止 Activity 是不一样的，因为旋转屏幕后，Activity 被销毁的同时会立刻创建新的 Activity 实例，这个时候 Activity 有机会再次立刻展示，所以系统要进行数据存储。这里可以简单地这么理解，系统只在 Activity 异常终止的时候才会调用 onSaveInstanceState 和 onRestoreInstanceState 来存储和恢复数据，其他情况不会触发这个过程。

2. 情况 2：资源内存不足导致低优先级的 Activity 被杀死

这种情况我们不好模拟，但是其数据存储和恢复过程和情况 1 完全一致。这里我们描述一下 Activity 的优先级情况。Activity 按照优先级从高到低，可以分为如下三种：

（1）前台 Activity——正在和用户交互的 Activity，优先级最高。

（2）可见但非前台 Activity——比如 Activity 中弹出了一个对话框，导致 Activity 可见但是位于后台无法和用户直接交互。

（3）后台 Activity——已经被暂停的 Activity，比如执行了 onStop，优先级最低。

当系统内存不足时，系统就会按照上述优先级去杀死目标 Activity 所在的进程，并在后续通过 onSaveInstanceState 和 onRestoreInstanceState 来存储和恢复数据。如果一个进程中没有四大组件在执行，那么这个进程将很快被系统杀死，因此，一些后台工作不适合脱离四大组件而独自运行在后台中，这样进程很容易被杀死。比较好的方法是将后台工作放入 Service 中从而保证进程有一定的优先级，这样就不会轻易地被系统杀死。

上面分析了系统的数据存储和恢复机制,我们知道,当系统配置发生改变后,Activity会被重新创建,那么有没有办法不重新创建呢?答案是有的,接下来我们就来分析这个问题。系统配置中有很多内容,如果当某项内容发生改变后,我们不想系统重新创建Activity,可以给Activity指定configChanges属性。比如不想让Activity在屏幕旋转的时候重新创建,就可以给configChanges属性添加orientation这个值,如下所示。

```
android:configChanges="orientation"
```

如果我们想指定多个值,可以用"|"连接起来,比如android:configChanges="orientation|keyboardHidden"。系统配置中所含的项目是非常多的,下面介绍每个项目的含义,如表1-1所示。

表1-1 configChanges 的项目和含义

项目	含义
mcc	SIM卡唯一标识IMSI(国际移动用户识别码)中的国家代码,由三位数字组成,中国为460。此项标识mcc代码发生了改变
mnc	SIM卡唯一标识IMSI(国际移动用户识别码)中的运营商代码,由两位数字组成,中国移动TD系统为00,中国联通为01,中国电信为03。此项标识mnc发生改变
locale	设备的本地位置发生了改变,一般指切换了系统语言
touchscreen	触摸屏发生了改变,这个很费解,正常情况下无法发生,可以忽略它
keyboard	键盘类型发生了改变,比如用户使用了外插键盘
keyboardHidden	键盘的可访问性发生了改变,比如用户调出了键盘
navigation	系统导航方式发生了改变,比如采用了轨迹球导航,这个有点费解,很难发生,可以忽略它
screenLayout	屏幕布局发生了改变,很可能是用户激活了另外一个显示设备
fontScale	系统字体缩放比例发生了改变,比如用户选择了一个新字号
uiMode	用户界面模式发生了改变,比如是否开启了夜间模式(API 8新添加)
orientation	屏幕方向发生了改变,这个是最常用的,比如旋转了手机屏幕
screenSize	当屏幕的尺寸信息发生了改变,当旋转设备屏幕时,屏幕尺寸会发生变化,这个选项比较特殊,它和编译选项有关,当编译选项中的minSdkVersion和targetSdkVersion均低于13时,此选项不会导致Activity重启,否则会导致Activity重启(API 13新添加)
smallestScreenSize	设备的物理屏幕尺寸发生改变,这个项目和屏幕的方向没关系,仅仅表示在实际的物理屏幕的尺寸改变的时候发生,比如用户切换到了外部的显示设备,这个选项和screenSize一样,当编译选项中的minSdkVersion和targetSdkVersion均低于13时,此选项不会导致Activity重启,否则会导致Activity重启(API 13新添加)
layoutDirection	当布局方向发生变化,这个属性用的比较少,正常情况下无须修改布局的layoutDirection属性(API 17新添加)

从表 1-1 可以知道，如果我们没有在 Activity 的 configChanges 属性中指定该选项的话，当配置发生改变后就会导致 Activity 重新创建。上面表格中的项目很多，但是我们常用的只有 locale、orientation 和 keyboardHidden 这三个选项，其他很少使用。需要注意的是 screenSize 和 smallestScreenSize，它们两个比较特殊，它们的行为和编译选项有关，但和运行环境无关。下面我们再看一个 demo，看看当我们指定了 configChanges 属性后，Activity 是否真的不会重新创建了。我们所要修改的代码很简单，只需要在 AndroidMenifest.xml 中加入 Activity 的声明即可，代码如下：

```xml
<uses-sdk
    android:minSdkVersion="8"
    android:targetSdkVersion="19" />
<activity
    android:name="com.ryg.chapter_1.MainActivity"
    android:configChanges="orientation|screenSize"
    android:label="@string/app_name" >
    <intent-filter>
        <action android:name="android.intent.action.MAIN" />
        <category android:name="android.intent.category.LAUNCHER" />
    </intent-filter>
</activity>

@Override
public void onConfigurationChanged(Configuration newConfig) {
    super.onConfigurationChanged(newConfig);
    Log.d(TAG, "onConfigurationChanged, newOrientation:" + newConfig.
    orientation);
}
```

需要说明的是，由于编译时笔者指定的 minSdkVersion 和 targetSdkVersion 有一个大于 13，所以为了防止旋转屏幕时 Activity 重启，除了 orientation，我们还要加上 screenSize，原因在上面的表格里已经说明了。其他代码还是不变，运行后看看 log，如图 1-6 所示。

Level	PID	TID	Application	Tag	Text
D	3565	3565	com.ryg.chapter_1	MainActivity	onConfigurationChanged, newOrientation:2
D	3565	3565	com.ryg.chapter_1	MainActivity	onConfigurationChanged, newOrientation:1
D	3565	3565	com.ryg.chapter_1	MainActivity	onConfigurationChanged, newOrientation:2

图 1-6　系统日志

由上面的日志可见，Activity 的确没有重新创建，并且也没有调用 onSaveInstanceState 和 onRestoreInstanceState 来存储和恢复数据，取而代之的是系统调用了 Activity 的 onConfigurationChanged 方法，这个时候我们就可以做一些自己的特殊处理了。

1.2 Activity 的启动模式

上一节介绍了 Activity 在标准情况下和异常情况下的生命周期，我们对 Activity 的生命周期应该有了深入的了解。除了 Activity 的生命周期外，Activity 的启动模式也是一个难点，原因是形形色色的启动模式和标志位实在是太容易被混淆了，但是 Activity 作为四大组件之首，它的的确确非常重要，有时候为了满足项目的特殊需求，就必须使用 Activity 的启动模式，所以我们必须要搞清楚它的启动模式和标志位，本节将会一一介绍。

1.2.1 Activity 的 LaunchMode

首先说一下 Activity 为什么需要启动模式。我们知道，在默认情况下，当我们多次启动同一个 Activity 的时候，系统会创建多个实例并把它们一一放入任务栈中，当我们单击 back 键，会发现这些 Activity 会一一回退。任务栈是一种"后进先出"的栈结构，这个比较好理解，每按一下 back 键就会有一个 Activity 出栈，直到栈空为止，当栈中无任何 Activity 的时候，系统就会回收这个任务栈。关于任务栈的系统工作原理，这里暂时不做说明，在后续章节会专门介绍任务栈。知道了 Activity 的默认启动模式以后，我们可能就会发现一个问题：多次启动同一个 Activity，系统重复创建多个实例，这样不是很傻吗？这样的确有点傻，Android 在设计的时候不可能不考虑到这个问题，所以它提供了启动模式来修改系统的默认行为。目前有四种启动模式：standard、singleTop、singleTask 和 singleInstance，下面先介绍各种启动模式的含义：

（1）standard：标准模式，这也是系统的默认模式。每次启动一个 Activity 都会重新创建一个新的实例，不管这个实例是否已经存在。被创建的实例的生命周期符合典型情况下 Activity 的生命周期，如上节描述，它的 onCreate、onStart、onResume 都会被调用。这是一种典型的多实例实现，一个任务栈中可以有多个实例，每个实例也可以属于不同的任务栈。在这种模式下，谁启动了这个 Activity，那么这个 Activity 就运行在启动它的那个 Activity 所在的栈中。比如 Activity A 启动了 Activity B（B 是标准模式），那么 B 就会进入到 A 所在的栈中。不知道读者是否注意到，当我们用 ApplicationContext 去启动 standard 模式的

Activity 的时候会报错，错误如下：

```
E/AndroidRuntime(674):    android.util.AndroidRuntimeException: Calling
startActivity from outside of an Activity context requires the FLAG_
ACTIVITY_NEW_TASK flag. Is this really what you want?
```

相信这句话读者一定不陌生，这是因为 standard 模式的 Activity 默认会进入启动它的 Activity 所属的任务栈中，但是由于非 Activity 类型的 Context（如 ApplicationContext）并没有所谓的任务栈，所以这就有问题了。解决这个问题的方法是为待启动 Activity 指定 FLAG_ACTIVITY_NEW_TASK 标记位，这样启动的时候就会为它创建一个新的任务栈，这个时候待启动 Activity 实际上是以 singleTask 模式启动的，读者可以仔细体会。

（2）singleTop：栈顶复用模式。在这种模式下，如果新 Activity 已经位于任务栈的栈顶，那么此 Activity 不会被重新创建，同时它的 onNewIntent 方法会被回调，通过此方法的参数我们可以取出当前请求的信息。需要注意的是，这个 Activity 的 onCreate、onStart 不会被系统调用，因为它并没有发生改变。如果新 Activity 的实例已存在但不是位于栈顶，那么新 Activity 仍然会重新重建。举个例子，假设目前栈内的情况为 ABCD，其中 ABCD 为四个 Activity，A 位于栈底，D 位于栈顶，这个时候假设要再次启动 D，如果 D 的启动模式为 singleTop，那么栈内的情况仍然为 ABCD；如果 D 的启动模式为 standard，那么由于 D 被重新创建，导致栈内的情况就变为 ABCDD。

（3）singleTask：栈内复用模式。这是一种单实例模式，在这种模式下，只要 Activity 在一个栈中存在，那么多次启动此 Activity 都不会重新创建实例，和 singleTop 一样，系统也会回调其 onNewIntent。具体一点，当一个具有 singleTask 模式的 Activity 请求启动后，比如 Activity A，系统首先会寻找是否存在 A 想要的任务栈，如果不存在，就重新创建一个任务栈，然后创建 A 的实例后把 A 放到栈中。如果存在 A 所需的任务栈，这时要看 A 是否在栈中有实例存在，如果有实例存在，那么系统就会把 A 调到栈顶并调用它的 onNewIntent 方法，如果实例不存在，就创建 A 的实例并把 A 压入栈中。举几个例子：

- 比如目前任务栈 S1 中的情况为 ABC，这个时候 Activity D 以 singleTask 模式请求启动，其所需要的任务栈为 S2，由于 S2 和 D 的实例均不存在，所以系统会先创建任务栈 S2，然后再创建 D 的实例并将其入栈到 S2。

- 另外一种情况，假设 D 所需的任务栈为 S1，其他情况如上面例子 1 所示，那么由于 S1 已经存在，所以系统会直接创建 D 的实例并将其入栈到 S1。

- 如果 D 所需的任务栈为 S1，并且当前任务栈 S1 的情况为 ADBC，根据栈内复用的原则，此时 D 不会重新创建，系统会把 D 切换到栈顶并调用其 onNewIntent 方法，同时由于 singleTask 默认具有 clearTop 的效果，会导致栈内所有在 D 上面的 Activity 全部出栈，于是最终 S1 中的情况为 AD。这一点比较特殊，在后面还会对此种情况详细地分析。

通过上述 3 个例子，读者应该能比较清晰地理解 singleTask 的含义了。

（4）singleInstance：单实例模式。这是一种加强的 singleTask 模式，它除了具有 singleTask 模式的所有特性外，还加强了一点，那就是具有此种模式的 Activity 只能单独地位于一个任务栈中，换句话说，比如 Activity A 是 singleInstance 模式，当 A 启动后，系统会为它创建一个新的任务栈，然后 A 独自在这个新的任务栈中，由于栈内复用的特性，后续的请求均不会创建新的 Activity，除非这个独特的任务栈被系统销毁了。

上面介绍了几种启动模式，这里需要指出一种情况，我们假设目前有 2 个任务栈，前台任务栈的情况为 AB，而后台任务栈的情况为 CD，这里假设 CD 的启动模式均为 singleTask。现在请求启动 D，那么整个后台任务栈都会被切换到前台，这个时候整个后退列表变成了 ABCD。当用户按 back 键的时候，列表中的 Activity 会一一出栈，如图 1-7 所示。如果不是请求启动 D 而是启动 C，那么情况就不一样了，请看图 1-8，具体原因在本节后面会再进行详细分析。

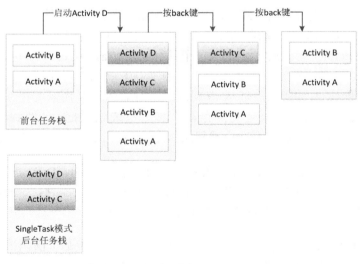

图 1-7 任务栈示例 1

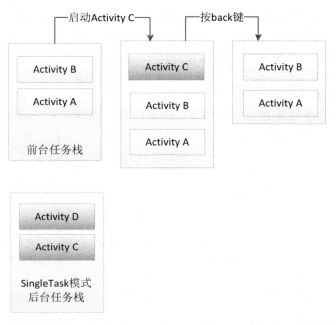

图 1-8　任务栈示例 2

另外一个问题是，在 singleTask 启动模式中，多次提到某个 Activity 所需的任务栈，什么是 Activity 所需要的任务栈呢？这要从一个参数说起：TaskAffinity，可以翻译为任务相关性。这个参数标识了一个 Activity 所需要的任务栈的名字，默认情况下，所有 Activity 所需的任务栈的名字为应用的包名。当然，我们可以为每个 Activity 都单独指定 TaskAffinity 属性，这个属性值必须不能和包名相同，否则就相当于没有指定。TaskAffinity 属性主要和 singleTask 启动模式或者 allowTaskReparenting 属性配对使用，在其他情况下没有意义。另外，任务栈分为前台任务栈和后台任务栈，后台任务栈中的 Activity 位于暂停状态，用户可以通过切换将后台任务栈再次调到前台。

当 TaskAffinity 和 singleTask 启动模式配对使用的时候，它是具有该模式的 Activity 的目前任务栈的名字，待启动的 Activity 会运行在名字和 TaskAffinity 相同的任务栈中。

当 TaskAffinity 和 allowTaskReparenting 结合的时候，这种情况比较复杂，会产生特殊的效果。当一个应用 A 启动了应用 B 的某个 Activity 后，如果这个 Activity 的 allowTaskReparenting 属性为 true 的话，那么当应用 B 被启动后，此 Activity 会直接从应用 A 的任务栈转移到应用 B 的任务栈中。这还是很抽象，再具体点，比如现在有 2 个应用 A

和 B，A 启动了 B 的一个 Activity C，然后按 Home 键回到桌面，然后再单击 B 的桌面图标，这个时候并不是启动了 B 的主 Activity，而是重新显示了已经被应用 A 启动的 Activity C，或者说，C 从 A 的任务栈转移到了 B 的任务栈中。可以这么理解，由于 A 启动了 C，这个时候 C 只能运行在 A 的任务栈中，但是 C 属于 B 应用，正常情况下，它的 TaskAffinity 值肯定不可能和 A 的任务栈相同（因为包名不同）。所以，当 B 被启动后，B 会创建自己的任务栈，这个时候系统发现 C 原本所想要的任务栈已经被创建了，所以就把 C 从 A 的任务栈中转移过来了。这种情况读者可以写个例子测试一下，这里就不做示例了。

如何给 Activity 指定启动模式呢？有两种方法，第一种是通过 AndroidMenifest 为 Activity 指定启动模式，如下所示。

```
<activity
    android:name="com.ryg.chapter_1.SecondActivity"
    android:configChanges="screenLayout"
    android:launchMode="singleTask"
    android:label="@string/app_name" />
```

另一种情况是通过在 Intent 中设置标志位来为 Activity 指定启动模式，比如：

```
Intent intent = new Intent();
intent.setClass(MainActivity.this, SecondActivity.class);
intent.addFlags(Intent.FLAG_ACTIVITY_NEW_TASK);
startActivity(intent);
```

这两种方式都可以为 Activity 指定启动模式，但是二者还是有区别的。首先，优先级上，第二种方式的优先级要高于第一种，当两种同时存在时，以第二种方式为准；其次，上述两种方式在限定范围上有所不同，比如，第一种方式无法直接为 Activity 设定 FLAG_ACTIVITY_CLEAR_TOP 标识，而第二种方式无法为 Activity 指定 singleInstance 模式。

关于 Intent 中为 Activity 指定的各种标记位，在下面的小节中会继续介绍。下面通过一个例子来体验启动模式的使用效果。还是前面的例子，这里我们把 MainActivity 的启动模式设为 singleTask，然后重复启动它，看看是否会重复创建，代码修改如下：

```
<activity
    android:name="com.ryg.chapter_1.MainActivity"
    android:configChanges="orientation|screenSize"
    android:label="@string/app_name"
```

```xml
            android:launchMode="singleTask" >
        <intent-filter>
            <action android:name="android.intent.action.MAIN" />

            <category android:name="android.intent.category.LAUNCHER" />
        </intent-filter>
</activity>
```

```java
@Override
protected void onNewIntent(Intent intent) {
    super.onNewIntent(intent);
    Log.d(TAG, "onNewIntent, time=" + intent.getLongExtra("time", 0));
}

findViewById(R.id.button1).setOnClickListener(new OnClickListener() {

    @Override
    public void onClick(View v) {
        Intent intent = new Intent();
        intent.setClass(MainActivity.this, MainActivity.class);
intent.putExtra("time", System.currentTimeMillis());
        startActivity(intent);
    }
});
```

根据上述修改，我们做如下操作，连续单击三次按钮启动 3 次 MainActivity，算上原本的 MainActvity 的实例，正常情况下，任务栈中应该有 4 个 MainActivity 的实例，但是我们为其指定了 singleTask 模式，现在来看一看到底有何不同。

执行 adb shell dumpsys activity 命令：

```
ACTIVITY MANAGER ACTIVITIES (dumpsys activity activities)
  Main stack:
    TaskRecord{41350dc8 #9 A com.ryg.chapter_1}
    Intent { cmp=com.ryg.chapter_1/.MainActivity (has extras) }
      Hist #1: ActivityRecord{412cc188 com.ryg.chapter_1/.MainActivity}
        Intent { act=android.intent.action.MAIN cat=[android.intent.
        category.LAUNCHER] flg=0x
0 cmp=com.ryg.chapter_1/.MainActivity bnds=[160,235][240,335] }
        ProcessRecord{411e6898 634:com.ryg.chapter_1/10052}
```

```
        TaskRecord{4125abc8 #2 A com.android.launcher}
        Intent { act=android.intent.action.MAIN cat=[android.intent.category.
        HOME] flg=0x10000000
m.android.launcher/com.android.launcher2.Launcher }
          Hist #0: ActivityRecord{412381f8 com.android.launcher/com.android.
          launcher2.Launcher}
            Intent { act=android.intent.action.MAIN cat=[android.intent.
            category.HOME] flg=0x1000
p=com.android.launcher/com.android.launcher2.Launcher }
            ProcessRecord{411d24c8 214:com.android.launcher/10013}

  Running activities (most recent first):
    TaskRecord{41350dc8 #9 A com.ryg.chapter_1}
      Run #1: ActivityRecord{412cc188 com.ryg.chapter_1/.MainActivity}
    TaskRecord{4125abc8 #2 A com.android.launcher}
      Run #0: ActivityRecord{412381f8 com.android.launcher/com.android.
      launcher2.Launcher}

  mResumedActivity: ActivityRecord{412cc188 com.ryg.chapter_1/.MainActivity}
  mFocusedActivity: ActivityRecord{412cc188 com.ryg.chapter_1/.MainActivity}

  Recent tasks:
   * Recent #0: TaskRecord{41350dc8 #9 A com.ryg.chapter_1}
   * Recent #1: TaskRecord{4125abc8 #2 A com.android.launcher}
   * Recent #2: TaskRecord{412b60a0 #5 A com.estrongs.android.pop.app.
     InstallMonitorActivity}
```

从上面导出的 Activity 信息可以看出，尽管启动了 4 次 MainActivity，但是它始终只有一个实例在任务栈中。从图 1-9 的 log 可以看出，Activity 的确没有重新创建，只是暂停了一下，然后调用了 onNewIntent，接着调用 onResume 就又继续了。

Level	PID	TID	Application	Tag	Text
D	755	755	com.ryg.chapter_1	MainActivity	onPause
D	755	755	com.ryg.chapter_1	MainActivity	onNewIntent, time=1422898165307
D	755	755	com.ryg.chapter_1	MainActivity	onResume
D	755	755	com.ryg.chapter_1	MainActivity	onPause
D	755	755	com.ryg.chapter_1	MainActivity	onNewIntent, time=1422898166173
D	755	755	com.ryg.chapter_1	MainActivity	onResume
D	.755	755	com.ryg.chapter_1	MainActivity	onPause
D	755	755	com.ryg.chapter_1	MainActivity	onNewIntent, time=1422898167429
D	755	755	com.ryg.chapter_1	MainActivity	onResume

图 1-9　系统日志

现在我们去掉 singleTask，再来对比一下，还是同样的操作，单击三次按钮启动 MainActivity 三次。

执行 adb shell dumpsys activity 命令：

```
ACTIVITY MANAGER ACTIVITIES (dumpsys activity activities)
  Main stack:
    TaskRecord{41325370 #17 A com.ryg.chapter_1}
    Intent { act=android.intent.action.MAIN cat=[android.intent.category.
    LAUNCHER] flg=0x100000
p=com.ryg.chapter_1/.MainActivity }
      Hist #4: ActivityRecord{41236968 com.ryg.chapter_1/.MainActivity}
        Intent { cmp=com.ryg.chapter_1/.MainActivity (has extras) }
        ProcessRecord{411e6898 803:com.ryg.chapter_1/10052}
      Hist #3: ActivityRecord{411f4b30 com.ryg.chapter_1/.MainActivity}
        Intent { cmp=com.ryg.chapter_1/.MainActivity (has extras) }
        ProcessRecord{411e6898 803:com.ryg.chapter_1/10052}
      Hist #2: ActivityRecord{411edcb8 com.ryg.chapter_1/.MainActivity}
        Intent { cmp=com.ryg.chapter_1/.MainActivity (has extras) }
        ProcessRecord{411e6898 803:com.ryg.chapter_1/10052}
      Hist #1: ActivityRecord{411e7588 com.ryg.chapter_1/.MainActivity}
        Intent { act=android.intent.action.MAIN cat=[android.intent.category.
        LAUNCHER] flg=0x10
0 cmp=com.ryg.chapter_1/.MainActivity }
        ProcessRecord{411e6898 803:com.ryg.chapter_1/10052}
    TaskRecord{4125abc8 #2 A com.android.launcher}
    Intent { act=android.intent.action.MAIN cat=[android.intent.category.
    HOME] flg=0x10000000 c
m.android.launcher/com.android.launcher2.Launcher }
      Hist #0: ActivityRecord{412381f8 com.android.launcher/com.android.
      launcher2.Launcher}
        Intent { act=android.intent.action.MAIN cat=[android.intent.cate-
        gory.HOME] flg=0x100000
p=com.android.launcher/com.android.launcher2.Launcher }
        ProcessRecord{411d24c8 214:com.android.launcher/10013}

  Running activities (most recent first):
    TaskRecord{41325370 #17 A com.ryg.chapter_1}
      Run #4: ActivityRecord{41236968 com.ryg.chapter_1/.MainActivity}
```

```
    Run #3: ActivityRecord{411f4b30 com.ryg.chapter_1/.MainActivity}
    Run #2: ActivityRecord{411edcb8 com.ryg.chapter_1/.MainActivity}
    Run #1: ActivityRecord{411e7588 com.ryg.chapter_1/.MainActivity}
  TaskRecord{4125abc8 #2 A com.android.launcher}
    Run #0: ActivityRecord{412381f8 com.android.launcher/com.android.
    launcher2.Launcher}

mResumedActivity: ActivityRecord{41236968 com.ryg.chapter_1/.MainAc-
tivity}
mFocusedActivity: ActivityRecord{41236968 com.ryg.chapter_1/.MainAc-
tivity}

Recent tasks:
* Recent #0: TaskRecord{41325370 #17 A com.ryg.chapter_1}
* Recent #1: TaskRecord{4125abc8 #2 A com.android.launcher}
* Recent #2: TaskRecord{412c8d58 #16 A com.estrongs.android.pop.app.
  InstallMonitorActivity}
```

上面的导出信息很多，我们可以有选择地看，比如就看 Running activities (most recent first)这一块，如下所示。

```
Running activities (most recent first):
  TaskRecord{41325370 #17 A com.ryg.chapter_1}
    Run #4: ActivityRecord{41236968 com.ryg.chapter_1/.MainActivity}
    Run #3: ActivityRecord{411f4b30 com.ryg.chapter_1/.MainActivity}
    Run #2: ActivityRecord{411edcb8 com.ryg.chapter_1/.MainActivity}
    Run #1: ActivityRecord{411e7588 com.ryg.chapter_1/.MainActivity}
  TaskRecord{4125abc8 #2 A com.android.launcher}
    Run #0: ActivityRecord{412381f8 com.android.launcher/com.android.
    launcher2.Launcher}
```

我们能够得出目前总共有 2 个任务栈，前台任务栈的 taskAffinity 值为 com.ryg.chapter_1，它里面有 4 个 Activity，后台任务栈的 taskAffinity 值为 com.android.launcher，它里面有 1 个 Activity，这个 Activity 就是桌面。通过这种方式来分析任务栈信息就清晰多了。

从上面的导出信息中可以看到，在任务栈中有 4 个 MainActivity，这也就验证了 Activity 的启动模式的工作方式。

上述四种启动模式，standard 和 singleTop 都比较好理解，singleInstance 由于其特殊性也好理解，但是关于 singleTask 有一种情况需要再说明一下。如图 1-7 所示，如果在 Activity B 中请求的不是 D 而是 C，那么情况如何呢？这里可以告诉读者的是，任务栈列表变成了 ABC，是不是很奇怪呢？Activity D 被直接出栈了。下面我们再用实例验证看看是不是这样。首先，还是使用上面的代码，但是我们做一下修改：

```xml
<activity
    android:name="com.ryg.chapter_1.MainActivity"
    android:configChanges="orientation|screenSize"
    android:label="@string/app_name"
    android:launchMode="standard" >
    <intent-filter>
        <action android:name="android.intent.action.MAIN" />
        <category android:name="android.intent.category.LAUNCHER" />
    </intent-filter>
</activity>
<activity
    android:name="com.ryg.chapter_1.SecondActivity"
    android:configChanges="screenLayout"
    android:label="@string/app_name"
    android:taskAffinity="com.ryg.task1"
    android:launchMode="singleTask" />

<activity
    android:name="com.ryg.chapter_1.ThirdActivity"
    android:configChanges="screenLayout"
    android:taskAffinity="com.ryg.task1"
    android:label="@string/app_name"
    android:launchMode="singleTask" />
```

我们将 SecondActivity 和 ThirdActivity 都设成 singleTask 并指定它们的 taskAffinity 属性为"com.ryg.task1"，注意这个 taskAffinity 属性的值为字符串，且中间必须含有包名分隔符"."。然后做如下操作，在 MainActivity 中单击按钮启动 SecondActivity，在 SecondActivity 中单击按钮启动 ThirdActivity，在 ThirdActivity 中单击按钮又启动 MainActivity，最后再在 MainActivity 中单击按钮启动 SecondActivity，现在按 2 次 back 键，然后看到的是哪个 Activity？答案是回到桌面。是不是有点摸不到头脑了？没关系，接下来我们分析这个问题。

首先，从理论上分析这个问题，先假设 MainActivity 为 A，SecondActivity 为 B，

ThirdActivity 为 C。我们知道 A 为 standard 模式，按照规定，A 的 taskAffinity 值继承自 Application 的 taskAffinity，而 Application 默认 taskAffinity 为包名，所以 A 的 taskAffinity 为包名。由于我们在 XML 中为 B 和 C 指定了 taskAffinity 和启动模式，所以 B 和 C 是 singleTask 模式且有相同的 taskAffinity 值"com.ryg.task1"。A 启动 B 的时候，按照 singleTask 的规则，这个时候需要为 B 重新创建一个任务栈"com.ryg.task1"。B 再启动 C，按照 singleTask 的规则，由于 C 所需的任务栈（和 B 为同一任务栈）已经被 B 创建，所以无须再创建新的任务栈，这个时候系统只是创建 C 的实例后将 C 入栈了。接着 C 再启动 A，A 是 standard 模式，所以系统会为它创建一个新的实例并将到加到启动它的那个 Activity 的任务栈，由于是 C 启动了 A，所以 A 会进入 C 的任务栈中并位于栈顶。这个时候已经有两个任务栈了，一个是名字为包名的任务栈，里面只有 A，另一个是名字为 "com.ryg.task1" 的任务栈，里面的 Activity 为 BCA。接下来，A 再启动 B，由于 B 是 singleTask，B 需要回到任务栈的栈顶，由于栈的工作模式为 "后进先出"，B 想要回到栈顶，只能是 CA 出栈。所以，到这里就很好理解了，如果再按 back 键，B 就出栈了，B 所在的任务栈已经不存在了，这个时候只能是回到后台任务栈并把 A 显示出来。注意这个 A 是后台任务栈的 A，不是 "com.ryg.task1" 任务栈的 A，接着再继续 back，就回到桌面了。分析到这里，我们得出一条结论，singleTask 模式的 Activity 切换到栈顶会导致在它之上的栈内的 Activity 出栈。

接着我们在实践中再次验证这个问题，还是采用 dumpsys 命令。我们省略中间的过程，直接看 C 启动 A 的那个状态，执行 adb shell dumpsys activity 命令，日志如下：

```
Running activities (most recent first):
TaskRecord{4132bd90 #12 A com.ryg.task1}
  Run #4: ActivityRecord{4133fd18 com.ryg.chapter_1/.MainActivity}
  Run #3: ActivityRecord{41349c58 com.ryg.chapter_1/.ThirdActivity}
  Run #2: ActivityRecord{4132bab0 com.ryg.chapter_1/.SecondActivity}
TaskRecord{4125a008 #11 A com.ryg.chapter_1}
  Run #1: ActivityRecord{41328c60 com.ryg.chapter_1/.MainActivity}
TaskRecord{41256440 #2 A com.android.launcher}
  Run #0: ActivityRecord{41231d30 com.android.launcher/com.android.launcher2.Launcher}
```

可以清楚地看到有 2 个任务栈，第一个（com.ryg.chapter_1）只有 A，第二个（com.ryg.task1）有 BCA，就如同我们上面分析的那样，然后再从 A 中启动 B，再看一下日志：

```
Running activities (most recent first):
TaskRecord{4132bd90 #12 A com.ryg.task1}
```

```
  Run #2: ActivityRecord{4132bab0 com.ryg.chapter_1/.SecondActivity}
TaskRecord{4125a008 #11 A com.ryg.chapter_1}
  Run #1: ActivityRecord{41328c60 com.ryg.chapter_1/.MainActivity}
TaskRecord{41256440 #2 A com.android.launcher}
  Run #0: ActivityRecord{41231d30 com.android.launcher/com.android.launch
er2.Launcher}
```

可以发现在任务栈 com.ryg.task1 中只剩下 B 了，C、A 都已经出栈了，这个时候再按 back 键，任务栈 com.ryg.chapter_1 中的 A 就显示出来了，如果再 back 就回到桌面了。分析到这里，相信读者对 Activity 的启动模式已经有很深入的理解了。下面介绍 Activity 中常用的标志位。

1.2.2 Activity 的 Flags

Activity 的 Flags 有很多，这里主要分析一些比较常用的标记位。标记位的作用很多，有的标记位可以设定 Activity 的启动模式，比如 FLAG_ACTIVITY_NEW_TASK 和 FLAG_ACTIVITY_SINGLE_TOP 等；还有的标记位可以影响 Activity 的运行状态，比如 FLAG_ACTIVITY_CLEAR_TOP 和 FLAG_ACTIVITY_EXCLUDE_FROM_RECENTS 等。下面主要介绍几个比较常用的标记位，剩下的标记位读者可以查看官方文档去了解，大部分情况下，我们不需要为 Activity 指定标记位，因此，对于标记位理解即可。在使用标记位的时候，要注意有些标记位是系统内部使用的，应用程序不需要去手动设置这些标记位以防出现问题。

FLAG_ACTIVITY_NEW_TASK

这个标记位的作用是为 Activity 指定"singleTask"启动模式，其效果和在 XML 中指定该启动模式相同。

FLAG_ACTIVITY_SINGLE_TOP

这个标记位的作用是为 Activity 指定"singleTop"启动模式，其效果和在 XML 中指定该启动模式相同。

FLAG_ACTIVITY_CLEAR_TOP

具有此标记位的 Activity，当它启动时，在同一个任务栈中所有位于它上面的 Activity 都要出栈。这个标记位一般会和 singleTask 启动模式一起出现,在这种情况下,被启动 Activity 的实例如果已经存在,那么系统就会调用它的 onNewIntent。如果被启动的 Activity 采用 standard 模式启动,那么它连同它之上的 Activity 都要出栈,系统会创建新的 Activity 实例并放入栈顶。通过 1.2.1 节中的分析可以知道,singleTask 启动模式默认就具有此标记位的效果。

FLAG_ACTIVITY_EXCLUDE_FROM_RECENTS

具有这个标记的 Activity 不会出现在历史 Activity 的列表中,当某些情况下我们不希望用户通过历史列表回到我们的 Activity 的时候这个标记比较有用。它等同于在 XML 中指定 Activity 的属性 android:excludeFromRecents="true"。

1.3　IntentFilter 的匹配规则

我们知道,启动 Activity 分为两种,显式调用和隐式调用。二者的区别这里就不多说了,显式调用需要明确地指定被启动对象的组件信息,包括包名和类名,而隐式调用则不需要明确指定组件信息。原则上一个 Intent 不应该既是显式调用又是隐式调用,如果二者共存的话以显式调用为主。显式调用很简单,这里主要介绍一下隐式调用。隐式调用需要 Intent 能够匹配目标组件的 IntentFilter 中所设置的过滤信息,如果不匹配将无法启动目标 Activity。IntentFilter 中的过滤信息有 action、category、data,下面是一个过滤规则的示例:

```xml
<activity
    android:name="com.ryg.chapter_1.ThirdActivity"
    android:configChanges="screenLayout"
    android:label="@string/app_name"
    android:launchMode="singleTask"
    android:taskAffinity="com.ryg.task1" >
    <intent-filter >
        <action android:name="com.ryg.charpter_1.c"/>
        <action android:name="com.ryg.charpter_1.d"/>
        <category android:name="com.ryg.category.c"/>
        <category android:name="com.ryg.category.d"/>
        <category android:name="android.intent.category.DEFAULT"/>
        <data android:mimeType="text/plain"/>
    </intent-filter>
</activity>
```

为了匹配过滤列表，需要同时匹配过滤列表中的 action、category、data 信息，否则匹配失败。一个过滤列表中的 action、category 和 data 可以有多个，所有的 action、category、data 分别构成不同类别，同一类别的信息共同约束当前类别的匹配过程。只有一个 Intent 同时匹配 action 类别、category 类别、data 类别才算完全匹配，只有完全匹配才能成功启动目标 Activity。另外一点，一个 Activity 中可以有多个 intent-filter，一个 Intent 只要能匹配任何一组 intent-filter 即可成功启动对应的 Activity，如下所示。

```xml
<activity android:name="ShareActivity">
    <!-- This activity handles "SEND" actions with text data -->
    <intent-filter>
        <action android:name="android.intent.action.SEND"/>
        <category android:name="android.intent.category.DEFAULT"/>
        <data android:mimeType="text/plain"/>
    </intent-filter>
    <!-- This activity also handles "SEND" and "SEND_MULTIPLE" with media
         data -->
    <intent-filter>
        <action android:name="android.intent.action.SEND"/>
        <action android:name="android.intent.action.SEND_MULTIPLE"/>
        <category android:name="android.intent.category.DEFAULT"/>
        <data android:mimeType="application/vnd.google.panorama360+jpg"/>
        <data android:mimeType="image/*"/>
        <data android:mimeType="video/*"/>
    </intent-filter>
</activity>
```

下面详细分析各种属性的匹配规则。

1. action 的匹配规则

action 是一个字符串，系统预定义了一些 action，同时我们也可以在应用中定义自己的 action。action 的匹配规则是 Intent 中的 action 必须能够和过滤规则中的 action 匹配，这里说的匹配是指 action 的字符串值完全一样。一个过滤规则中可以有多个 action，那么只要 Intent 中的 action 能够和过滤规则中的任何一个 action 相同即可匹配成功。针对上面的过滤规则，只要我们的 Intent 中 action 值为 "com.ryg.charpter_1.c" 或者 "com.ryg.charpter_1.d" 都能成功匹配。需要注意的是，Intent 中如果没有指定 action，那么匹配失败。总结一下，action 的匹配要求 Intent 中的 action 存在且必须和过滤规则中的其中一个 action 相同，这里

需要注意它和 category 匹配规则的不同。另外，action 区分大小写，大小写不同字符串相同的 action 会匹配失败。

2. category 的匹配规则

category 是一个字符串，系统预定义了一些 category，同时我们也可以在应用中定义自己的 category。category 的匹配规则和 action 不同，它要求 Intent 中如果含有 category，那么所有的 category 都必须和过滤规则中的其中一个 category 相同。换句话说，Intent 中如果出现了 category，不管有几个 category，对于每个 category 来说，它必须是过滤规则中已经定义了的 category。当然，Intent 中可以没有 category，如果没有 category 的话，按照上面的描述，这个 Intent 仍然可以匹配成功。这里要注意下它和 action 匹配过程的不同，action 是要求 Intent 中必须有一个 action 且必须能够和过滤规则中的某个 action 相同，而 category 要求 Intent 可以没有 category，但是如果你一旦有 category，不管有几个，每个都要能够和过滤规则中的任何一个 category 相同。为了匹配前面的过滤规则中的 category，我们可以写出下面的 Intent，intent.addcategory ("com.ryg.category.c")或者 Intent. addcategory ("com.ryg.category.d")亦或者不设置 category。为什么不设置 category 也可以匹配呢？原因是系统在调用 startActivity 或者 startActivityForResult 的时候会默认为 Intent 加上 "android.intent.category.DEFAULT" 这个 category，所以这个 category 就可以匹配前面的过滤规则中的第三个 category。同时，为了我们的 activity 能够接收隐式调用，就必须在 intent-filter 中指定 "android.intent.category.DEFAULT" 这个 category，原因刚才已经说明了。

3. data 的匹配规则

data 的匹配规则和 action 类似，如果过滤规则中定义了 data，那么 Intent 中必须也要定义可匹配的 data。在介绍 data 的匹配规则之前，我们需要先了解一下 data 的结构，因为 data 稍微有些复杂。

data 的语法如下所示。

```
<data android:scheme="string"
    android:host="string"
    android:port="string"
    android:path="string"
    android:pathPattern="string"
    android:pathPrefix="string"
    android:mimeType="string" />
```

data 由两部分组成，mimeType 和 URI。mimeType 指媒体类型，比如 image/jpeg、audio/mpeg4-generic 和 video/*等，可以表示图片、文本、视频等不同的媒体格式，而 URI 中包含的数据就比较多了，下面是 URI 的结构：

```
<scheme>://<host>:<port>/[<path>|<pathPrefix>|<pathPattern>]
```

这里再给几个实际的例子就比较好理解了，如下所示。

```
content://com.example.project:200/folder/subfolder/etc
http://www.baidu.com:80/search/info
```

看了上面的两个示例应该就瞬间明白了，没错，就是这么简单。不过下面还是要介绍一下每个数据的含义。

Scheme：URI 的模式，比如 http、file、content 等，如果 URI 中没有指定 scheme，那么整个 URI 的其他参数无效，这也意味着 URI 是无效的。

Host：URI 的主机名，比如 www.baidu.com，如果 host 未指定，那么整个 URI 中的其他参数无效，这也意味着 URI 是无效的。

Port：URI 中的端口号，比如 80，仅当 URI 中指定了 scheme 和 host 参数的时候 port 参数才是有意义的。

Path、pathPattern 和 pathPrefix：这三个参数表述路径信息，其中 path 表示完整的路径信息；pathPattern 也表示完整的路径信息，但是它里面可以包含通配符"*"，"*"表示 0 个或多个任意字符，需要注意的是，由于正则表达式的规范，如果想表示真实的字符串，那么"*"要写成"*"，"\"要写成"\\\\"；pathPrefix 表示路径的前缀信息。

介绍完 data 的数据格式后，我们要说一下 data 的匹配规则了。前面说到，data 的匹配规则和 action 类似，它也要求 Intent 中必须含有 data 数据，并且 data 数据能够完全匹配过滤规则中的某一个 data.这里的完全匹配是指过滤规则中出现的 data 部分也出现在了 Intent 中的 data 中。下面分情况说明。

（1）如下过滤规则：

```
<intent-filter>
    <data android:mimeType="image/*" />
```

```
    ...
</intent-filter>
```

这种规则指定了媒体类型为所有类型的图片,那么 Intent 中的 mimeType 属性必须为 "image/*" 才能匹配,这种情况下虽然过滤规则没有指定 URI,但是却有默认值,URI 的默认值为 content 和 file。也就是说,虽然没有指定 URI,但是 Intent 中的 URI 部分的 schema 必须为 content 或者 file 才能匹配,这点是需要尤其注意的。为了匹配(1)中规则,我们可以写出如下示例:

```
intent.setDataAndType(Uri.parse("file://abc"),"image/png")。
```

另外,如果要为 Intent 指定完整的 data,必须要调用 setDataAndType 方法,不能先调用 setData 再调用 setType,因为这两个方法彼此会清除对方的值,这个看源码就很容易理解,比如 setData:

```
public Intent setData(Uri data) {
    mData = data;
    mType = null;
    return this;
}
```

可以发现,setData 会把 mimeType 置为 null,同理 setType 也会把 URI 置为 null。

(2)如下过滤规则:

```
<intent-filter>
    <data android:mimeType="video/mpeg" android:scheme="http" ... />
    <data android:mimeType="audio/mpeg" android:scheme="http" ... />
    ...
</intent-filter>
```

这种规则指定了两组 data 规则,且每个 data 都指定了完整的属性值,既有 URI 又有 mimeType。为了匹配(2)中规则,我们可以写出如下示例:

```
intent.setDataAndType(Uri.parse("http://abc"),"video/mpeg")
```

或者

```
intent. setDataAndType (Uri.parse("http://abc"),"audio/mpeg")
```

通过上面两个示例，读者应该已经明白了 data 的匹配规则，关于 data 还有一个特殊情况需要说明下，这也是它和 action 不同的地方，如下两种特殊的写法，它们的作用是一样的：

```
<intent-filter ...>
    <data android:scheme="file" android:host="www.baidu.com" />
    ...
</intent-filter>

<intent-filter ...>
    <data android:scheme="file" />
    <data android:host="www.baidu.com" />
    ...
</intent-filter>
```

到这里我们已经把 IntentFilter 的过滤规则都讲解了一遍，还记得本节前面给出的一个 intent-filter 的示例吗？现在我们给出完全匹配它的 Intent：

```
Intent intent = new Intent("com.ryg.charpter_1.c");
intent.addCategory("com.ryg.category.c");
intent.setDataAndType(Uri.parse("file://abc"), "text/plain");
startActivity(intent);
```

还记得 URI 的 schema 是有默认值的吗？如果把上面的 intent.setDataAndType(Uri.parse("file://abc"), "text/plain") 这句改成 intent.setDataAndType(Uri.parse("http://abc"), "text/plain")，打开 Activity 的时候就会报错，提示无法找到 Activity，如图 1-10 所示。另外一点，Intent-filter 的匹配规则对于 Service 和 BroadcastReceiver 也是同样的道理，不过系统对于 Service 的建议是尽量使用显式调用方式来启动服务。

图 1-10　系统日志

最后，当我们通过隐式方式启动一个 Activity 的时候，可以做一下判断，看是否有 Activity 能够匹配我们的隐式 Intent，如果不做判断就有可能出现上述的错误了。判断方法有两种：采用 PackageManager 的 resolveActivity 方法或者 Intent 的 resolveActivity 方法，如果它们找不到匹配的 Activity 就会返回 null，我们通过判断返回值就可以规避上述错误了。另外，PackageManager 还提供了 queryIntentActivities 方法，这个方法和 resolveActivity 方法不同的是：它不是返回最佳匹配的 Activity 信息而是返回所有成功匹配的 Activity 信息。我们看一下 queryIntentActivities 和 resolveActivity 的方法原型：

```
public abstract List<ResolveInfo> queryIntentActivities(Intent intent, int flags);
public abstract ResolveInfo resolveActivity(Intent intent, int flags);
```

上述两个方法的第一个参数比较好理解，第二个参数需要注意，我们要使用 MATCH_DEFAULT_ONLY 这个标记位，这个标记位的含义是仅仅匹配那些在 intent-filter 中声明了 <category android:name="android.intent.category.DEFAULT"/>这个 category 的 Activity。使用这个标记位的意义在于，只要上述两个方法不返回 null，那么 startActivity 一定可以成功。如果不用这个标记位，就可以把 intent-filter 中 category 不含 DEFAULT 的那些 Activity 给匹配出来，从而导致 startActivity 可能失败。因为不含有 DEFAULT 这个 category 的 Activity 是无法接收隐式 Intent 的。在 action 和 category 中，有一类 action 和 category 比较重要，它们是：

```
<action android:name="android.intent.action.MAIN" />
<category android:name="android.intent.category.LAUNCHER" />
```

这二者共同作用是用来标明这是一个入口 Activity 并且会出现在系统的应用列表中，少了任何一个都没有实际意义，也无法出现在系统的应用列表中，也就是二者缺一不可。另外，针对 Service 和 BroadcastReceiver，PackageManager 同样提供了类似的方法去获取成功匹配的组件信息。

第 2 章　IPC 机制

本章主要讲解 Android 中的 IPC 机制。首先介绍 Android 中的多进程概念以及多进程开发模式中常见的注意事项，接着介绍 Android 中的序列化机制和 Binder，然后详细介绍 Bundle、文件共享、AIDL、Messenger、ContentProvider 和 Socket 等进程间通信的方式。为了更好地使用 AIDL 来进行进程间通信，本章还引入了 Binder 连接池的概念。最后，本章讲解各种进程间通信方式的优缺点和适用场景。通过本章，可以让读者对 Android 中的 IPC 机制和多进程开发模式有深入的理解。

2.1　Android IPC 简介

IPC 是 Inter-Process Communication 的缩写，含义为进程间通信或者跨进程通信，是指两个进程之间进行数据交换的过程。说起进程间通信，我们首先要理解什么是进程，什么是线程，进程和线程是截然不同的概念。按照操作系统中的描述，线程是 CPU 调度的最小单元，同时线程是一种有限的系统资源。而进程一般指一个执行单元，在 PC 和移动设备上指一个程序或者一个应用。一个进程可以包含多个线程，因此进程和线程是包含与被包含的关系。最简单的情况下，一个进程中可以只有一个线程，即主线程，在 Android 里面主线程也叫 UI 线程，在 UI 线程里才能操作界面元素。很多时候，一个进程中需要执行大量耗时的任务，如果这些任务放在主线程中去执行就会造成界面无法响应，严重影响用户体验，这种情况在 PC 系统和移动系统中都存在，在 Android 中有一个特殊的名字叫做 ANR（Application Not Responding），即应用无响应。解决这个问题就需要用到线程，把一些耗时的任务放在线程中即可。

IPC 不是 Android 中所独有的，任何一个操作系统都需要有相应的 IPC 机制，比如 Windows 上可以通过剪贴板、管道和邮槽等来进行进程间通信；Linux 上可以通过命名管道、共享内存、信号量等来进行进程间通信。可以看到不同的操作系统平台有着不同的进程间通信方式，对于 Android 来说，它是一种基于 Linux 内核的移动操作系统，它的进程间通信方式并不能完全继承自 Linux，相反，它有自己的进程间通信方式。在 Android 中最有特色的进程间通信方式就是 Binder 了，通过 Binder 可以轻松地实现进程间通信。除了 Binder，Android 还支持 Socket，通过 Socket 也可以实现任意两个终端之间的通信，当然同一个设备上的两个进程通过 Socket 通信自然也是可以的。

说到 IPC 的使用场景就必须提到多进程，只有面对多进程这种场景下，才需要考虑进程间通信。这个是很好理解的，如果只有一个进程在运行，又何谈多进程呢？多进程的情况分为两种。第一种情况是一个应用因为某些原因自身需要采用多进程模式来实现，至于原因，可能有很多，比如有些模块由于特殊原因需要运行在单独的进程中，又或者为了加大一个应用可使用的内存所以需要通过多进程来获取多份内存空间。Android 对单个应用所使用的最大内存做了限制，早期的一些版本可能是 16MB，不同设备有不同的大小。另一种情况是当前应用需要向其他应用获取数据，由于是两个应用，所以必须采用跨进程的方式来获取所需的数据，甚至我们通过系统提供的 ContentProvider 去查询数据的时候，其实也是一种进程间通信，只不过通信细节被系统内部屏蔽了，我们无法感知而已。后续章节会详细介绍 ContentProvider 的底层实现，这里就先不做详细介绍了。总之，不管由于何种原因，我们采用了多进程的设计方法，那么应用中就必须妥善地处理进程间通信的各种问题。

2.2　Android 中的多进程模式

在正式介绍进程间通信之前，我们必须先要理解 Android 中的多进程模式。通过给四大组件指定 android:process 属性，我们可以轻易地开启多进程模式，这看起来很简单，但是实际使用过程中却暗藏杀机，多进程远远没有我们想的那么简单，有时候我们通过多进程得到的好处甚至都不足以弥补使用多进程所带来的代码层面的负面影响。下面会详细分析这些问题。

2.2.1　开启多进程模式

正常情况下，在 Android 中多进程是指一个应用中存在多个进程的情况，因此这里不讨论两个应用之间的多进程情况。首先，在 Android 中使用多进程只有一种方法，那就是给四大组件（Activity、Service、Receiver、ContentProvider）在 AndroidMenifest 中指定 android:process 属性，除此之外没有其他办法，也就是说我们无法给一个线程或者一个实体类指定其运行时所在的进程。其实还有另一种非常规的多进程方法，那就是通过 JNI 在 native 层去 fork 一个新的进程，但是这种方法属于特殊情况，也不是常用的创建多进程的方式，因此我们暂时不考虑这种方式。下面是一个示例，描述了如何在 Android 中创建多进程：

```xml
<activity
    android:name="com.ryg.chapter_2.MainActivity"
    android:configChanges="orientation|screenSize"
    android:label="@string/app_name"
    android:launchMode="standard" >
    <intent-filter>
        <action android:name="android.intent.action.MAIN" />
        <category android:name="android.intent.category.LAUNCHER" />
    </intent-filter>
</activity>
<activity
    android:name="com.ryg.chapter_2.SecondActivity"
    android:configChanges="screenLayout"
    android:label="@string/app_name"
    android:process=":remote" />
<activity
    android:name="com.ryg.chapter_2.ThirdActivity"
    android:configChanges="screenLayout"
    android:label="@string/app_name"
    android:process="com.ryg.chapter_2.remote" />
```

上面的示例分别为 SecondActivity 和 ThirdActivity 指定了 process 属性，并且它们的属性值不同，这意味着当前应用又增加了两个新进程。假设当前应用的包名为"com.ryg.chapter_2"，当 SecondActivity 启动时，系统会为它创建一个单独的进程，进程名为"com.ryg.chapter_2:remote"；当 ThirdActivity 启动时，系统也会为它创建一个单独的进程，进程名为"com.ryg.chapter_2.remote"。同时入口 Activity 是 MainActivity，没有为它指定 process 属性，那么它运行在默认进程中，默认进程的进程名是包名。下面我们运行一下看看效果，如图 2-1 所示。进程列表末尾存在 3 个进程，进程 id 分别为 645、659、672，这说明我们的应用成功地使用了多进程技术，是不是很简单呢？这只是开始，实际使用中多进程是有很多问题需要处理的。

avd4.0.3 [emulator-5554]	Online	avd4.0.3
system_process	77	8600
com.android.systemui	169	8601
com.android.inputmethod.latin	187	8602
com.android.phone	200	8603
com.android.launcher	214	8604
com.android.settings	241	8605
android.process.acore	263	8606
com.android.calendar	309	8607
com.android.deskclock	349	8610
com.android.providers.calendar	363	8612
com.android.defcontainer	393	8615
android.process.media	421	8618
com.android.exchange	440	8619
com.android.email	454	8622
com.android.mms	476	8623
com.svox.pico	526	8625
com.estrongs.android.pop	539	8628
com.android.quicksearchbox	553	8629
com.android.keychain	613	8633
com.ryg.chapter_2	645	8632
com.ryg.chapter_2:remote	659	8634
com.ryg.chapter_2.remote	672	8635

图 2-1　系统进程列表

　　除了在 Eclipse 的 DDMS 视图中查看进程信息，还可以用 shell 来查看，命令为：adb shell ps 或者 adb shell ps | grep com.ryg.chapter_2。其中 com.ryg.chapter_2 是包名，如图 2-2 所示，通过 ps 命令也可以查看一个包名中当前所存在的进程信息。

```
renyugang@dell ~
$ adb shell ps | grep com.ryg.chapter_2
app_53    645   36   113324 30112 ffffffff 400113c0 S com.ryg.chapter_2
app_53    659   36   113328 29536 ffffffff 400113c0 S com.ryg.chapter_2:remote
app_53    672   36   114968 30168 ffffffff 400113c0 S com.ryg.chapter_2.remote
```

图 2-2　通过 ps 命令来查看进程信息

　　不知道读者朋友有没有注意到，SecondActivity 和 ThirdActivity 的 android:process 属性分别为 ":remote" 和 "com.ryg.chapter_2.remote"，那么这两种方式有区别吗？其实是有区别的，区别有两方面：首先，":" 的含义是指要在当前的进程名前面附加上当前的包名，这是一种简写的方法，对于 SecondActivity 来说，它完整的进程名为 com.ryg.chapter_2:remote，这一点通过图 2-1 和 2-2 中的进程信息也能看出来，而对于 ThirdActivity 中的声明方式，它是一种完整的命名方式，不会附加包名信息；其次，进程名以"："开头的进程属于当前应用的私有进程，其他应用的组件不可以和它跑在同一个进程中，而进程名不以"："开头的进程属于全局进程，其他应用通过 ShareUID 方式可以和它跑在同一个进程中。

我们知道 Android 系统会为每个应用分配一个唯一的 UID，具有相同 UID 的应用才能共享数据。这里要说明的是，两个应用通过 ShareUID 跑在同一个进程中是有要求的，需要这两个应用有相同的 ShareUID 并且签名相同才可以。在这种情况下，它们可以互相访问对方的私有数据，比如 data 目录、组件信息等，不管它们是否跑在同一个进程中。当然如果它们跑在同一个进程中，那么除了能共享 data 目录、组件信息，还可以共享内存数据，或者说它们看起来就像是一个应用的两个部分。

2.2.2 多进程模式的运行机制

如果用一句话来形容多进程，那笔者只能这样说："当应用开启了多进程以后，各种奇怪的现象都出现了"。为什么这么说呢？这是有原因的。大部分人都认为开启多进程是很简单的事情，只需要给四大组件指定 android:process 属性即可。比如说在实际的产品开发中，可能会有多进程的需求，需要把某些组件放在单独的进程中去运行，很多人都会觉得这不很简单吗？然后迅速地给那些组件指定了 android:process 属性，然后编译运行，发现"正常地运行起来了"。这里笔者想说的是，那是真的正常地运行起来了吗？现在先不置可否，下面先给举个例子，然后引入本节的话题。还是本章刚开始说的那个例子，其中 SecondActivity 通过指定 android:process 属性从而使其运行在一个独立的进程中，这里做了一些改动，我们新建了一个类，叫做 UserManager，这个类中有一个 public 的静态成员变量，如下所示。

```
public class UserManager {
    public static int sUserId = 1;
}
```

然后在 MainActivity 的 onCreate 中我们把这个 sUserId 重新赋值为 2，打印出这个静态变量的值后再启动 SecondActivity，在 SecondActivity 中我们再打印一下 sUserId 的值。按照正常的逻辑，静态变量是可以在所有的地方共享的，并且一处有修改处处都会同步，图 2-3 是运行时所打印的日志，我们看一下结果如何。

看了图 2-3 中的日志，发现结果和我们想的完全不一致，正常情况下 SecondActivity 中打印的 sUserId 的值应该是 2 才对，但是从日志上看它竟然还是 1，可是我们的确已经在 MainActivity 中把 sUserId 重新赋值为 2 了。看到这里，大家应该明白了这就是多进程所带来的问题，多进程绝非只是仅仅指定一个 android:process 属性那么简单。

Level	PID	TID	Application	Tag	Text
D	686	686	com.ryg.chapter_2	MainActivity	UserManage.sUserId=2
D	722	722	com.ryg.chapter_2:remote	SecondActivity	onCreate
D	722	722	com.ryg.chapter_2:remote	SecondActivity	UserManage.sUserId=1

图 2-3　系统日志

上述问题出现的原因是 SecondActivity 运行在一个单独的进程中，我们知道 Android 为每一个应用分配了一个独立的虚拟机，或者说为每个进程都分配一个独立的虚拟机，不同的虚拟机在内存分配上有不同的地址空间，这就导致在不同的虚拟机中访问同一个类的对象会产生多份副本。拿我们这个例子来说，在进程 com.ryg.chapter_2 和进程 com.ryg.chapter_2:remote 中都存在一个 UserManager 类，并且这两个类是互不干扰的，在一个进程中修改 sUserId 的值只会影响当前进程，对其他进程不会造成任何影响，这样我们就可以理解为什么在 MainActivity 中修改了 sUserId 的值，但是在 SecondActivity 中 sUserId 的值却没有发生改变这个现象。

所有运行在不同进程中的四大组件，只要它们之间需要通过内存来共享数据，都会共享失败，这也是多进程所带来的主要影响。正常情况下，四大组件中间不可能不通过一些中间层来共享数据，那么通过简单地指定进程名来开启多进程都会无法正确运行。当然，特殊情况下，某些组件之间不需要共享数据，这个时候可以直接指定 android:process 属性来开启多进程，但是这种场景是不常见的，几乎所有情况都需要共享数据。

一般来说，使用多进程会造成如下几方面的问题：

（1）静态成员和单例模式完全失效。

（2）线程同步机制完全失效。

（3）SharedPreferences 的可靠性下降。

（4）Application 会多次创建。

第 1 个问题在上面已经进行了分析。第 2 个问题本质上和第一个问题是类似的，既然都不是一块内存了，那么不管是锁对象还是锁全局类都无法保证线程同步，因为不同进程锁的不是同一个对象。第 3 个问题是因为 SharedPreferences 不支持两个进程同时去执行写操作，否则会导致一定几率的数据丢失，这是因为 SharedPreferences 底层是通过读/写 XML 文件来实现的，并发写显然是可能出问题的，甚至并发读/写都有可能出问题。第 4 个问题也是显而易见的,当一个组件跑在一个新的进程中的时候,由于系统要在创建新的进程同时分配独立的虚拟机,所以这个过程其实就是启动一个应用的过程。因此,相当于系统又把这个应用重新启动了一遍,既然重新启动了,那么自然会创建新的 Application。这个问题其实可以这么理解,运行在同一个进程中的组件是属于同一个虚拟机和同一个 Application 的,同理,运行在不同进程中的组件是属于两个不同的虚拟机和 Application 的。为了更加清晰地展示这一点,下面我们来做一个测试,首先在 Application 的 onCreate 方法中打印出当前进程的名字,然后连续启动三个同一个应用内但属于不同进程的 Activity,按照期望,Application 的 onCreate 应该执行三次并打印出三次进程名不同的 log,代码如下所示。

```java
public class MyApplication extends Application {

    private static final String TAG = "MyApplication";

    @Override
    public void onCreate() {
        super.onCreate();
        String processName = MyUtils.getProcessName(getApplicationContext(),
        Process.myPid());
        Log.d(TAG, "application start, process name:" + processName);
    }

}
```

运行后看一下 log,如图 2-4 所示。通过 log 可以看出,Application 执行了三次 onCreate,并且每次的进程名称和进程 id 都不一样,它们的进程名和我们为 Activity 指定的 android:process 属性一致。这也就证实了在多进程模式中,不同进程的组件的确会拥有独立的虚拟机、Application 以及内存空间,这会给实际的开发带来很多困扰,是尤其需要注意的。或者我们也可以这么理解同一个应用间的多进程:它就相当于两个不同的应用采用了 SharedUID 的模式,这样能够更加直接地理解多进程模式的本质。

Level	PID	Tag	Text
D	28299	MyApplication	application start, process name:com.ryg.chapter_2
D	28358	MyApplication	application start, process name:com.ryg.chapter_2:remote
D	28385	MyApplication	application start, process name:com.ryg.chapter_2.remote

图 2-4 系统日志

本节我们分析了多进程所带来的问题，但是我们不能因为多进程有很多问题就不去正视它。为了解决这个问题，系统提供了很多跨进程通信方法，虽然说不能直接地共享内存，但是通过跨进程通信我们还是可以实现数据交互。实现跨进程通信的方式很多，比如通过Intent 来传递数据，共享文件和 SharedPreferences，基于 Binder 的 Messenger 和 AIDL 以及Socket 等，但是为了更好地理解各种 IPC 方式，我们需要先熟悉一些基础概念，比如序列化相关的 Serializable 和 Parcelable 接口，以及 Binder 的概念，熟悉完这些基础概念以后，再去理解各种 IPC 方式就比较简单了。

2.3 IPC 基础概念介绍

本节主要介绍 IPC 中的一些基础概念，主要包含三方面内容：Serializable 接口、Parcelable 接口以及 Binder，只有熟悉这三方面的内容后，我们才能更好地理解跨进程通信的各种方式。Serializable 和 Parcelable 接口可以完成对象的序列化过程，当我们需要通过Intent 和 Binder 传输数据时就需要使用 Parcelable 或者 Serializable。还有的时候我们需要把对象持久化到存储设备上或者通过网络传输给其他客户端，这个时候也需要使用Serializable 来完成对象的持久化，下面先介绍如何使用 Serializable 来完成对象的序列化。

2.3.1 Serializable 接口

Serializable 是 Java 所提供的一个序列化接口，它是一个空接口，为对象提供标准的序列化和反序列化操作。使用 Serializable 来实现序列化相当简单，只需要在类的声明中指定一个类似下面的标识即可自动实现默认的序列化过程。

```
private static final long serialVersionUID = 8711368828010083044L
```

在 Android 中也提供了新的序列化方式，那就是 Parcelable 接口，使用 Parcelable 来实现对象的序列化，其过程要稍微复杂一些，本节先介绍 Serializable 接口。上面提到，想让一个对象实现序列化，只需要这个类实现 Serializable 接口并声明一个 serialVersionUID 即可，实际上，甚至这个 serialVersionUID 也不是必需的，我们不声明这个 serialVersionUID 同样也可以实现序列化，但是这将会对反序列化过程产生影响，具体什么影响后面再介绍。User 类就是一个实现了 Serializable 接口的类，它是可以被序列化和反序列化的，如下所示。

```
public class User implements Serializable {
```

```
    private static final long serialVersionUID = 5190671237212957731L;

    public int userId;
    public String userName;
    public boolean isMale;
    ...
}
```

通过 Serializable 方式来实现对象的序列化，实现起来非常简单，几乎所有工作都被系统自动完成了。如何进行对象的序列化和反序列化也非常简单，只需要采用 ObjectOutputStream 和 ObjectInputStream 即可轻松实现。下面举个简单的例子。

```
//序列化过程
User user = new User(0, "jake", true);
ObjectOutputStream out = new ObjectOutputStream(
        new FileOutputStream("cache.txt"));
out.writeObject(user);
out.close();

//反序列化过程
ObjectInputStream in = new ObjectInputStream(
        new FileInputStream("cache.txt"));
User newUser = (User) in.readObject();
in.close();
```

上述代码演示了采用 Serializable 方式序列化对象的典型过程，很简单，只需要把实现了 Serializable 接口的 User 对象写到文件中就可以快速恢复了，恢复后的对象 newUser 和 user 的内容完全一样，但是两者并不是同一个对象。

刚开始提到，即使不指定 serialVersionUID 也可以实现序列化，那到底要不要指定呢？如果指定的话，serialVersionUID 后面那一长串数字又是什么含义呢？我们要明白，系统既然提供了这个 serialVersionUID，那么它必须是有用的。这个 serialVersionUID 是用来辅助序列化和反序列化过程的，原则上序列化后的数据中的 serialVersionUID 只有和当前类的 serialVersionUID 相同才能够正常地被反序列化。serialVersionUID 的详细工作机制是这样的：序列化的时候系统会把当前类的 serialVersionUID 写入序列化的文件中（也可能是其他中介），当反序列化的时候系统会去检测文件中的 serialVersionUID，看它是否和当前类的 serialVersionUID 一致，如果一致就说明序列化的类的版本和当前类的版本是相同的，这个时候可以成功反序列化；否则就说明当前类和序列化的类相比发生了某些变换，比如成员变量的数量、类型可能发生了改变，这个时候是无法正常反序列化的，因此会报如下错误：

```
java.io.InvalidClassException: Main; local class incompatible: stream
classdesc serialVersionUID = 8711368828010083044, local class serial-
VersionUID = 8711368828010083043.
```

一般来说，我们应该手动指定 serialVersionUID 的值，比如 1L，也可以让 Eclipse 根据当前类的结构自动去生成它的 hash 值，这样序列化和反序列化时两者的 serialVersionUID 是相同的，因此可以正常进行反序列化。如果不手动指定 serialVersionUID 的值，反序列化时当前类有所改变，比如增加或者删除了某些成员变量，那么系统就会重新计算当前类的 hash 值并把它赋值给 serialVersionUID，这个时候当前类的 serialVersionUID 就和序列化的数据中的 serialVersionUID 不一致，于是反序列化失败，程序就会出现 crash。所以，我们可以明显感觉到 serialVersionUID 的作用，当我们手动指定了它以后，就可以在很大程度上避免反序列化过程的失败。比如当版本升级后，我们可能删除了某个成员变量也可能增加了一些新的成员变量，这个时候我们的反向序列化过程仍然能够成功，程序仍然能够最大限度地恢复数据，相反，如果不指定 serialVersionUID 的话，程序则会挂掉。当然我们还要考虑另外一种情况，如果类结构发生了非常规性改变，比如修改了类名，修改了成员变量的类型，这个时候尽管 serialVersionUID 验证通过了，但是反序列化过程还是会失败，因为类结构有了毁灭性的改变，根本无法从老版本的数据中还原出一个新的类结构的对象。

根据上面的分析，我们可以知道，给 serialVersionUID 指定为 1L 或者采用 Eclipse 根据当前类结构去生成的 hash 值，这两者并没有本质区别，效果完全一样。以下两点需要特别提一下，首先静态成员变量属于类不属于对象，所以不会参与序列化过程；其次用 transient 关键字标记的成员变量不参与序列化过程。

另外，系统的默认序列化过程也是可以改变的，通过实现如下两个方法即可重写系统默认的序列化和反序列化过程，具体怎么去重写这两个方法就是很简单的事了，这里就不再详细介绍了，毕竟这不是本章的重点，而且大部分情况下我们不需要重写这两个方法。

```
private void writeObject(java.io.ObjectOutputStream out)
  throws IOException {
 // write 'this' to 'out'...
}

private void readObject(java.io.ObjectInputStream in)
   throws IOException, ClassNotFoundException {
```

```
    // populate the fields of 'this' from the data in 'in'...
}
```

2.3.2　Parcelable 接口

上一节我们介绍了通过 Serializable 方式来实现序列化的方法，本节接着介绍另一种序列化方式：Parcelable。Parcelable 也是一个接口，只要实现这个接口，一个类的对象就可以实现序列化并可以通过 Intent 和 Binder 传递。下面的示例是一个典型的用法。

```
public class User implements Parcelable {

    public int userId;
    public String userName;
    public boolean isMale;

    public Book book;

    public User(int userId, String userName, boolean isMale) {
        this.userId = userId;
        this.userName = userName;
        this.isMale = isMale;
    }

    public int describeContents() {
        return 0;
    }

    public void writeToParcel(Parcel out, int flags) {
        out.writeInt(userId);
        out.writeString(userName);
        out.writeInt(isMale ? 1 : 0);
        out.writeParcelable(book, 0);
    }

    public static final Parcelable.Creator<User> CREATOR = new Parcelable.Creator<User>() {
        public User createFromParcel(Parcel in) {
            return new User(in);
        }
```

```
        public User[] newArray(int size) {
            return new User[size];
        }
    };

    private User(Parcel in) {
        userId = in.readInt();
        userName = in.readString();
        isMale = in.readInt() == 1;
        book = in.readParcelable(Thread.currentThread().getContextClass-
        Loader());
    }
}
```

这里先说一下 Parcel，Parcel 内部包装了可序列化的数据，可以在 Binder 中自由传输。从上述代码中可以看出，在序列化过程中需要实现的功能有序列化、反序列化和内容描述。序列化功能由 writeToParcel 方法来完成，最终是通过 Parcel 中的一系列 write 方法来完成的；反序列化功能由 CREATOR 来完成，其内部标明了如何创建序列化对象和数组，并通过 Parcel 的一系列 read 方法来完成反序列化过程；内容描述功能由 describeContents 方法来完成，几乎在所有情况下这个方法都应该返回 0，仅当当前对象中存在文件描述符时，此方法返回 1。需要注意的是，在 User(Parcel in) 方法中，由于 book 是另一个可序列化对象，所以它的反序列化过程需要传递当前线程的上下文类加载器，否则会报无法找到类的错误。详细的方法说明请参看表 2-1。

表 2-1 Parcelable 的方法说明

方　　法	功　　能	标　记　位
createFromParcel(Parcel in)	从序列化后的对象中创建原始对象	
newArray(int size)	创建指定长度的原始对象数组	
User(Parcel in)	从序列化后的对象中创建原始对象	
writeToParcel (Parcel out, int flags)	将当前对象写入序列化结构中，其中 flags 标识有两种值：0 或者 1（参见右侧标记位）。为 1 时标识当前对象需要作为返回值返回，不能立即释放资源，几乎所有情况都为 0	PARCELABLE_WRITE_RETURN_VALUE
describeContents	返回当前对象的内容描述。如果含有文件描述符，返回 1（参见右侧标记位），否则返回 0，几乎所有情况都返回 0	CONTENTS_FILE_DESCRIPTOR

系统已经为我们提供了许多实现了 Parcelable 接口的类，它们都是可以直接序列化的，

比如 Intent、Bundle、Bitmap 等，同时 List 和 Map 也可以序列化，前提是它们里面的每个元素都是可序列化的。

既然 Parcelable 和 Serializable 都能实现序列化并且都可用于 Intent 间的数据传递，那么二者该如何选取呢？Serializable 是 Java 中的序列化接口，其使用起来简单但是开销很大，序列化和反序列化过程需要大量 I/O 操作。而 Parcelable 是 Android 中的序列化方式，因此更适合用在 Android 平台上，它的缺点就是使用起来稍微麻烦点，但是它的效率很高，这是 Android 推荐的序列化方式，因此我们要首选 Parcelable。Parcelable 主要用在内存序列化上，通过 Parcelable 将对象序列化到存储设备中或者将对象序列化后通过网络传输也都是可以的，但是这个过程会稍显复杂，因此在这两种情况下建议大家使用 Serializable。以上就是 Parcelable 和 Serializable 的区别。

2.3.3　Binder

Binder 是一个很深入的话题，笔者也看过一些别人写的 Binder 相关的文章，发现很少有人能把它介绍清楚，不是深入代码细节不能自拔，就是长篇大论不知所云，看完后都是晕晕的感觉。所以，本节笔者不打算深入探讨 Binder 的底层细节，因为 Binder 太复杂了。本节的侧重点是介绍 Binder 的使用以及上层原理，为接下来的几节内容做铺垫。

直观来说，Binder 是 Android 中的一个类，它实现了 IBinder 接口。从 IPC 角度来说，Binder 是 Android 中的一种跨进程通信方式，Binder 还可以理解为一种虚拟的物理设备，它的设备驱动是/dev/binder，该通信方式在 Linux 中没有；从 Android Framework 角度来说，Binder 是 ServiceManager 连接各种 Manager（ActivityManager、WindowManager，等等）和相应 ManagerService 的桥梁；从 Android 应用层来说，Binder 是客户端和服务端进行通信的媒介，当 bindService 的时候，服务端会返回一个包含了服务端业务调用的 Binder 对象，通过这个 Binder 对象，客户端就可以获取服务端提供的服务或者数据，这里的服务包括普通服务和基于 AIDL 的服务。

Android 开发中，Binder 主要用在 Service 中，包括 AIDL 和 Messenger，其中普通 Service 中的 Binder 不涉及进程间通信，所以较为简单，无法触及 Binder 的核心，而 Messenger 的底层其实是 AIDL，所以这里选择用 AIDL 来分析 Binder 的工作机制。为了分析 Binder 的工作机制，我们需要新建一个 AIDL 示例，SDK 会自动为我们生产 AIDL 所对应的 Binder 类，然后我们就可以分析 Binder 的工作过程。还是采用本章开始时用的例子，新建 Java 包 com.ryg.chapter_2.aidl，然后新建三个文件 Book.java、Book.aidl 和 IBookManager.aidl，代码如下所示。

```java
//Book.java
package com.ryg.chapter_2.aidl;

import android.os.Parcel;
import android.os.Parcelable;

public class Book implements Parcelable {
    public int bookId;
    public String bookName;

    public Book(int bookId, String bookName) {
        this.bookId = bookId;
        this.bookName = bookName;
    }

    public int describeContents() {
        return 0;
    }

    public void writeToParcel(Parcel out, int flags) {
        out.writeInt(bookId);
        out.writeString(bookName);
    }

    public static final Parcelable.Creator<Book> CREATOR = new Parcelable.Creator<Book>() {
        public Book createFromParcel(Parcel in) {
            return new Book(in);
        }

        public Book[] newArray(int size) {
            return new Book[size];
        }
    };

    private Book(Parcel in) {
        bookId = in.readInt();
        bookName = in.readString();
    }
}
```

```
//Book.aidl
package com.ryg.chapter_2.aidl;

parcelable Book;

// IBookManager.aidl
package com.ryg.chapter_2.aidl;

import com.ryg.chapter_2.aidl.Book;
interface IBookManager {
    List<Book> getBookList();
    void addBook(in Book book);
}
```

上面三个文件中，Book.java 是一个表示图书信息的类，它实现了 Parcelable 接口。Book.aidl 是 Book 类在 AIDL 中的声明。IBookManager.aidl 是我们定义的一个接口，里面有两个方法：getBookList 和 addBook，其中 getBookList 用于从远程服务端获取图书列表，而 addBook 用于往图书列表中添加一本书，当然这两个方法主要是示例用，不一定要有实际意义。我们可以看到，尽管 Book 类已经和 IBookManager 位于相同的包中，但是在 IBookManager 中仍然要导入 Book 类，这就是 AIDL 的特殊之处。下面我们先看一下系统为 IBookManager.aidl 生成的 Binder 类，在 gen 目录下的 com.ryg.chapter_2.aidl 包中有一个 IBookManager.java 的类，这就是我们要找的类。接下来我们需要根据这个系统生成的 Binder 类来分析 Binder 的工作原理，代码如下：

```
//IBookManager.java
/*
 * This file is auto-generated.  DO NOT MODIFY.
 * Original file: E:\\workspace\\Chapter_2\\src\\com\\ryg\\chapter_2\\
   aidl\\IBookManager.aidl
 */
package com.ryg.chapter_2.aidl;

public interface IBookManager extends android.os.IInterface {

    /** Local-side IPC implementation stub class. */
    public static abstract class Stub extends android.os.Binder implements
            com.ryg.chapter_2.aidl.IBookManager {
```

```java
private static final java.lang.String DESCRIPTOR = "com.ryg.chapter_2.aidl.IBookManager";

/** Construct the stub at attach it to the interface. */
public Stub() {
    this.attachInterface(this, DESCRIPTOR);
}

/**
 * Cast an IBinder object into an com.ryg.chapter_2.aidl.IBookManager
 * interface, generating a proxy if needed.
 */
public static com.ryg.chapter_2.aidl.IBookManager asInterface(
        android.os.IBinder obj) {
    if ((obj == null)) {
        return null;
    }
    android.os.IInterface iin = obj.queryLocalInterface(DESCRIPTOR);
    if (((iin != null) && (iin instanceof com.ryg.chapter_2.aidl.IBookManager))) {
        return ((com.ryg.chapter_2.aidl.IBookManager) iin);
    }
    return new com.ryg.chapter_2.aidl.IBookManager.Stub.Proxy(obj);
}

@Override
public android.os.IBinder asBinder() {
    return this;
}

@Override
public boolean onTransact(int code, android.os.Parcel data,
        android.os.Parcel reply, int flags) throws android.os.RemoteException {
    switch (code) {
    case INTERFACE_TRANSACTION: {
        reply.writeString(DESCRIPTOR);
        return true;
    }
    case TRANSACTION_getBookList: {
```

```java
            data.enforceInterface(DESCRIPTOR);
            java.util.List<com.ryg.chapter_2.aidl.Book> _result = this.
            getBookList();
            reply.writeNoException();
            reply.writeTypedList(_result);
            return true;
        }
        case TRANSACTION_addBook: {
            data.enforceInterface(DESCRIPTOR);
            com.ryg.chapter_2.aidl.Book _arg0;
            if ((0 != data.readInt())) {
                _arg0 = com.ryg.chapter_2.aidl.Book.CREATOR.create-
                FromParcel(data);
            } else {
                _arg0 = null;
            }
            this.addBook(_arg0);
            reply.writeNoException();
            return true;
        }
    }
    return super.onTransact(code, data, reply, flags);
}

private static class Proxy implements com.ryg.chapter_2.aidl.
IBookManager {
    private android.os.IBinder mRemote;

    Proxy(android.os.IBinder remote) {
        mRemote = remote;
    }

    @Override
    public android.os.IBinder asBinder() {
        return mRemote;
    }

    public java.lang.String getInterfaceDescriptor() {
        return DESCRIPTOR;
    }
```

```java
@Override
public java.util.List<com.ryg.chapter_2.aidl.Book> getBookList()
        throws android.os.RemoteException {
    android.os.Parcel _data = android.os.Parcel.obtain();
    android.os.Parcel _reply = android.os.Parcel.obtain();
    java.util.List<com.ryg.chapter_2.aidl.Book> _result;
    try {
        _data.writeInterfaceToken(DESCRIPTOR);
        mRemote.transact(Stub.TRANSACTION_getBookList, _data,
                _reply, 0);
        _reply.readException();
        _result = _reply
                .createTypedArrayList(com.ryg.chapter_2.aidl.
                Book.CREATOR);
    } finally {
        _reply.recycle();
        _data.recycle();
    }
    return _result;
}

@Override
public void addBook(com.ryg.chapter_2.aidl.Book book)
        throws android.os.RemoteException {
    android.os.Parcel _data = android.os.Parcel.obtain();
    android.os.Parcel _reply = android.os.Parcel.obtain();
    try {
        _data.writeInterfaceToken(DESCRIPTOR);
        if ((book != null)) {
            _data.writeInt(1);
            book.writeToParcel(_data, 0);
        } else {
            _data.writeInt(0);
        }
        mRemote.transact(Stub.TRANSACTION_addBook,_data,_reply,0);
        _reply.readException();
    } finally {
        _reply.recycle();
        _data.recycle();
```

```
            }
        }
    }

    static final int TRANSACTION_getBookList = (android.os.IBinder.
    FIRST_CALL_TRANSACTION + 0);
    static final int TRANSACTION_addBook = (android.os.IBinder.FIRST_
    CALL_TRANSACTION + 1);
}

public java.util.List<com.ryg.chapter_2.aidl.Book> getBookList()
        throws android.os.RemoteException;

public void addBook(com.ryg.chapter_2.aidl.Book book)
        throws android.os.RemoteException;
}
```

上述代码是系统生成的，为了方便查看笔者稍微做了一下格式上的调整。在 gen 目录下，可以看到根据 IBookManager.aidl 系统为我们生成了 IBookManager.java 这个类，它继承了 IInterface 这个接口，同时它自己也还是个接口，所有可以在 Binder 中传输的接口都需要继承 IInterface 接口。这个类刚开始看起来逻辑混乱，但是实际上还是很清晰的，通过它我们可以清楚地了解到 Binder 的工作机制。这个类的结构其实很简单，首先，它声明了两个方法 getBookList 和 addBook，显然这就是我们在 IBookManager.aidl 中所声明的方法，同时它还声明了两个整型的 id 分别用于标识这两个方法，这两个 id 用于标识在 transact 过程中客户端所请求的到底是哪个方法。接着，它声明了一个内部类 Stub，这个 Stub 就是一个 Binder 类，当客户端和服务端都位于同一个进程时，方法调用不会走跨进程的 transact 过程，而当两者位于不同进程时，方法调用需要走 transact 过程，这个逻辑由 Stub 的内部代理类 Proxy 来完成。这么来看，IBookManager 这个接口的确很简单，但是我们也应该认识到，这个接口的核心实现就是它的内部类 Stub 和 Stub 的内部代理类 Proxy，下面详细介绍针对这两个类的每个方法的含义。

DESCRIPTOR

Binder 的唯一标识，一般用当前 Binder 的类名表示，比如本例中的 "com.ryg.chapter_2.aidl.IBookManager"。

asInterface(android.os.IBinder obj)

用于将服务端的 Binder 对象转换成客户端所需的 AIDL 接口类型的对象,这种转换过程是区分进程的,如果客户端和服务端位于同一进程,那么此方法返回的就是服务端的 Stub 对象本身,否则返回的是系统封装后的 Stub.proxy 对象。

asBinder

此方法用于返回当前 Binder 对象。

onTransact

这个方法运行在服务端中的 Binder 线程池中,当客户端发起跨进程请求时,远程请求会通过系统底层封装后交由此方法来处理。该方法的原型为 public Boolean onTransact (int code, android.os.Parcel data, android.os.Parcel reply, int flags)。服务端通过 code 可以确定客户端所请求的目标方法是什么,接着从 data 中取出目标方法所需的参数(如果目标方法有参数的话),然后执行目标方法。当目标方法执行完毕后,就向 reply 中写入返回值(如果目标方法有返回值的话),onTransact 方法的执行过程就是这样的。需要注意的是,如果此方法返回 false,那么客户端的请求会失败,因此我们可以利用这个特性来做权限验证,毕竟我们也不希望随便一个进程都能远程调用我们的服务。

Proxy#getBookList

这个方法运行在客户端,当客户端远程调用此方法时,它的内部实现是这样的:首先创建该方法所需要的输入型 Parcel 对象_data、输出型 Parcel 对象_reply 和返回值对象 List;然后把该方法的参数信息写入_data 中(如果有参数的话);接着调用 transact 方法来发起 RPC(远程过程调用)请求,同时当前线程挂起;然后服务端的 onTransact 方法会被调用,直到 RPC 过程返回后,当前线程继续执行,并从_reply 中取出 RPC 过程的返回结果;最后返回_reply 中的数据。

Proxy#addBook

这个方法运行在客户端,它的执行过程和 getBookList 是一样的,addBook 没有返回值,所以它不需要从_reply 中取出返回值。

通过上面的分析,读者应该已经了解了 Binder 的工作机制,但是有两点还是需要额外说明一下:首先,当客户端发起远程请求时,由于当前线程会被挂起直至服务端进程返回数据，所以如果一个远程方法是很耗时的，那么不能在 UI 线程中发起此远程请求；其次，由于服务端的 Binder 方法运行在 Binder 的线程池中，所以 Binder 方法不管是否耗时都应该采用同步的方式去实现，因为它已经运行在一个线程中了。为了更好地说明 Binder，下面给出一个 Binder 的工作机制图，如图 2-5 所示。

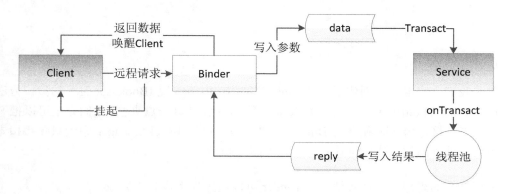

图 2-5　Binder 的工作机制

从上述分析过程来看，我们完全可以不提供 AIDL 文件即可实现 Binder，之所以提供 AIDL 文件，是为了方便系统为我们生成代码。系统根据 AIDL 文件生成 Java 文件的格式是固定的，我们可以抛开 AIDL 文件直接写一个 Binder 出来，接下来我们就介绍如何手动写一个 Binder。还是上面的例子，但是这次我们不提供 AIDL 文件。参考上面系统自动生成的 IBookManager.java 这个类的代码，可以发现这个类是相当有规律的，根据它的特点，我们完全可以自己写一个和它一模一样的类出来，然后这个不借助 AIDL 文件的 Binder 就完成了。但是我们发现系统生成的类看起来结构不清晰，我们想试着对它进行结构上的调整，可以发现这个类主要由两部分组成，首先它本身是一个 Binder 的接口（继承了 IInterface），其次它的内部有个 Stub 类，这个类就是个 Binder。还记得我们怎么写一个 Binder 的服务端吗？代码如下所示。

```
private final IBookManager.Stub mBinder = new IBookManager.Stub() {
    @Override
    public List<Book> getBookList() throws RemoteException {
        synchronized (mBookList) {
            return mBookList;
        }
    }
```

```
    @Override
    public void addBook(Book book) throws RemoteException {
        synchronized (mBookList) {
            if (!mBookList.contains(book)) {
                mBookList.add(book);
            }
        }
    }
}
```

首先我们会实现一个创建了一个 Stub 对象并在内部实现 IBookManager 的接口方法，然后在 Service 的 onBind 中返回这个 Stub 对象。因此，从这一点来看，我们完全可以把 Stub 类提取出来直接作为一个独立的 Binder 类来实现，这样 IBookManager 中就只剩接口本身了，通过这种分离的方式可以让它的结构变得清晰点。

根据上面的思想，手动实现一个 Binder 可以通过如下步骤来完成：

（1）声明一个 AIDL 性质的接口，只需要继承 IInterface 接口即可，IInterface 接口中只有一个 asBinder 方法。这个接口的实现如下：

```
public interface IBookManager extends IInterface {

    static final String DESCRIPTOR = "com.ryg.chapter_2.manualbinder.IBookManager";

    static final int TRANSACTION_getBookList = IBinder.FIRST_CALL_TRANSACTION + 0;
    static final int TRANSACTION_addBook = IBinder.FIRST_CALL_TRANSACTION + 1;

    public List<Book> getBookList() throws RemoteException;

    public void addBook(Book book) throws RemoteException;
}
```

可以看到，在接口中声明了一个 Binder 描述符和另外两个 id，这两个 id 分别表示的是 getBookList 和 addBook 方法，这段代码原本也是系统生成的，我们仿照系统生成的规则去手动书写这部分代码。如果我们有三个方法，应该怎么做呢？很显然，我们要再声明一个

id，然后按照固定模式声明这个新方法即可，这个比较好理解，不再多说。

（2）实现 Stub 类和 Stub 类中的 Proxy 代理类，这段代码我们可以自己写，但是写出来后会发现和系统自动生成的代码是一样的，因此这个 Stub 类我们只需要参考系统生成的代码即可，只是结构上需要做一下调整，调整后的代码如下所示。

```java
public class BookManagerImpl extends Binder implements IBookManager {

    /** Construct the stub at attach it to the interface. */
    public BookManagerImpl() {
        this.attachInterface(this, DESCRIPTOR);
    }

    /**
     * Cast an IBinder object into an IBookManager interface, generating a proxy
     * if needed.
     */
    public static IBookManager asInterface(IBinder obj) {
        if ((obj == null)) {
            return null;
        }
        android.os.IInterface iin = obj.queryLocalInterface(DESCRIPTOR);
        if (((iin != null) && (iin instanceof IBookManager))) {
            return ((IBookManager) iin);
        }
        return new BookManagerImpl.Proxy(obj);
    }

    @Override
    public IBinder asBinder() {
        return this;
    }

    @Override
    public boolean onTransact(int code, Parcel data, Parcel reply, int flags)
            throws RemoteException {
        switch (code) {
        case INTERFACE_TRANSACTION: {
```

```java
            reply.writeString(DESCRIPTOR);
            return true;
        }
        case TRANSACTION_getBookList: {
            data.enforceInterface(DESCRIPTOR);
            List<Book> result = this.getBookList();
            reply.writeNoException();
            reply.writeTypedList(result);
            return true;
        }
        case TRANSACTION_addBook: {
            data.enforceInterface(DESCRIPTOR);
            Book arg0;
            if ((0 != data.readInt())) {
                arg0 = Book.CREATOR.createFromParcel(data);
            } else {
                arg0 = null;
            }
            this.addBook(arg0);
            reply.writeNoException();
            return true;
        }
    }
    return super.onTransact(code, data, reply, flags);
}

@Override
public List<Book> getBookList() throws RemoteException {
    // TODO 待实现
    return null;
}

@Override
public void addBook(Book book) throws RemoteException {
    // TODO 待实现
}

private static class Proxy implements IBookManager {
    private IBinder mRemote;
```

```java
Proxy(IBinder remote) {
    mRemote = remote;
}

@Override
public IBinder asBinder() {
    return mRemote;
}

public java.lang.String getInterfaceDescriptor() {
    return DESCRIPTOR;
}

@Override
public List<Book> getBookList() throws RemoteException {
    Parcel data = Parcel.obtain();
    Parcel reply = Parcel.obtain();
    List<Book> result;
    try {
        data.writeInterfaceToken(DESCRIPTOR);
        mRemote.transact(TRANSACTION_getBookList, data, reply, 0);
        reply.readException();
        result = reply.createTypedArrayList(Book.CREATOR);
    } finally {
        reply.recycle();
        data.recycle();
    }
    return result;
}

@Override
public void addBook(Book book) throws RemoteException {
    Parcel data = Parcel.obtain();
    Parcel reply = Parcel.obtain();
    try {
        data.writeInterfaceToken(DESCRIPTOR);
        if ((book != null)) {
            data.writeInt(1);
            book.writeToParcel(data, 0);
        } else {
```

```
            data.writeInt(0);
        }
        mRemote.transact(TRANSACTION_addBook, data, reply, 0);
        reply.readException();
    } finally {
        reply.recycle();
        data.recycle();
    }
  }
}
```

通过将上述代码和系统生成的代码对比，可以发现简直是一模一样的。也许有人会问：既然和系统生成的一模一样，那我们为什么要手动去写呢？我们在实际开发中完全可以通过 AIDL 文件让系统去自动生成，手动去写的意义在于可以让我们更加理解 Binder 的工作原理，同时也提供了一种不通过 AIDL 文件来实现 Binder 的新方式。也就是说，AIDL 文件并不是实现 Binder 的必需品。如果是我们手写的 Binder，那么在服务端只需要创建一个 BookManagerImpl 的对象并在 Service 的 onBind 方法中返回即可。最后，是否手动实现 Binder 没有本质区别，二者的工作原理完全一样，AIDL 文件的本质是系统为我们提供了一种快速实现 Binder 的工具，仅此而已。

接下来，我们介绍 Binder 的两个很重要的方法 linkToDeath 和 unlinkToDeath。我们知道，Binder 运行在服务端进程，如果服务端进程由于某种原因异常终止，这个时候我们到服务端的 Binder 连接断裂（称之为 Binder 死亡），会导致我们的远程调用失败。更为关键的是，如果我们不知道 Binder 连接已经断裂，那么客户端的功能就会受到影响。为了解决这个问题，Binder 中提供了两个配对的方法 linkToDeath 和 unlinkToDeath，通过 linkToDeath 我们可以给 Binder 设置一个死亡代理，当 Binder 死亡时，我们就会收到通知，这个时候我们就可以重新发起连接请求从而恢复连接。那么到底如何给 Binder 设置死亡代理呢？也很简单。

首先，声明一个 DeathRecipient 对象。DeathRecipient 是一个接口，其内部只有一个方法 binderDied，我们需要实现这个方法，当 Binder 死亡的时候，系统就会回调 binderDied 方法，然后我们就可以移出之前绑定的 binder 代理并重新绑定远程服务：

```
private IBinder.DeathRecipient mDeathRecipient = new IBinder.Death-
Recipient() {
```

```
    @Override
    public void binderDied() {
        if (mBookManager == null)
            return;
        mBookManager.asBinder().unlinkToDeath(mDeathRecipient, 0);
        mBookManager = null;
        // TODO: 这里重新绑定远程 Service
    }
};
```

其次，在客户端绑定远程服务成功后，给 binder 设置死亡代理：

```
mService = IMessageBoxManager.Stub.asInterface(binder);
binder.linkToDeath(mDeathRecipient, 0);
```

其中 linkToDeath 的第二个参数是个标记位，我们直接设为 0 即可。经过上面两个步骤，就给我们的 Binder 设置了死亡代理，当 Binder 死亡的时候我们就可以收到通知了。另外，通过 Binder 的方法 isBinderAlive 也可以判断 Binder 是否死亡。

到这里，IPC 的基础知识就介绍完毕了，下面开始进入正题，直面形形色色的进程间通信方式。

2.4　Android 中的 IPC 方式

在上节中，我们介绍了 IPC 的几个基础知识：序列化和 Binder，本节开始详细分析各种跨进程通信方式。具体方式有很多，比如可以通过在 Intent 中附加 extras 来传递信息，或者通过共享文件的方式来共享数据，还可以采用 Binder 方式来跨进程通信，另外，ContentProvider 天生就是支持跨进程访问的，因此我们也可以采用它来进行 IPC。此外，通过网络通信也是可以实现数据传递的，所以 Socket 也可以实现 IPC。上述所说的各种方法都能实现 IPC，它们在使用方法和侧重点上都有很大的区别，下面会一一进行展开。

2.4.1　使用 Bundle

我们知道，四大组件中的三大组件（Activity、Service、Receiver）都是支持在 Intent 中传递 Bundle 数据的，由于 Bundle 实现了 Parcelable 接口，所以它可以方便地在不同的进程间传输。基于这一点，当我们在一个进程中启动了另一个进程的 Activity、Service 和 Receiver，我们就可以在 Bundle 中附加我们需要传输给远程进程的信息并通过 Intent 发送出去。当然，我们传输的数据必须能够被序列化，比如基本类型、实现了 Parcelable 接口的对象、实现了 Serializable 接口的对象以及一些 Android 支持的特殊对象，具体内容可以看 Bundle 这个类，就可以看到所有它支持的类型。Bundle 不支持的类型我们无法通过它在进程间传递数据，这个很简单，就不再详细介绍了。这是一种最简单的进程间通信方式，

除了直接传递数据这种典型的使用场景，它还有一种特殊的使用场景。比如 A 进程正在进行一个计算，计算完成后它要启动 B 进程的一个组件并把计算结果传递给 B 进程，可是遗憾的是这个计算结果不支持放入 Bundle 中，因此无法通过 Intent 来传输，这个时候如果我们用其他 IPC 方式就会略显复杂。可以考虑如下方式：我们通过 Intent 启动进程 B 的一个 Service 组件（比如 IntentService），让 Service 在后台进行计算，计算完毕后再启动 B 进程中真正要启动的目标组件，由于 Service 也运行在 B 进程中，所以目标组件就可以直接获取计算结果，这样一来就轻松解决了跨进程的问题。这种方式的核心思想在于将原本需要在 A 进程的计算任务转移到 B 进程的后台 Service 中去执行，这样就成功地避免了进程间通信问题，而且只用了很小的代价。

2.4.2 使用文件共享

共享文件也是一种不错的进程间通信方式，两个进程通过读/写同一个文件来交换数据，比如 A 进程把数据写入文件，B 进程通过读取这个文件来获取数据。我们知道，在 Windows 上，一个文件如果被加了排斥锁将会导致其他线程无法对其进行访问，包括读和写，而由于 Android 系统基于 Linux，使得其并发读/写文件可以没有限制地进行，甚至两个线程同时对同一个文件进行写操作都是允许的，尽管这可能出问题。通过文件交换数据很好使用，除了可以交换一些文本信息外，我们还可以序列化一个对象到文件系统中的同时从另一个进程中恢复这个对象，下面就展示这种使用方法。

还是本章刚开始的那个例子，这次我们在 MainActivity 的 onResume 中序列化一个 User 对象到 sd 卡上的一个文件里，然后在 SecondActivity 的 onResume 中去反序列化，我们期望在 SecondActivity 中能够正确地恢复 User 对象的值。关键代码如下：

```
//在MainActivity中的修改
private void persistToFile() {
    new Thread(new Runnable() {
```

```java
        @Override
        public void run() {
            User user = new User(1, "hello world", false);
            File dir = new File(MyConstants.CHAPTER_2_PATH);
            if (!dir.exists()) {
                dir.mkdirs();
            }
            File cachedFile = new File(MyConstants.CACHE_FILE_PATH);
            ObjectOutputStream objectOutputStream = null;
            try {
                objectOutputStream = new ObjectOutputStream(
                        new FileOutputStream(cachedFile));
                objectOutputStream.writeObject(user);
                Log.d(TAG, "persist user:" + user);
            } catch (IOException e) {
                e.printStackTrace();
            } finally {
                MyUtils.close(objectOutputStream);
            }
        }
    }).start();
}
//SecondActivity 中的修改
private void recoverFromFile() {
    new Thread(new Runnable() {

        @Override
        public void run() {
            User user = null;
            File cachedFile = new File(MyConstants.CACHE_FILE_PATH);
            if (cachedFile.exists()) {
                ObjectInputStream objectInputStream = null;
                try {
                    objectInputStream = new ObjectInputStream(
                            new FileInputStream(cachedFile));
                    user = (User) objectInputStream.readObject();
                    Log.d(TAG, "recover user:" + user);
                } catch (IOException e) {
                    e.printStackTrace();
                } catch (ClassNotFoundException e) {
                    e.printStackTrace();
```

```
            } finally {
                MyUtils.close(objectInputStream);
            }
        }
    }
}).start();
```

下面看一下 log，很显然，在 SecondActivity 中成功地从文件从恢复了之前存储的 User 对象的内容，这里之所以说内容，是因为反序列化得到的对象只是在内容上和序列化之前的对象是一样的，但它们本质上还是两个对象。

```
D/MainActivity(10744): persist user:User:{userId:1, userName:hello world,
isMale:false}, with child:{null}
D/SecondActivity(10877): recover user:User:{userId:1, userName:hello world,
isMale:false}, with child:{null}
```

通过文件共享这种方式来共享数据对文件格式是没有具体要求的，比如可以是文本文件，也可以是 XML 文件，只要读/写双方约定数据格式即可。通过文件共享的方式也是有局限性的，比如并发读/写的问题，像上面的那个例子，如果并发读/写，那么我们读出的内容就有可能不是最新的，如果是并发写的话那就更严重了。因此我们要尽量避免并发写这种情况的发生或者考虑使用线程同步来限制多个线程的写操作。通过上面的分析，我们可以知道，文件共享方式适合在对数据同步要求不高的进程之间进行通信，并且要妥善处理并发读/写的问题。

当然，SharedPreferences 是个特例，众所周知，SharedPreferences 是 Android 中提供的轻量级存储方案，它通过键值对的方式来存储数据，在底层实现上它采用 XML 文件来存储键值对，每个应用的 SharedPreferences 文件都可以在当前包所在的 data 目录下查看到。一般来说，它的目录位于/data/data/package name/shared_prefs 目录下，其中 package name 表示的是当前应用的包名。从本质上来说，SharedPreferences 也属于文件的一种，但是由于系统对它的读/写有一定的缓存策略，即在内存中会有一份 SharedPreferences 文件的缓存，因此在多进程模式下，系统对它的读/写就变得不可靠，当面对高并发的读/写访问，Sharedpreferences 有很大几率会丢失数据，因此，不建议在进程间通信中使用 SharedPreferences。

2.4.3 使用 Messenger

Messenger 可以翻译为信使，顾名思义，通过它可以在不同进程中传递 Message 对象，在 Message 中放入我们需要传递的数据，就可以轻松地实现数据的进程间传递了。Messenger 是一种轻量级的 IPC 方案，它的底层实现是 AIDL，为什么这么说呢，我们大致看一下 Messenger 这个类的构造方法就明白了。下面是 Messenger 的两个构造方法，从构造方法的实现上我们可以明显看出 AIDL 的痕迹，不管是 IMessenger 还是 Stub.asInterface，这种使用方法都表明它的底层是 AIDL。

```
public Messenger(Handler target) {
    mTarget = target.getIMessenger();
}

public Messenger(IBinder target) {
    mTarget = IMessenger.Stub.asInterface(target);
}
```

Messenger 的使用方法很简单，它对 AIDL 做了封装，使得我们可以更简便地进行进程间通信。同时，由于它一次处理一个请求，因此在服务端我们不用考虑线程同步的问题，这是因为服务端中不存在并发执行的情形。实现一个 Messenger 有如下几个步骤，分为服务端和客户端。

1. 服务端进程

首先，我们需要在服务端创建一个 Service 来处理客户端的连接请求，同时创建一个 Handler 并通过它来创建一个 Messenger 对象，然后在 Service 的 onBind 中返回这个 Messenger 对象底层的 Binder 即可。

2. 客户端进程

客户端进程中，首先要绑定服务端的 Service，绑定成功后用服务端返回的 IBinder 对象创建一个 Messenger，通过这个 Messenger 就可以向服务端发送消息了，发消息类型为 Message 对象。如果需要服务端能够回应客户端，就和服务端一样，我们还需要创建一个 Handler 并创建一个新的 Messenger，并把这个 Messenger 对象通过 Message 的 replyTo 参数传递给服务端，服务端通过这个 replyTo 参数就可以回应客户端。这听起来可能还是有点抽象，不过看了下面的两个例子，读者肯定就都明白了。首先，我们来看一个简单点的例子，在这个例子中服务端无法回应客户端。

首先看服务端的代码，这是服务端的典型代码，可以看到 MessengerHandler 用来处理客户端发送的消息，并从消息中取出客户端发来的文本信息。而 mMessenger 是一个 Messenger 对象，它和 MessengerHandler 相关联，并在 onBind 方法中返回它里面的 Binder 对象，可以看出，这里 Messenger 的作用是将客户端发送的消息传递给 MessengerHandler 处理。

```java
public class MessengerService extends Service {

    private static final String TAG = "MessengerService";

    private static class MessengerHandler extends Handler {
        @Override
        public void handleMessage(Message msg) {
            switch (msg.what) {
            case MyConstants.MSG_FROM_CLIENT:
                Log.i(TAG, "receive msg from Client:" + msg.getData().
                getString("msg"));
                break;
            default:
                super.handleMessage(msg);
            }
        }
    }

    private final Messenger mMessenger = new Messenger(new Messenger-
    Handler());

    @Override
    public IBinder onBind(Intent intent) {
        return mMessenger.getBinder();
    }

}
```

然后，注册 service，让其运行在单独的进程中：

```xml
<service
    android:name="com.ryg.chapter_2.messenger.MessengerService"
    android:process=":remote" >
```

接下来再看看客户端的实现，客户端的实现也比较简单，首先需要绑定远程进程的

MessengerService，绑定成功后，根据服务端返回的 binder 对象创建 Messenger 对象并使用此对象向服务端发送消息。下面的代码在 Bundle 中向服务端发送了一句话，在上面的服务端代码中会打印出这句话。

```java
public class MessengerActivity extends Activity {

    private static final String TAG = " MessengerActivity";

    private Messenger mService;

    private ServiceConnection mConnection = new ServiceConnection() {
        public void onServiceConnected(ComponentName className, IBinder service) {
            mService = new Messenger(service);
            Message msg = Message.obtain(null, MyConstants.MSG_FROM_CLIENT);
            Bundle data = new Bundle();
            data.putString("msg", "hello, this is client.");
            msg.setData(data);
            try {
                mService.send(msg);
            } catch (RemoteException e) {
                e.printStackTrace();
            }
        }

        public void onServiceDisconnected(ComponentName className) {
        }
    };

    @Override
    protected void onCreate(Bundle savedInstanceState) {
        super.onCreate(savedInstanceState);
        setContentView(R.layout.activity_messenger);
        Intent intent = new Intent(this, MessengerService.class);
        bindService(intent, mConnection, Context.BIND_AUTO_CREATE);
    }

    @Override
    protected void onDestroy() {
```

```
            unbindService(mConnection);
            super.onDestroy();
        }
    }
```

最后，我们运行程序，看一下 log，很显然，服务端成功收到了客户端所发来的问候语："hello, this is client."。

```
I/MessengerService( 1037): receive msg from Client:hello, this is client.
```

通过上面的例子可以看出，在 Messenger 中进行数据传递必须将数据放入 Message 中，而 Messenger 和 Message 都实现了 Parcelable 接口，因此可以跨进程传输。简单来说，Message 中所支持的数据类型就是 Messenger 所支持的传输类型。实际上，通过 Messenger 来传输 Message，Message 中能使用的载体只有 what、arg1、arg2、Bundle 以及 replyTo。Message 中的另一个字段 object 在同一个进程中是很实用的，但是在进程间通信的时候，在 Android 2.2 以前 object 字段不支持跨进程传输，即便是 2.2 以后，也仅仅是系统提供的实现了 Parcelable 接口的对象才能通过它来传输。这就意味着我们自定义的 Parcelable 对象是无法通过 object 字段来传输的，读者可以试一下。非系统的 Parcelable 对象的确无法通过 object 字段来传输，这也导致了 object 字段的实用性大大降低，所幸我们还有 Bundle，Bundle 中可以支持大量的数据类型。

上面的例子演示了如何在服务端接收客户端中发送的消息，但是有时候我们还需要能回应客户端，下面就介绍如何实现这种效果。还是采用上面的例子，但是稍微做一下修改，每当客户端发来一条消息，服务端就会自动回复一条"嗯，你的消息我已经收到，稍后会回复你。"，这很类似邮箱的自动回复功能。

首先看服务端的修改，服务端只需要修改 MessengerHandler，当收到消息后，会立即回复一条消息给客户端。

```
private static class MessengerHandler extends Handler {
    @Override
    public void handleMessage(Message msg) {
        switch (msg.what) {
        case MyConstants.MSG_FROM_CLIENT:
            Log.i(TAG, "receive msg from Client:" + msg.getData().getString
                ("msg"));
            Messenger client = msg.replyTo;
```

```
        Message relpyMessage = Message.obtain(null, MyConstants.MSG_
FROM_SERVICE);
        Bundle bundle = new Bundle();
        bundle.putString("reply", "嗯，你的消息我已经收到，稍后会回复你。");
        relpyMessage.setData(bundle);
        try {
            client.send(relpyMessage);
        } catch (RemoteException e) {
            e.printStackTrace();
        }
        break;
    default:
        super.handleMessage(msg);
    }
  }
}
```

接着再看客户端的修改，为了接收服务端的回复，客户端也需要准备一个接收消息的 Messenger 和 Handler，如下所示。

```
private Messenger mGetReplyMessenger = new Messenger(new Messenger-
Handler());

private static class MessengerHandler extends Handler {
    @Override
    public void handleMessage(Message msg) {
        switch (msg.what) {
        case MyConstants.MSG_FROM_SERVICE:
            Log.i(TAG, "receive msg from Service:" + msg.getData().
            getString("reply"));
            break;
        default:
            super.handleMessage(msg);
        }
    }
}
```

除了上述修改，还有很关键的一点，当客户端发送消息的时候，需要把接收服务端回复的 Messenger 通过 Message 的 replyTo 参数传递给服务端，如下所示。

```
mService = new Messenger(service);
Message msg = Message.obtain(null, MyConstants.MSG_FROM_CLIENT);
Bundle data = new Bundle();
data.putString("msg", "hello, this is client.");
msg.setData(data);
//注意下面这句
msg.replyTo = mGetReplyMessenger;
try {
    mService.send(msg);
} catch (RemoteException e) {
    e.printStackTrace();
}
```

通过上述修改，我们再运行程序，然后看一下 log，很显然，客户端收到了服务端的回复 "嗯，你的消息我已经收到，稍后会回复你。"，这说明我们的功能已经完成。

```
I/MessengerService( 1419): receive msg from Client:hello, this is client.
I/MessengerActivity( 1404): receive msg from Service:嗯，你的消息我已经收到，
稍后会回复你。
```

到这里，我们已经把采用 Messenger 进行进程间通信的方法都介绍完了，读者可以试着通过 Messenger 来实现更复杂的跨进程通信功能。下面给出一张 Messenger 的工作原理图以方便读者更好地理解 Messenger，如图 2-6 所示。

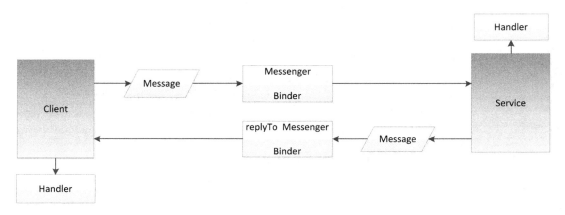

图 2-6　Messenger 的工作原理

关于进程间通信，可能有的读者会觉得笔者提供的示例都是针对同一个应用的，有没

有针对不同应用的？是这样的，之所以选择在同一个应用内进行进程间通信，是因为操作起来比较方便，但是效果和在两个应用间进行进程间通信是一样的。在本章刚开始就说过，同一个应用的不同组件，如果它们运行在不同进程中，那么和它们分别属于两个应用没有本质区别，关于这点需要深刻理解，因为这是理解进程间通信的基础。

2.4.4 使用 AIDL

上一节我们介绍了使用 Messenger 来进行进程间通信的方法，可以发现，Messenger 是以串行的方式处理客户端发来的消息，如果大量的消息同时发送到服务端，服务端仍然只能一个个处理，如果有大量的并发请求，那么用 Messenger 就不太合适了。同时，Messenger 的作用主要是为了传递消息，很多时候我们可能需要跨进程调用服务端的方法，这种情形用 Messenger 就无法做到了，但是我们可以使用 AIDL 来实现跨进程的方法调用。AIDL 也是 Messenger 的底层实现，因此 Messenger 本质上也是 AIDL，只不过系统为我们做了封装从而方便上层的调用而已。在上一节中，我们介绍了 Binder 的概念，大家对 Binder 也有了一定的了解，在 Binder 的基础上我们可以更加容易地理解 AIDL。这里先介绍使用 AIDL 来进行进程间通信的流程，分为服务端和客户端两个方面。

1．服务端

服务端首先要创建一个 Service 用来监听客户端的连接请求，然后创建一个 AIDL 文件，将暴露给客户端的接口在这个 AIDL 文件中声明，最后在 Service 中实现这个 AIDL 接口即可。

2．客户端

客户端所要做事情就稍微简单一些，首先需要绑定服务端的 Service，绑定成功后，将服务端返回的 Binder 对象转成 AIDL 接口所属的类型，接着就可以调用 AIDL 中的方法了。

上面描述的只是一个感性的过程，AIDL 的实现过程远不止这么简单，接下来会对其中的细节和难点进行详细介绍，并完善我们在 Binder 那一节所提供的的实例。

3．AIDL 接口的创建

首先看 AIDL 接口的创建，如下所示，我们创建了一个后缀为 AIDL 的文件，在里面声明了一个接口和两个接口方法。

```
// IBookManager.aidl
package com.ryg.chapter_2.aidl;

import com.ryg.chapter_2.aidl.Book;

interface IBookManager {
    List<Book> getBookList();
    void addBook(in Book book);
}
```

在 AIDL 文件中，并不是所有的数据类型都是可以使用的，那么到底 AIDL 文件支持哪些数据类型呢？如下所示。

- 基本数据类型（int、long、char、boolean、double 等）；
- String 和 CharSequence；
- List：只支持 ArrayList，里面每个元素都必须能够被 AIDL 支持；
- Map：只支持 HashMap，里面的每个元素都必须被 AIDL 支持，包括 key 和 value；
- Parcelable：所有实现了 Parcelable 接口的对象；
- AIDL：所有的 AIDL 接口本身也可以在 AIDL 文件中使用。

以上 6 种数据类型就是 AIDL 所支持的所有类型，其中自定义的 Parcelable 对象和 AIDL 对象必须要显式 import 进来，不管它们是否和当前的 AIDL 文件位于同一个包内。比如 IBookManager.aidl 这个文件，里面用到了 Book 这个类，这个类实现了 Parcelable 接口并且和 IBookManager.aidl 位于同一个包中，但是遵守 AIDL 的规范，我们仍然需要显式地 import 进来：import com.ryg.chapter_2.aidl.Book。AIDL 中会大量使用到 Parcelable，至于如何使用 Parcelable 接口来序列化对象，在本章的前面已经介绍过，这里就不再赘述。

另外一个需要注意的地方是，如果 AIDL 文件中用到了自定义的 Parcelable 对象，那么必须新建一个和它同名的 AIDL 文件，并在其中声明它为 Parcelable 类型。在上面的 IBookManager.aidl 中，我们用到了 Book 这个类，所以，我们必须要创建 Book.aidl，然后在里面添加如下内容：

```
package com.ryg.chapter_2.aidl;
parcelable Book;
```

我们需要注意，AIDL 中每个实现了 Parcelable 接口的类都需要按照上面那种方式去创建相应的 AIDL 文件并声明那个类为 parcelable。除此之外，AIDL 中除了基本数据类型，其他类型的参数必须标上方向：in、out 或者 inout，in 表示输入型参数，out 表示输出型参数，inout 表示输入输出型参数，至于它们具体的区别，这个就不说了。我们要根据实际需要去指定参数类型，不能一概使用 out 或者 inout，因为这在底层实现是有开销的。最后，AIDL 接口中只支持方法，不支持声明静态常量，这一点区别于传统的接口。

为了方便 AIDL 的开发，建议把所有和 AIDL 相关的类和文件全部放入同一个包中，这样做的好处是，当客户端是另外一个应用时，我们可以直接把整个包复制到客户端工程中，对于本例来说，就是要把 com.ryg.chapter_2.aidl 这个包和包中的文件原封不动地复制到客户端中。如果 AIDL 相关的文件位于不同的包中时，那么就需要把这些包一一复制到客户端工程中，这样操作起来比较麻烦而且也容易出错。需要注意的是，AIDL 的包结构在服务端和客户端要保持一致，否则运行会出错，这是因为客户端需要反序列化服务端中和 AIDL 接口相关的所有类，如果类的完整路径不一样的话，就无法成功反序列化，程序也就无法正常运行。为了方便演示，本章的所有示例都是在同一个工程中进行的，但是读者要理解，一个工程和两个工程的多进程本质是一样的，两个工程的情况，除了需要复制 AIDL 接口所相关的包到客户端，其他完全一样，读者可以自行试验。

4．远程服务端 Service 的实现

上面讲述了如何定义 AIDL 接口，接下来我们就需要实现这个接口了。我们先创建一个 Service，称为 BookManagerService，代码如下：

```
public class BookManagerService extends Service {

    private static final String TAG = "BMS";

    private CopyOnWriteArrayList<Book> mBookList = new CopyOnWriteArray-
    List<Book>();

    private Binder mBinder = new IBookManager.Stub() {

        @Override
        public List<Book> getBookList() throws RemoteException {
```

```java
            return mBookList;
        }

        @Override
        public void addBook(Book book) throws RemoteException {
            mBookList.add(book);
        }
    };

    @Override
    public void onCreate() {
        super.onCreate();
        mBookList.add(new Book(1, "Android"));
        mBookList.add(new Book(2, "Ios"));
    }

    @Override
    public IBinder onBind(Intent intent) {
        return mBinder;
    }
}
```

上面是一个服务端 Service 的典型实现，首先在 onCreate 中初始化添加了两本图书的信息，然后创建了一个 Binder 对象并在 onBind 中返回它，这个对象继承自 IBookManager.Stub 并实现了它内部的 AIDL 方法，这个过程在 Binder 那一节已经介绍过了，这里就不多说了。这里主要看 getBookList 和 addBook 这两个 AIDL 方法的实现，实现过程也比较简单，注意这里采用了 CopyOnWriteArrayList，这个 CopyOnWriteArrayList 支持并发读/写。在前面我们提到，AIDL 方法是在服务端的 Binder 线程池中执行的，因此当多个客户端同时连接的时候，会存在多个线程同时访问的情形，所以我们要在 AIDL 方法中处理线程同步，而我们这里直接使用 CopyOnWriteArrayList 来进行自动的线程同步。

前面我们提到，AIDL 中能够使用的 List 只有 ArrayList，但是我们这里却使用了 CopyOnWriteArrayList（注意它不是继承自 ArrayList），为什么能够正常工作呢？这是因为 AIDL 中所支持的是抽象的 List，而 List 只是一个接口，因此虽然服务端返回的是 CopyOnWriteArrayList，但是在 Binder 中会按照 List 的规范去访问数据并最终形成一个新的 ArrayList 传递给客户端。所以，我们在服务端采用 CopyOnWriteArrayList 是完全可以的。

和此类似的还有 ConcurrentHashMap，读者可以体会一下这种转换情形。然后我们需要在 XML 中注册这个 Service，如下所示。注意 BookManagerService 是运行在独立进程中的，它和客户端的 Activity 不在同一个进程中，这样就构成了进程间通信的场景。

```
<service
    android:name=".aidl.BookManagerService"
    android:process=":remote" >
</service>
```

5．客户端的实现

客户端的实现就比较简单了，首先要绑定远程服务，绑定成功后将服务端返回的 Binder 对象转换成 AIDL 接口，然后就可以通过这个接口去调用服务端的远程方法了，代码如下所示。

```java
public class BookManagerActivity extends Activity {

    private static final String TAG = "BookManagerActivity";

    private ServiceConnection mConnection = new ServiceConnection() {
        public void onServiceConnected(ComponentName className, IBinder service) {
            IBookManager bookManager = IBookManager.Stub.asInterface(service);
            try {
                List<Book> list = bookManager.getBookList();
                Log.i(TAG, "query book list, list type:" + list.getClass().getCanonicalName());
                Log.i(TAG, "query book list:" + list.toString());
            } catch (RemoteException e) {
                e.printStackTrace();
            }
        }

        public void onServiceDisconnected(ComponentName className) {
        }
    };

    @Override
```

```
    protected void onCreate(Bundle savedInstanceState) {
        super.onCreate(savedInstanceState);
        setContentView(R.layout.activity_book_manager);
        Intent intent = new Intent(this, BookManagerService.class);
        bindService(intent, mConnection, Context.BIND_AUTO_CREATE);
    }

    @Override
    protected void onDestroy() {
        unbindService(mConnection);
        super.onDestroy();
    }
}
```

绑定成功以后，会通过 bookManager 去调用 getBookList 方法，然后打印出所获取的图书信息。需要注意的是，服务端的方法有可能需要很久才能执行完毕，这个时候下面的代码就会导致 ANR，这一点是需要注意的，后面会再介绍这种情况，之所以先这么写是为了让读者更好地了解 AIDL 的实现步骤。

接着在 XML 中注册此 Activity，运行程序，log 如下所示。

```
I/BookManagerActivity(3047): query book list, list type:java.util.ArrayList
I/BookManagerActivity(3047): query book list:[[bookId:1, bookName:Android],
[bookId:2, bokName:Ios]]
```

可以发现，虽然我们在服务端返回的是 CopyOnWriteArrayList 类型，但是客户端收到的仍然是 ArrayList 类型，这也证实了我们在前面所做的分析。第二行 log 表明客户端成功地得到了服务端的图书列表信息。

这就是一次完完整整的使用 AIDL 进行 IPC 的过程，到这里相信读者对 AIDL 应该有了一个整体的认识了，但是还没完，AIDL 的复杂性远不止这些，下面继续介绍 AIDL 中常见的一些难点。

我们接着再调用一下另外一个接口 addBook，我们在客户端给服务端添加一本书，然后再获取一次，看程序是否能够正常工作。还是上面的代码，客户端在服务连接后，在 onServiceConnected 中做如下改动：

```
    public void onServiceConnected(ComponentName className, IBinder service) {
```

```java
        IBookManager bookManager = IBookManager.Stub.asInterface(service);
        try {
            List<Book> list = bookManager.getBookList();
            Log.i(TAG, "query book list:" + list.toString());
            Book newBook = new Book(3, "Android开发艺术探索");
            bookManager.addBook(newBook);
            Log.i(TAG, "add book:" + newBook);
            List<Book> newList = bookManager.getBookList();
            Log.i(TAG, "query book list:" + newList.toString());
        } catch (RemoteException e) {
            e.printStackTrace();
        }
    }
}
```

运行后我们再看一下 log，很显然，我们成功地向服务端添加了一本书 "Android 开发艺术探索"。

```
I/BookManagerActivity( 3148): query book list:[[bookId:1, bookName:Android],
[bookId:2, bookName:Ios]]
I/BookManagerActivity( 3148): add book:[bookId:3, bookName:Android开发艺术探索]
I/BookManagerActivity( 3148): query book list:[[bookId:1, bookName:Android],
[bookId:2, bookName:Ios], [bookId:3, bookName:Android开发艺术探索]]
```

现在我们考虑一种情况，假设有一种需求：用户不想时不时地去查询图书列表了，太累了，于是，他去问图书馆，"当有新书时能不能把书的信息告诉我呢？"。大家应该明白了，这就是一种典型的观察者模式，每个感兴趣的用户都观察新书，当新书到的时候，图书馆就通知每一个对这本书感兴趣的用户，这种模式在实际开发中用得很多，下面我们就来模拟这种情形。首先，我们需要提供一个 AIDL 接口，每个用户都需要实现这个接口并且向图书馆申请新书的提醒功能，当然用户也可以随时取消这种提醒。之所以选择 AIDL 接口而不是普通接口，是因为 AIDL 中无法使用普通接口。这里我们创建一个 IOnNewBookArrivedListener.aidl 文件，我们所期望的情况是：当服务端有新书到来时，就会通知每一个已经申请提醒功能的用户。从程序上来说就是调用所有 IOnNewBookArrivedListener 对象中的 onNewBookArrived 方法，并把新书的对象通过参数传递给客户端，内容如下所示。

```
package com.ryg.chapter_2.aidl;
import com.ryg.chapter_2.aidl.Book;
```

```
interface IOnNewBookArrivedListener {
    void onNewBookArrived(in Book newBook);
}
```

除了要新加一个 AIDL 接口，还需要在原有的接口中添加两个新方法，代码如下所示。

```
package com.ryg.chapter_2.aidl;

import com.ryg.chapter_2.aidl.Book;
import com.ryg.chapter_2.aidl.IOnNewBookArrivedListener;

interface IBookManager {
    List<Book> getBookList();
    void addBook(in Book book);
    void registerListener(IOnNewBookArrivedListener listener);
    void unregisterListener(IOnNewBookArrivedListener listener);
}
```

接着，服务端中 Service 的实现也要稍微修改一下，主要是 Service 中 IBookManager.Stub 的实现，因为我们在 IBookManager 新加了两个方法，所以在 IBookManager.Stub 中也要实现这两个方法。同时，在 BookManagerService 中还开启了一个线程，每隔 5s 就向书库中增加一本新书并通知所有感兴趣的用户，整个代码如下所示。

```
public class BookManagerService extends Service {

    private static final String TAG = "BMS";

    private AtomicBoolean mIsServiceDestoryed = new AtomicBoolean(false);

    private CopyOnWriteArrayList<Book> mBookList = new CopyOnWriteArrayList<Book>();
    private CopyOnWriteArrayList<IOnNewBookArrivedListener> mListenerList
        = new CopyOnWriteArrayList<IOnNewBookArrivedListener>();

    private Binder mBinder = new IBookManager.Stub() {

        @Override
        public List<Book> getBookList() throws RemoteException {
```

```java
        return mBookList;
    }

    @Override
    public void addBook(Book book) throws RemoteException {
        mBookList.add(book);
    }

    @Override
    public void registerListener(IOnNewBookArrivedListener listener)
            throws RemoteException {
        if (!mListenerList.contains(listener)) {
            mListenerList.add(listener);
        } else {
            Log.d(TAG, "already exists.");
        }
        Log.d(TAG, "registerListener, size:" + mListenerList.size());
    }

    @Override
    public void unregisterListener(IOnNewBookArrivedListener listener)
            throws RemoteException {
        if (mListenerList.contains(listener)) {
            mListenerList.remove(listener);
            Log.d(TAG, "unregister listener succeed.");
        } else {
            Log.d(TAG, "not found, can not unregister.");
        }
        Log.d(TAG, "unregisterListener, current size:" + mListenerList.
        size());
    };

};

@Override
public void onCreate() {
    super.onCreate();
    mBookList.add(new Book(1, "Android"));
    mBookList.add(new Book(2, "Ios"));
    new Thread(new ServiceWorker()).start();
```

```java
    }

    @Override
    public IBinder onBind(Intent intent) {
        return mBinder;
    }

    @Override
    public void onDestroy() {
        mIsServiceDestoryed.set(true);
        super.onDestroy();
    }

    private void onNewBookArrived(Book book) throws RemoteException {
        mBookList.add(book);
        Log.d(TAG, "onNewBookArrived, notify listeners:" + mListenerList.
        size());
        for (int i = 0; i < mListenerList.size(); i++) {
            IOnNewBookArrivedListener listener = mListenerList.get(i);
            Log.d(TAG, "onNewBookArrived, notify listener:" + listener);
            listener.onNewBookArrived(book);
        }
    }

    private class ServiceWorker implements Runnable {
        @Override
        public void run() {
            // do background processing here.....
            while (!mIsServiceDestoryed.get()) {
                try {
                    Thread.sleep(5000);
                } catch (InterruptedException e) {
                    e.printStackTrace();
                }
                int bookId = mBookList.size() + 1;
                Book newBook = new Book(bookId, "new book#" + bookId);
                try {
                    onNewBookArrived(newBook);
                } catch (RemoteException e) {
```

```
                e.printStackTrace();
            }
        }
    }
}
```

最后，我们还需要修改一下客户端的代码，主要有两方面：首先客户端要注册 IOnNewBookArrivedListener 到远程服务端，这样当有新书时服务端才能通知当前客户端，同时我们要在 Activity 退出时解除这个注册；另一方面，当有新书时，服务端会回调客户端的 IOnNewBookArrivedListener 对象中的 onNewBookArrived 方法，但是这个方法是在客户端的 Binder 线程池中执行的，因此，为了便于进行 UI 操作，我们需要有一个 Handler 可以将其切换到客户端的主线程中去执行，这个原理在 Binder 中已经做了分析，这里就不多说了。客户端的代码修改如下：

```
public class BookManagerActivity extends Activity {

    private static final String TAG = "BookManagerActivity";
    private static final int MESSAGE_NEW_BOOK_ARRIVED = 1;

    private IBookManager mRemoteBookManager;

    private Handler mHandler = new Handler() {
        @Override
        public void handleMessage(Message msg) {
            switch (msg.what) {
            case MESSAGE_NEW_BOOK_ARRIVED:
                Log.d(TAG, "receive new book :" + msg.obj);
                break;
            default:
                super.handleMessage(msg);
            }
        }
    };

    private ServiceConnection mConnection = new ServiceConnection() {
        public void onServiceConnected(ComponentName className, IBinder
        service) {
```

```java
            IBookManager bookManager = IBookManager.Stub.asInterface
            (service);
            try {
                mRemoteBookManager = bookManager;
                List<Book> list = bookManager.getBookList();
                Log.i(TAG, "query book list, list type:"
                        + list.getClass().getCanonicalName());
                Log.i(TAG, "query book list:" + list.toString());
                Book newBook = new Book(3, "Android进阶");
                bookManager.addBook(newBook);
                Log.i(TAG, "add book:" + newBook);
                List<Book> newList = bookManager.getBookList();
                Log.i(TAG, "query book list:" + newList.toString());
                bookManager.registerListener(mOnNewBookArrivedListener);
            } catch (RemoteException e) {
                e.printStackTrace();
            }
        }

        public void onServiceDisconnected(ComponentName className) {
            mRemoteBookManager = null;
            Log.e(TAG, "binder died.");
        }
    };

    private IOnNewBookArrivedListener mOnNewBookArrivedListener = new
    IOnNewBookArrivedListener.Stub() {

        @Override
        public void onNewBookArrived(Book newBook) throws RemoteException {
            mHandler.obtainMessage(MESSAGE_NEW_BOOK_ARRIVED, newBook)
                    .sendToTarget();
        }
    };

    @Override
    protected void onCreate(Bundle savedInstanceState) {
        super.onCreate(savedInstanceState);
        setContentView(R.layout.activity_book_manager);
        Intent intent = new Intent(this, BookManagerService.class);
```

```
        bindService(intent, mConnection, Context.BIND_AUTO_CREATE);
    }

    @Override
    protected void onDestroy() {
        if (mRemoteBookManager != null
                && mRemoteBookManager.asBinder().isBinderAlive()) {
            try {
                Log.i(TAG, "unregister listener:" + mOnNewBookArrived-
                  Listener);
                mRemoteBookManager
                        .unregisterListener(mOnNewBookArrivedListener);
            } catch (RemoteException e) {
                e.printStackTrace();
            }
        }
        unbindService(mConnection);
        super.onDestroy();
    }

}
```

运行程序，看一下 log，从 log 中可以看出，客户端的确收到了服务端每 5s 一次的新书推送，我们的功能也就实现了。

```
D/BMS(3414):onNewBookArrived, notify listener:com.ryg.chapter_2.aidl.
IOnNewBookArrivedListener$Stub$Proxy@4052a648
D/BookManagerActivity( 3385): receive new book :[bookId:4, bookName:new
book#4]
D/BMS(3414):onNewBookArrived, notify listener:com.ryg.chapter_2.aidl.
IOnNewBookArrivedListener$Stub$Proxy@4052a648
D/BookManagerActivity( 3385): receive new book :[bookId:5, bookName:new
book#5]
```

如果你以为到这里 AIDL 的介绍就结束了，那你就错了，之前就说过，AIDL 远不止这么简单，目前还有一些难点是我们还没有涉及的，接下来将继续为读者介绍。

从上面的代码可以看出，当 BookManagerActivity 关闭时，我们会在 onDestroy 中去解除已经注册到服务端的 listener，这就相当于我们不想再接收图书馆的新书提醒了，所以我们可以随时取消这个提醒服务。按 back 键退出 BookManagerActivity，下面是打印出的 log。

```
I/BookManagerActivity(5642): unregister listener:com.ryg.chapter_2.aidl.
BookManagerActivity$3@405284c8
D/BMS(5650): not found, can not unregister.
D/BMS(5650): unregisterListener, current size:1
```

从上面的 log 可以看出，程序没有像我们所预期的那样执行。在解注册的过程中，服务端竟然无法找到我们之前注册的那个 listener，在客户端我们注册和解注册时明明传递的是同一个 listener 啊！最终，服务端由于无法找到要解除的 listener 而宣告解注册失败！这当然不是我们想要的结果，但是仔细想想，好像这种方式的确无法完成解注册。其实，这是必然的，这种解注册的处理方式在日常开发过程中时常使用到，但是放到多进程中却无法奏效，因为 Binder 会把客户端传递过来的对象重新转化并生成一个新的对象。虽然我们在注册和解注册过程中使用的是同一个客户端对象，但是通过 Binder 传递到服务端后，却会产生两个全新的对象。别忘了对象是不能跨进程直接传输的，对象的跨进程传输本质上都是反序列化的过程，这就是为什么 AIDL 中的自定义对象都必须要实现 Parcelable 接口的原因。那么到底我们该怎么做才能实现解注册功能呢？答案是使用 RemoteCallbackList，这看起来很抽象，不过没关系，请看接下来的详细分析。

RemoteCallbackList 是系统专门提供的用于删除跨进程 listener 的接口。RemoteCallbackList 是一个泛型，支持管理任意的 AIDL 接口，这点从它的声明就可以看出，因为所有的 AIDL 接口都继承自 IInterface 接口，读者还有印象吗？

```
public class RemoteCallbackList<E extends IInterface>
```

它的工作原理很简单，在它的内部有一个 Map 结构专门用来保存所有的 AIDL 回调，这个 Map 的 key 是 IBinder 类型，value 是 Callback 类型，如下所示。

```
ArrayMap<IBinder, Callback> mCallbacks = new ArrayMap<IBinder, Callback>();
```

其中 Callback 中封装了真正的远程 listener。当客户端注册 listener 的时候，它会把这个 listener 的信息存入 mCallbacks 中，其中 key 和 value 分别通过下面的方式获得：

```
IBinder key= listener.asBinder()
Callback value = new Callback(listener, cookie)
```

到这里，读者应该都明白了，虽然说多次跨进程传输客户端的同一个对象会在服务端生成不同的对象，但是这些新生成的对象有一个共同点，那就是它们底层的 Binder 对象是同一个，利用这个特性，就可以实现上面我们无法实现的功能。当客户端解注册的时候，我们只要遍历服务端所有的 listener，找出那个和解注册 listener 具有相同 Binder 对象的服务端 listener 并把它删掉即可，这就是 RemoteCallbackList 为我们做的事情。同时 RemoteCallbackList 还有一个很有用的功能，那就是当客户端进程终止后，它能够自动移除客户端所注册的 listener。另外，RemoteCallbackList 内部自动实现了线程同步的功能，所以我们使用它来注册和解注册时，不需要做额外的线程同步工作。由此可见，RemoteCallbackList 的确是个很有价值的类，下面就演示如何使用它来完成解注册。

RemoteCallbackList 使用起来很简单，我们要对 BookManagerService 做一些修改，首先要创建一个 RemoteCallbackList 对象来替代之前的 CopyOnWriteArrayList，如下所示。

```
private RemoteCallbackList<IOnNewBookArrivedListener> mListenerList = new
RemoteCallbackList<IOnNewBookArrivedListener>();
```

然后修改 registerListener 和 unregisterListener 这两个接口的实现，如下所示。

```
@Override
public void registerListener(IOnNewBookArrivedListener listener)
        throws RemoteException {
    mListenerList.register(listener);
}

@Override
public void unregisterListener(IOnNewBookArrivedListener listener)
        throws RemoteException {
    mListenerList.unregister(listener);
};
```

怎么样？是不是用起来很简单，接着要修改 onNewBookArrived 方法，当有新书时，我们就要通知所有已注册的 listener，如下所示。

```
private void onNewBookArrived(Book book) throws RemoteException {
    mBookList.add(book);
    final int N = mListenerList.beginBroadcast();
    for (int i = 0; i < N; i++) {
        IOnNewBookArrivedListener l = mListenerList.getBroadcastItem(i);
```

```
        if (l != null) {
            try {
                l.onNewBookArrived(book);
            } catch (RemoteException e) {
                e.printStackTrace();
            }
        }
    }
    mListenerList.finishBroadcast();
}
```

BookManagerService 的修改已经完毕了,为了方便我们验证程序的功能,我们还需要添加一些 log,在注册和解注册后我们分别打印出所有 listener 的数量。如果程序正常工作的话,那么注册之后 listener 总数量是 1,解注册之后总数量应该是 0,我们再次运行一下程序,看是否如此。从下面的 log 来看,很显然,使用 RemoteCallbackList 的确可以完成跨进程的解注册功能。

```
I/BookManagerActivity(8419): register listener:com.ryg.chapter_2.aidl.
BookManagerActivity$3@40537610
D/BMS(8427): registerListener, current size:1
I/BookManagerActivity(8419): unregister listener:com.ryg.chapter_2.aidl.
BookManagerActivity$3@40537610
D/BMS(8427): unregister success.
D/BMS(8427): unregisterListener, current size:0
```

使用 RemoteCallbackList,有一点需要注意,我们无法像操作 List 一样去操作它,尽管它的名字中也带个 List,但是它并不是一个 List。遍历 RemoteCallbackList,必须要按照下面的方式进行,其中 beginBroadcast 和 finishBroadcast 必须要配对使用,哪怕我们仅仅是想要获取 RemoteCallbackList 中的元素个数,这是必须要注意的地方。

```
final int N = mListenerList.beginBroadcast();
for (int i = 0; i < N; i++) {
    IOnNewBookArrivedListener l = mListenerList.getBroadcastItem(i);
    if (l != null) {
        //TODO handle l
    }
}
mListenerList.finishBroadcast();
```

到这里，AIDL 的基本使用方法已经介绍完了，但是有几点还需要再次说明一下。我们知道，客户端调用远程服务的方法，被调用的方法运行在服务端的 Binder 线程池中，同时客户端线程会被挂起，这个时候如果服务端方法执行比较耗时，就会导致客户端线程长时间地阻塞在这里，而如果这个客户端线程是 UI 线程的话，就会导致客户端 ANR，这当然不是我们想要看到的。因此，如果我们明确知道某个远程方法是耗时的，那么就要避免在客户端的 UI 线程中去访问远程方法。由于客户端的 onServiceConnected 和 onServiceDisconnected 方法都运行在 UI 线程中，所以也不可以在它们里面直接调用服务端的耗时方法，这点要尤其注意。另外，由于服务端的方法本身就运行在服务端的 Binder 线程池中，所以服务端方法本身就可以执行大量耗时操作，这个时候切记不要在服务端方法中开线程去进行异步任务，除非你明确知道自己在干什么，否则不建议这么做。下面我们稍微改造一下服务端的 getBookList 方法，我们假定这个方法是耗时的，那么服务端可以这么实现：

```
@Override
public List<Book> getBookList() throws RemoteException {
    SystemClock.sleep(5000);
    return mBookList;
}
```

然后在客户端中放一个按钮，单击它的时候就会调用服务端的 getBookList 方法，可以预知，连续单击几次，客户端就 ANR 了，如图 2-7 所示，感兴趣读者可以自行试一下。

图 2-7　UI 线程中调用远程耗时方法导致的 ANR

避免出现上述这种 ANR 其实很简单，我们只需要把调用放在非 UI 线程即可，如下所示。

```
public void onButton1Click(View view) {
    Toast.makeText(this, "click button1", Toast.LENGTH_SHORT).show();
    new Thread(new Runnable() {

        @Override
        public void run() {
            if (mRemoteBookManager != null) {
                try {
                    List<Book> newList = mRemoteBookManager.getBookList();
                } catch (RemoteException e) {
                    e.printStackTrace();
                }
            }
        }
    }).start();
}
```

同理，当远程服务端需要调用客户端的 listener 中的方法时，被调用的方法也运行在 Binder 线程池中，只不过是客户端的线程池。所以，我们同样不可以在服务端中调用客户端的耗时方法。比如针对 BookManagerService 的 onNewBookArrived 方法，如下所示。在它内部调用了客户端的 IOnNewBookArrivedListener 中的 onNewBookArrived 方法，如果客户端的这个 onNewBookArrived 方法比较耗时的话，那么请确保 BookManagerService 中的 onNewBookArrived 运行在非 UI 线程中，否则将导致服务端无法响应。

```
private void onNewBookArrived(Book book) throws RemoteException {
    mBookList.add(book);
    Log.d(TAG, "onNewBookArrived, notify listeners:" + mListenerList.size());
    for (int i = 0; i < mListenerList.size(); i++) {
        IOnNewBookArrivedListener listener = mListenerList.get(i);
        Log.d(TAG, "onNewBookArrived, notify listener:" + listener);
        listener.onNewBookArrived(book);
    }
}
```

另外，由于客户端的 IOnNewBookArrivedListener 中的 onNewBookArrived 方法运行在

客户端的 Binder 线程池中，所以不能在它里面去访问 UI 相关的内容，如果要访问 UI，请使用 Handler 切换到 UI 线程，这一点在前面的代码实例中已经有所体现，这里就不再详细描述了。

为了程序的健壮性，我们还需要做一件事。Binder 是可能意外死亡的，这往往是由于服务端进程意外停止了，这时我们需要重新连接服务。有两种方法，第一种方法是给 Binder 设置 DeathRecipient 监听，当 Binder 死亡时，我们会收到 binderDied 方法的回调，在 binderDied 方法中我们可以重连远程服务，具体方法在 Binder 那一节已经介绍过了，这里就不再详细描述了。另一种方法是在 onServiceDisconnected 中重连远程服务。这两种方法我们可以随便选择一种来使用，它们的区别在于：onServiceDisconnected 在客户端的 UI 线程中被回调，而 binderDied 在客户端的 Binder 线程池中被回调。也就是说，在 binderDied 方法中我们不能访问 UI，这就是它们的区别。下面验证一下二者之间的区别，首先我们通过 DDMS 杀死服务端进程，接着在这两个方法中打印出当前线程的名称，如下所示。

```
D/BookManagerActivity(13652): onServiceDisconnected. tname:main
D/BookManagerActivity(13652): binder died. tname:Binder Thread #2
```

从上面的 log 和图 2-8 我们可以看到，onServiceDisconnected 运行在 main 线程中，即 UI 线程，而 binderDied 运行在 "Binder Thread #2" 这个线程中，很显然，它是 Binder 线程池中的一个线程。

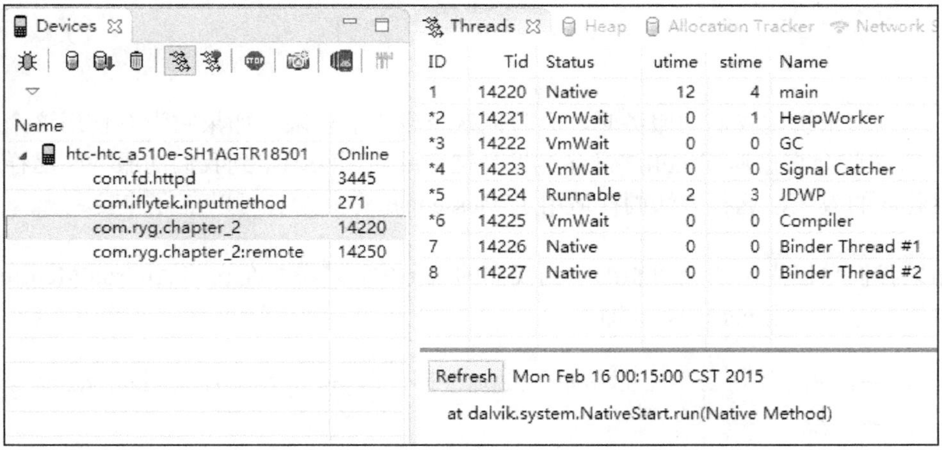

图 2-8　DDMS 中的线程信息

到此为止，我们已经对 AIDL 有了一个系统性的认识，但是还差最后一步：如何在 AIDL

中使用权限验证功能。默认情况下，我们的远程服务任何人都可以连接，但这应该不是我们愿意看到的，所以我们必须给服务加入权限验证功能，权限验证失败则无法调用服务中的方法。在 AIDL 中进行权限验证，这里介绍两种常用的方法。

第一种方法，我们可以在 onBind 中进行验证，验证不通过就直接返回 null，这样验证失败的客户端直接无法绑定服务，至于验证方式可以有多种，比如使用 permission 验证。使用这种验证方式，我们要先在 AndroidMenifest 中声明所需的权限，比如：

```xml
<permission
    android:name="com.ryg.chapter_2.permission.ACCESS_BOOK_SERVICE"
    android:protectionLevel="normal" />
```

关于 permission 的定义方式请读者查看相关资料，这里就不详细展开了，毕竟本节的主要内容是介绍 AIDL。定义了权限以后，就可以在 BookManagerService 的 onBind 方法中做权限验证了，如下所示。

```java
public IBinder onBind(Intent intent) {
    int check = checkCallingOrSelfPermission("com.ryg.chapter_2.permission.ACCESS_BOOK_SERVICE");
    if (check == PackageManager.PERMISSION_DENIED) {
        return null;
    }
    return mBinder;
}
```

一个应用来绑定我们的服务时，会验证这个应用的权限，如果它没有使用这个权限，onBind 方法就会直接返回 null，最终结果是这个应用无法绑定到我们的服务，这样就达到了权限验证的效果，这种方法同样适用于 Messenger 中，读者可以自行扩展。

如果我们自己内部的应用想绑定到我们的服务中，只需要在它的 AndroidMenifest 文件中采用如下方式使用 permission 即可。

```xml
<uses-permission android:name="com.ryg.chapter_2.permission.ACCESS_BOOK_SERVICE" />
```

第二种方法，我们可以在服务端的 onTransact 方法中进行权限验证，如果验证失败就直接返回 false，这样服务端就不会终止执行 AIDL 中的方法从而达到保护服务端的效果。至于具体的验证方式有很多，可以采用 permission 验证，具体实现方式和第一种方法一样。

还可以采用 Uid 和 Pid 来做验证，通过 getCallingUid 和 getCallingPid 可以拿到客户端所属应用的 Uid 和 Pid，通过这两个参数我们可以做一些验证工作，比如验证包名。在下面的代码中，既验证了 permission，又验证了包名。一个应用如果想远程调用服务中的方法，首先要使用我们的自定义权限"com.ryg.chapter_2.permission.ACCESS_BOOK_SERVICE"，其次包名必须以"com.ryg"开始，否则调用服务端的方法会失败。

```
public boolean onTransact(int code, Parcel data, Parcel reply, int flags)
        throws RemoteException {
    int check = checkCallingOrSelfPermission("com.ryg.chapter_2.permission.
ACCESS_BOOK_SERVICE");
    if (check == PackageManager.PERMISSION_DENIED) {
        return false;
    }

    String packageName = null;
    String[] packages = getPackageManager().getPackagesForUid(getCalling-
Uid());
    if (packages != null && packages.length > 0) {
        packageName = packages[0];
    }
    if (!packageName.startsWith("com.ryg")) {
        return false;
    }

    return super.onTransact(code, data, reply, flags);
}
```

上面介绍了两种 AIDL 中常用的权限验证方法，但是肯定还有其他方法可以做权限验证，比如为 Service 指定 android:permission 属性等，这里就不一一进行介绍了。到这里为止，本节的内容就全部结束了，读者应该对 AIDL 的使用过程有很深入的理解了，接下来会介绍另一个 IPC 方式，那就是使用 ContentProvider。

2.4.5 使用 ContentProvider

ContentProvider 是 Android 中提供的专门用于不同应用间进行数据共享的方式，从这一点来看，它天生就适合进程间通信。和 Messenger 一样，ContentProvider 的底层实现同样也是 Binder，由此可见，Binder 在 Android 系统中是何等的重要。虽然 ContentProvider 的底层实现是 Binder，但是它的使用过程要比 AIDL 简单许多，这是因为系统已经为我们做了封装，使得我们无须关心底层细节即可轻松实现 IPC。ContentProvider 虽然使用起来很简单，包括自己创建一个 ContentProvider 也不是什么难事，尽管如此，它的细节还是相当多，比如 CRUD 操作、防止 SQL 注入和权限控制等。由于章节主题限制，在本节中，笔者暂时不对 ContentProvider 的使用细节以及工作机制进行详细分析，而是为读者介绍采用 ContentProvider 进行跨进程通信的主要流程，至于使用细节和内部工作机制会在后续章节进行详细分析。

系统预置了许多 ContentProvider，比如通讯录信息、日程表信息等，要跨进程访问这些信息，只需要通过 ContentResolver 的 query、update、insert 和 delete 方法即可。在本节中，我们来实现一个自定义的 ContentProvider，并演示如何在其他应用中获取 ContentProvider 中的数据从而实现进程间通信这一目的。首先，我们创建一个 ContentProvider，名字就叫 BookProvider。创建一个自定义的 ContentProvider 很简单，只需要继承 ContentProvider 类并实现六个抽象方法即可：onCreate、query、update、insert、delete 和 getType。这六个抽象方法都很好理解，onCreate 代表 ContentProvider 的创建，一般来说我们需要做一些初始化工作；getType 用来返回一个 Uri 请求所对应的 MIME 类型（媒体类型），比如图片、视频等，这个媒体类型还是有点复杂的，如果我们的应用不关注这个选项，可以直接在这个方法中返回 null 或者 "*/*"；剩下的四个方法对应于 CRUD 操作，即实现对数据表的增删改查功能。根据 Binder 的工作原理，我们知道这六个方法均运行在 ContentProvider 的进程中，除了 onCreate 由系统回调并运行在主线程里，其他五个方法均由外界回调并运行在 Binder 线程池中，这一点在接下来的例子中可以再次证明。

ContentProvider 主要以表格的形式来组织数据，并且可以包含多个表，对于每个表格来说，它们都具有行和列的层次性，行往往对应一条记录，而列对应一条记录中的一个字段，这点和数据库很类似。除了表格的形式，ContentProvider 还支持文件数据，比如图片、视频等。文件数据和表格数据的结构不同，因此处理这类数据时可以在 ContentProvider 中返回文件的句柄给外界从而让外界来访问 ContentProvider 中的文件信息。Android 系统所提供的 MediaStore 功能就是文件类型的 ContentProvider，详细实现可以参考 MediaStore。另外，虽然 ContentProvider 的底层数据看起来很像一个 SQLite 数据库，但是 ContentProvider 对底层的数据存储方式没有任何要求，我们既可以使用 SQLite 数据库，也可以使用普通的文件，甚至可以采用内存中的一个对象来进行数据的存储，这一点在后续的章节中会再次介绍，所以这里不再深入了。

下面看一个最简单的示例，它演示了 ContentProvider 的工作工程。首先创建一个 BookProvider 类，它继承自 ContentProvider 并实现了 ContentProvider 的六个必须需要实现的抽象方法。在下面的代码中，我们什么都没干，尽管如此，这个 BookProvider 也是可以工作的，只是它无法向外界提供有效的数据而已。

```java
// ContentProvider.java
public class BookProvider extends ContentProvider {

    private static final String TAG = "BookProvider";

    @Override
    public boolean onCreate() {
        Log.d(TAG, "onCreate, current thread:" + Thread.currentThread().
        getName());
        return false;
    }

    @Override
    public Cursor query(Uri uri, String[] projection, String selection,
            String[] selectionArgs, String sortOrder) {
        Log.d(TAG, "query, current thread:" + Thread.currentThread().
        getName());
        return null;
    }

    @Override
    public String getType(Uri uri) {
        Log.d(TAG, "getType");
        return null;
    }

    @Override
    public Uri insert(Uri uri, ContentValues values) {
        Log.d(TAG, "insert");
        return null;
    }

    @Override
    public int delete(Uri uri, String selection, String[] selectionArgs) {
        Log.d(TAG, "delete");
        return 0;
    }

    @Override
    public int update(Uri uri, ContentValues values, String selection,
```

```
        String[] selectionArgs) {
    Log.d(TAG, "update");
    return 0;
}

}
```

接着我们需要注册这个 BookProvider，如下所示。其中 android:authorities 是 ContentProvider 的唯一标识，通过这个属性外部应用就可以访问我们的 BookProvider，因此，android:authorities 必须是唯一的，这里建议读者在命名的时候加上包名前缀。为了演示进程间通信，我们让 BookProvider 运行在独立的进程中并给它添加了权限，这样外界应用如果想访问 BookProvider，就必须声明"com.ryg.PROVIDER"这个权限。ContentProvider 的权限还可以细分为读权限和写权限，分别对应 android:readPermission 和 android:writePermission 属性，如果分别声明了读权限和写权限，那么外界应用也必须依次声明相应的权限才可以进行读/写操作，否则外界应用会异常终止。关于权限这一块，请读者自行查阅相关资料，本章不进行详细介绍。

```
<provider
    android:name=".provider.BookProvider"
    android:authorities="com.ryg.chapter_2.book.provider"
    android:permission="com.ryg.PROVIDER"
    android:process=":provider" >
</provider>
```

注册了 ContentProvider 以后，我们就可以在外部应用中访问它了。为了方便演示，这里仍然选择在同一个应用的其他进程中去访问这个 BookProvider，至于在单独的应用中去访问这个 BookProvider，和同一个应用中访问的效果是一样的，读者可以自行试一下（注意要声明对应权限）。

```
//ProviderActivity.java
public class ProviderActivity extends Activity {

    @Override
    protected void onCreate(Bundle savedInstanceState) {
        super.onCreate(savedInstanceState);
        setContentView(R.layout.activity_provider);
        Uri uri = Uri.parse("content://com.ryg.chapter_2.book.provider");
        getContentResolver().query(uri, null, null, null, null);
```

```
            getContentResolver().query(uri, null, null, null, null);
            getContentResolver().query(uri, null, null, null, null);
    }
}
```

在上面的代码中,我们通过 ContentResolver 对象的 query 方法去查询 BookProvider 中的数据,其中"content://com.ryg.chapter_2.book.provider"唯一标识了 BookProvider,而这个标识正是我们前面为 BookProvider 的 android:authorities 属性所指定的值。我们运行后看一下 log。从下面 log 可以看出,BookProvider 中的 query 方法被调用了三次,并且这三次调用不在同一个线程中。可以看出,它们运行在一个 Binder 线程中,前面提到 update、insert 和 delete 方法同样也运行在 Binder 线程中。另外,onCreate 运行在 main 线程中,也就是 UI 线程,所以我们不能在 onCreate 中做耗时操作。

```
D/BookProvider(2091): onCreate, current thread:main
D/BookProvider(2091): query, current thread:Binder Thread #2
D/BookProvider(2091): query, current thread:Binder Thread #1
D/BookProvider(2091): query, current thread:Binder Thread #2
D/MyApplication(2091): application start, process name:com.ryg.chapter_
2:provider
```

到这里,整个 ContentProvider 的流程我们已经跑通了,虽然 ContentProvider 中没有返回任何数据。接下来,在上面的基础上,我们继续完善 BookProvider,从而使其能够对外部应用提供数据。继续本章提到的那个例子,现在我们要提供一个 BookProvider,外部应用可以通过 BookProvider 来访问图书信息,为了更好地演示 ContentProvider 的使用,用户还可以通过 BookProvider 访问到用户信息。为了完成上述功能,我们需要一个数据库来管理图书和用户信息,这个数据库不难实现,代码如下:

```
// DbOpenHelper.java
public class DbOpenHelper extends SQLiteOpenHelper {

    private static final String DB_NAME = "book_provider.db";
    public static final String BOOK_TABLE_NAME = "book";
    public static final String USER_TALBE_NAME = "user";

    private static final int DB_VERSION = 1;

    // 图书和用户信息表
    private String CREATE_BOOK_TABLE = "CREATE TABLE IF NOT EXISTS "
```

```
        + BOOK_TABLE_NAME + "(_id INTEGER PRIMARY KEY," + "name TEXT)";

private String CREATE_USER_TABLE = "CREATE TABLE IF NOT EXISTS "
        + USER_TALBE_NAME + "(_id INTEGER PRIMARY KEY," + "name TEXT," +
        "sex INT)";

public DbOpenHelper(Context context) {
    super(context, DB_NAME, null, DB_VERSION);
}

@Override
public void onCreate(SQLiteDatabase db) {
    db.execSQL(CREATE_BOOK_TABLE);
    db.execSQL(CREATE_USER_TABLE);
}

@Override
public void onUpgrade(SQLiteDatabase db, int oldVersion, int newVersion){
    // TODO ignored
}

}
```

上述代码是一个最简单的数据库的实现，我们借助 SQLiteOpenHelper 来管理数据库的创建、升级和降级。下面我们就要通过 BookProvider 向外界提供上述数据库中的信息了。我们知道，ContentProvider 通过 Uri 来区分外界要访问的的数据集合，在本例中支持外界对 BookProvider 中的 book 表和 user 表进行访问，为了知道外界要访问的是哪个表，我们需要为它们定义单独的 Uri 和 Uri_Code，并将 Uri 和对应的 Uri_Code 相关联，我们可以使用 UriMatcher 的 addURI 方法将 Uri 和 Uri_Code 关联到一起。这样，当外界请求访问 BookProvider 时，我们就可以根据请求的 Uri 来得到 Uri_Code，有了 Uri_Code 我们就可以知道外界想要访问哪个表，然后就可以进行相应的数据操作了，具体代码如下所示。

```
public class BookProvider extends ContentProvider {

    private static final String TAG = "BookProvider";

    public static final String AUTHORITY = "com.ryg.chapter_2.book.
    provider";
```

```java
    public static final Uri BOOK_CONTENT_URI = Uri.parse("content://"
            + AUTHORITY + "/book");
    public static final Uri USER_CONTENT_URI = Uri.parse("content://"
            + AUTHORITY + "/user");

    public static final int BOOK_URI_CODE = 0;
    public static final int USER_URI_CODE = 1;
    private static final UriMatcher sUriMatcher = new UriMatcher(
            UriMatcher.NO_MATCH);

    static {
        sUriMatcher.addURI(AUTHORITY, "book", BOOK_URI_CODE);
        sUriMatcher.addURI(AUTHORITY, "user", USER_URI_CODE);
    }
    ...
}
```

从上面代码可以看出，我们分别为 book 表和 user 表指定了 Uri，分别为"content://com.ryg.chapter_2.book.provider/book"和"content://com.ryg.chapter_2.book.provider/user"，这两个 Uri 所关联的 Uri_Code 分别为 0 和 1。这个关联过程是通过下面的语句来完成的：

```java
sUriMatcher.addURI(AUTHORITY, "book", BOOK_URI_CODE);
sUriMatcher.addURI(AUTHORITY, "user", USER_URI_CODE);
```

将 Uri 和 Uri_Code 管理以后，我们就可以通过如下方式来获取外界所要访问的数据源，根据 Uri 先取出 Uri_Code，根据 Uri_Code 再得到数据表的名称，知道了外界要访问的表，接下来就可以响应外界的增删改查请求了。

```java
private String getTableName(Uri uri) {
    String tableName = null;
    switch (sUriMatcher.match(uri)) {
    case BOOK_URI_CODE:
        tableName = DbOpenHelper.BOOK_TABLE_NAME;
        break;
    case USER_URI_CODE:
        tableName = DbOpenHelper.USER_TALBE_NAME;
        break;
    default:break;
```

```
    }

    return tableName;
}
```

接着，我们就可以实现 query、update、insert、delete 方法了。如下是 query 方法的实现，首先我们要从 Uri 中取出外界要访问的表的名称，然后根据外界传递的查询参数就可以进行数据库的查询操作了，这个过程比较简单。

```
@Override
public Cursor query(Uri uri, String[] projection, String selection,
        String[] selectionArgs, String sortOrder) {
    Log.d(TAG, "query, current thread:" + Thread.currentThread().getName());
    String table = getTableName(uri);
    if (table == null) {
        throw new IllegalArgumentException("Unsupported URI: " + uri);
    }
    return mDb.query(table, projection, selection, selectionArgs, null,
        null, sortOrder, null);
}
```

另外三个方法的实现思想和 query 是类似的，只有一点不同，那就是 update、insert 和 delete 方法会引起数据源的改变，这个时候我们需要通过 ContentResolver 的 notifyChange 方法来通知外界当前 ContentProvider 中的数据已经发生改变。要观察一个 ContentProvider 中的数据改变情况，可以通过 ContentResolver 的 registerContentObserver 方法来注册观察者，通过 unregisterContentObserver 方法来解除观察者。对于这三个方法，这里不再详细解释了，BookProvider 的完整代码如下：

```
public class BookProvider extends ContentProvider {

    private static final String TAG = "BookProvider";

    public static final String AUTHORITY = "com.ryg.chapter_2.book.provider";

    public static final Uri BOOK_CONTENT_URI = Uri.parse("content://"
            + AUTHORITY + "/book");
    public static final Uri USER_CONTENT_URI = Uri.parse("content://"
            + AUTHORITY + "/user");
```

```java
    public static final int BOOK_URI_CODE = 0;
    public static final int USER_URI_CODE = 1;
    private static final UriMatcher sUriMatcher = new UriMatcher(
            UriMatcher.NO_MATCH);

    static {
        sUriMatcher.addURI(AUTHORITY, "book", BOOK_URI_CODE);
        sUriMatcher.addURI(AUTHORITY, "user", USER_URI_CODE);
    }

    private Context mContext;
    private SQLiteDatabase mDb;

    @Override
    public boolean onCreate() {
        Log.d(TAG, "onCreate, current thread:"
                + Thread.currentThread().getName());
        mContext = getContext();
//ContentProvider 创建时，初始化数据库。注意：这里仅仅是为了演示，实际使用中不推荐在主线程中进行耗时的数据库操作
        initProviderData();
        return true;
    }

    private void initProviderData() {
        mDb = new DbOpenHelper(mContext).getWritableDatabase();
        mDb.execSQL("delete from " + DbOpenHelper.BOOK_TABLE_NAME);
        mDb.execSQL("delete from " + DbOpenHelper.USER_TALBE_NAME);
        mDb.execSQL("insert into book values(3,'Android');");
        mDb.execSQL("insert into book values(4,'Ios');");
        mDb.execSQL("insert into book values(5,'Html5');");
        mDb.execSQL("insert into user values(1,'jake',1);");
        mDb.execSQL("insert into user values(2,'jasmine',0);");
    }

    @Override
    public Cursor query(Uri uri, String[] projection, String selection,
            String[] selectionArgs, String sortOrder) {
```

```java
        Log.d(TAG, "query, current thread:" + Thread.currentThread().
    getName());
    String table = getTableName(uri);
    if (table == null) {
        throw new IllegalArgumentException("Unsupported URI: " + uri);
    }
    return mDb.query(table, projection, selection, selectionArgs, null,
    null, sortOrder, null);
}

@Override
public String getType(Uri uri) {
    Log.d(TAG, "getType");
    return null;
}

@Override
public Uri insert(Uri uri, ContentValues values) {
    Log.d(TAG, "insert");
    String table = getTableName(uri);
    if (table == null) {
        throw new IllegalArgumentException("Unsupported URI: " + uri);
    }
    mDb.insert(table, null, values);
    mContext.getContentResolver().notifyChange(uri, null);
    return uri;
}

@Override
public int delete(Uri uri, String selection, String[] selectionArgs) {
    Log.d(TAG, "delete");
    String table = getTableName(uri);
    if (table == null) {
        throw new IllegalArgumentException("Unsupported URI: " + uri);
    }
    int count = mDb.delete(table, selection, selectionArgs);
    if (count > 0) {
        getContext().getContentResolver().notifyChange(uri, null);
    }
    return count;
```

```
}

@Override
public int update(Uri uri, ContentValues values, String selection,
        String[] selectionArgs) {
    Log.d(TAG, "update");
    String table = getTableName(uri);
    if (table == null) {
        throw new IllegalArgumentException("Unsupported URI: " + uri);
    }
    int row = mDb.update(table, values, selection, selectionArgs);
    if (row > 0) {
        getContext().getContentResolver().notifyChange(uri, null);
    }
    return row;
}

private String getTableName(Uri uri) {
    String tableName = null;
    switch (sUriMatcher.match(uri)) {
    case BOOK_URI_CODE:
        tableName = DbOpenHelper.BOOK_TABLE_NAME;
        break;
    case USER_URI_CODE:
        tableName = DbOpenHelper.USER_TALBE_NAME;
        break;
        default:break;
    }

    return tableName;
}
}
```

需要注意的是，query、update、insert、delete 四大方法是存在多线程并发访问的，因此方法内部要做好线程同步。在本例中，由于采用的是 SQLite 并且只有一个 SQLiteDatabase 的连接，所以可以正确应对多线程的情况。具体原因是 SQLiteDatabase 内部对数据库的操作是有同步处理的，但是如果通过多个 SQLiteDatabase 对象来操作数据库就无法保证线程同步，因为 SQLiteDatabase 对象之间无法进行线程同步。如果 ContentProvider 的底层数据集是一块内存的话，比如是 List，在这种情况下同 List 的遍历、插入、删除操作就需要进

行线程同步,否则就会引起并发错误,这点是尤其需要注意的。到这里 BookProvider 已经实现完成了,接着我们在外部访问一下它,看看是否能够正常工作。

```java
public class ProviderActivity extends Activity {
    private static final String TAG = "ProviderActivity";

    @Override
    protected void onCreate(Bundle savedInstanceState) {
        super.onCreate(savedInstanceState);
         ...

        Uri bookUri = Uri.parse("content://com.ryg.chapter_2.book.provider/book");
        ContentValues values = new ContentValues();
        values.put("_id", 6);
        values.put("name", "程序设计的艺术");
        getContentResolver().insert(bookUri, values);
        Cursor bookCursor = getContentResolver().query(bookUri, new String[]{"_id", "name"}, null, null, null);
        while (bookCursor.moveToNext()) {
            Book book = new Book();
            book.bookId = bookCursor.getInt(0);
            book.bookName = bookCursor.getString(1);
            Log.d(TAG, "query book:" + book.toString());
        }
        bookCursor.close();

        Uri userUri = Uri.parse("content://com.ryg.chapter_2.book.provider/user");
        Cursor userCursor = getContentResolver().query(userUri, new String[]{"_id", "name", "sex"}, null, null, null);
        while (userCursor.moveToNext()) {
            User user = new User();
            user.userId = userCursor.getInt(0);
            user.userName = userCursor.getString(1);
            user.isMale = userCursor.getInt(2) == 1;
            Log.d(TAG, "query user:" + user.toString());
        }
```

```
        userCursor.close();
    }
}
```

默认情况下，BookProvider 的数据库中有三本书和两个用户，在上面的代码中，我们首先添加一本书："程序设计的艺术"。接着查询所有的图书，这个时候应该查询出四本书，因为我们刚刚添加了一本。然后查询所有的用户，这个时候应该查询出两个用户。是不是这样呢？我们运行一下程序，看一下 log。

```
D/BookProvider( 1127): insert
D/BookProvider( 1127): query, current thread:Binder Thread #1
D/ProviderActivity( 1114): query book:[bookId:3, bookName:Android]
D/ProviderActivity( 1114): query book:[bookId:4, bookName:Ios]
D/ProviderActivity( 1114): query book:[bookId:5, bookName:Html5]
D/ProviderActivity( 1114): query book:[bookId:6, bookName:程序设计的艺术]
D/MyApplication( 1127): application start, process name:com.ryg.chapter_
2:provider
D/BookProvider( 1127): query, current thread:Binder Thread #3
D/ProviderActivity( 1114): query user:User:{userId:1, userName:jake,
isMale:true}, with child:{null}
D/ProviderActivity( 1114): query user:User:{userId:2, userName:jasmine,
isMale:false}, with child:{null}
```

从上述 log 可以看到，我们的确查询到了 4 本书和 2 个用户，这说明 BookProvider 已经能够正确地处理外部的请求了，读者可以自行验证一下 update 和 delete 操作，这里就不再验证了。同时，由于 ProviderActivity 和 BookProvider 运行在两个不同的进程中，因此，这也构成了进程间的通信。ContentProvider 除了支持对数据源的增删改查这四个操作，还支持自定义调用，这个过程是通过 ContentResolver 的 Call 方法和 ContentProvider 的 Call 方法来完成的。关于使用 ContentProvider 来进行 IPC 就介绍到这里，ContentProvider 本身还有一些细节这里并没有介绍，读者可以自行了解，本章侧重的是各种进程间通信的方法以及它们的区别，因此针对某种特定的方法可能不会介绍得面面俱到。另外，ContentProvider 在后续章节还会有进一步的讲解，主要包括细节问题和工作原理，读者可以阅读后面的相应章节。

2.4.6　使用 Socket

在本节中，我们通过 Socket 来实现进程间的通信。Socket 也称为"套接字"，是网络通信中的概念，它分为流式套接字和用户数据报套接字两种，分别对应于网络的传输控制层中的 TCP 和 UDP 协议。TCP 协议是面向连接的协议，提供稳定的双向通信功能，TCP

连接的建立需要经过"三次握手"才能完成,为了提供稳定的数据传输功能,其本身提供了超时重传机制,因此具有很高的稳定性;而 UDP 是无连接的,提供不稳定的单向通信功能,当然 UDP 也可以实现双向通信功能。在性能上,UDP 具有更好的效率,其缺点是不保证数据一定能够正确传输,尤其是在网络拥塞的情况下。关于 TCP 和 UDP 的介绍就这么多,更详细的资料请查看相关网络资料。接下来我们演示一个跨进程的聊天程序,两个进程可以通过 Socket 来实现信息的传输,Socket 本身可以支持传输任意字节流,这里为了简单起见,仅仅传输文本信息,很显然,这是一种 IPC 方式。

使用 Socket 来进行通信,有两点需要注意,首先需要声明权限:

```
<uses-permission android:name="android.permission.INTERNET" />
<uses-permission android:name="android.permission.ACCESS_NETWORK_STATE" />
```

其次要注意不能在主线程中访问网络,因为这会导致我们的程序无法在 Android 4.0 及其以上的设备中运行,会抛出如下异常:android.os.NetworkOnMainThreadException。而且进行网络操作很可能是耗时的,如果放在主线程中会影响程序的响应效率,从这方面来说,也不应该在主线程中访问网络。下面就开始设计我们的聊天室程序了,比较简单,首先在远程 Service 建立一个 TCP 服务,然后在主界面中连接 TCP 服务,连接上了以后,就可以给服务端发消息。对于我们发送的每一条文本消息,服务端都会随机地回应我们一句话。为了更好地展示 Socket 的工作机制,在服务端我们做了处理,使其能够和多个客户端同时建立连接并响应。

先看一下服务端的设计,当 Service 启动时,会在线程中建立 TCP 服务,这里监听的是 8688 端口,然后就可以等待客户端的连接请求。当有客户端连接时,就会生成一个新的 Socket,通过每次新创建的 Socket 就可以分别和不同的客户端通信了。服务端每收到一次客户端的消息就会随机回复一句话给客户端。当客户端断开连接时,服务端这边也会相应的关闭对应 Socket 并结束通话线程,这点是如何做到的呢?方法有很多,这里是通过判断服务端输入流的返回值来确定的,当客户端断开连接后,服务端这边的输入流会返回 null,这个时候我们就知道客户端退出了。服务端的代码如下所示。

```java
public class TCPServerService extends Service {

    private boolean mIsServiceDestoryed = false;
    private String[] mDefinedMessages = new String[] {
            "你好啊,哈哈",
            "请问你叫什么名字呀?",
```

```
            "今天北京天气不错啊, shy",
            "你知道吗？我可是可以和多个人同时聊天的哦",
            "给你讲个笑话吧：据说爱笑的人运气不会太差，不知道真假。"
    };

    @Override
    public void onCreate() {
        new Thread(new TcpServer()).start();
        super.onCreate();
    }

    @Override
    public IBinder onBind(Intent intent) {
        return null;
    }

    @Override
    public void onDestroy() {
        mIsServiceDestoryed = true;
        super.onDestroy();
    }

    private class TcpServer implements Runnable {

        @SuppressWarnings("resource")
        @Override
        public void run() {
            ServerSocket serverSocket = null;
            try {
                //监听本地 8688 端口
                serverSocket = new ServerSocket(8688);
            } catch (IOException e) {
                System.err.println("establish tcp server failed, port:8688");
                e.printStackTrace();
                return;
            }

            while (!mIsServiceDestoryed) {
                try {
                    // 接收客户端请求
```

```java
            final Socket client = serverSocket.accept();
            System.out.println("accept");
            new Thread() {
                @Override
                public void run() {
                    try {
                        responseClient(client);
                    } catch (IOException e) {
                        e.printStackTrace();
                    }
                };
            }.start();

        } catch (IOException e) {
            e.printStackTrace();
        }
    }
}

private void responseClient(Socket client) throws IOException {
    // 用于接收客户端消息
    BufferedReader in = new BufferedReader(new InputStreamReader(
            client.getInputStream()));
    // 用于向客户端发送消息
    PrintWriter out = new PrintWriter(new BufferedWriter(
            new OutputStreamWriter(client.getOutputStream())), true);
    out.println("欢迎来到聊天室!");
    while (!mIsServiceDestoryed) {
        String str = in.readLine();
        System.out.println("msg from client:" + str);
        if (str == null) {
            //客户端断开连接
            break;
        }
        int i = new Random().nextInt(mDefinedMessages.length);
        String msg = mDefinedMessages[i];
        out.println(msg);
        System.out.println("send :" + msg);
    }
```

```
        System.out.println("client quit.");
        // 关闭流
        MyUtils.close(out);
        MyUtils.close(in);
        client.close();
    }
}
```

接着看一下客户端，客户端 Activity 启动时，会在 onCreate 中开启一个线程去连接服务端 Socket，至于为什么用线程在前面已经做了介绍。为了确定能够连接成功，这里采用了超时重连的策略，每次连接失败后都会重新建立尝试建立连接。当然为了降低重试机制的开销，我们加入了休眠机制，即每次重试的时间间隔为 1000 毫秒。

```
Socket socket = null;
while (socket == null) {
    try {
        socket = new Socket("localhost", 8688);
        mClientSocket = socket;
        mPrintWriter = new PrintWriter(new BufferedWriter(
                new OutputStreamWriter(socket.getOutputStream())), true);
        mHandler.sendEmptyMessage(MESSAGE_SOCKET_CONNECTED);
        System.out.println("connect server success.");
    } catch (IOException e) {
        SystemClock.sleep(1000);
        System.out.println("connect tcp server failed, retry...");
    }
}
```

服务端连接成功以后，就可以和服务端进行通信了。下面的代码在线程中通过 while 循环不断地去读取服务端发送过来的消息，同时当 Activity 退出时，就退出循环并终止线程。

```
// 接收服务器端的消息
BufferedReader br = new BufferedReader(new InputStreamReader(
        socket.getInputStream()));
while (!TCPClientActivity.this.isFinishing()) {
    String msg = br.readLine();
    System.out.println("receive :" + msg);
    if (msg != null) {
```

```java
            String time = formatDateTime(System.currentTimeMillis());
            final String showedMsg = "server " + time + ":" + msg
                + "\n";
            mHandler.obtainMessage(MESSAGE_RECEIVE_NEW_MSG, showedMsg)
                .sendToTarget();
        }
    }
```

同时，当 Activity 退出时，还要关闭当前的 Socket，如下所示。

```java
@Override
protected void onDestroy() {
    if (mClientSocket != null) {
        try {
            mClientSocket.shutdownInput();
            mClientSocket.close();
        } catch (IOException e) {
            e.printStackTrace();
        }
    }
    super.onDestroy();
}
```

接着是发送消息的过程，这个就很简单了，这里不再详细说明。客户端的完整代码如下：

```java
public class TCPClientActivity extends Activity implements OnClickListener {

    private static final int MESSAGE_RECEIVE_NEW_MSG = 1;
    private static final int MESSAGE_SOCKET_CONNECTED = 2;

    private Button mSendButton;
    private TextView mMessageTextView;
    private EditText mMessageEditText;

    private PrintWriter mPrintWriter;
    private Socket mClientSocket;

    @SuppressLint("HandlerLeak")
    private Handler mHandler = new Handler() {
```

```java
        @Override
        public void handleMessage(Message msg) {
            switch (msg.what) {
            case MESSAGE_RECEIVE_NEW_MSG: {
                mMessageTextView.setText(mMessageTextView.getText()
                        + (String) msg.obj);
                break;
            }
            case MESSAGE_SOCKET_CONNECTED: {
                mSendButton.setEnabled(true);
                break;
            }
            default:
                break;
            }
        }
    };

    @Override
    protected void onCreate(Bundle savedInstanceState) {
        super.onCreate(savedInstanceState);
        setContentView(R.layout.activity_tcpclient);
        mMessageTextView = (TextView) findViewById(R.id.msg_container);
        mSendButton = (Button) findViewById(R.id.send);
        mSendButton.setOnClickListener(this);
        mMessageEditText = (EditText) findViewById(R.id.msg);
        Intent service = new Intent(this, TCPServerService.class);
        startService(service);
        new Thread() {
            @Override
            public void run() {
                connectTCPServer();
            }
        }.start();
    }

    @Override
    protected void onDestroy() {
        if (mClientSocket != null) {
```

```java
            try {
                mClientSocket.shutdownInput();
                mClientSocket.close();
            } catch (IOException e) {
                e.printStackTrace();
            }
        }
        super.onDestroy();
    }

    @Override
    public void onClick(View v) {
        if (v == mSendButton) {
            final String msg = mMessageEditText.getText().toString();
            if (!TextUtils.isEmpty(msg) && mPrintWriter != null) {
                mPrintWriter.println(msg);
                mMessageEditText.setText("");
                String time = formatDateTime(System.currentTimeMillis());
                final String showedMsg = "self " + time + ":" + msg + "\n";
                mMessageTextView.setText(mMessageTextView.getText() + showedMsg);
            }
        }
    }

    @SuppressLint("SimpleDateFormat")
    private String formatDateTime(long time) {
        return new SimpleDateFormat("(HH:mm:ss)").format(new Date(time));
    }

    private void connectTCPServer() {
        Socket socket = null;
        while (socket == null) {
            try {
                socket = new Socket("localhost", 8688);
                mClientSocket = socket;
                mPrintWriter = new PrintWriter(new BufferedWriter(
                        new OutputStreamWriter(socket.getOutputStream())),
                        true);
```

```java
            mHandler.sendEmptyMessage(MESSAGE_SOCKET_CONNECTED);
            System.out.println("connect server success");
        } catch (IOException e) {
            SystemClock.sleep(1000);
            System.out.println("connect tcp server failed, retry...");
        }
    }

    try {
        // 接收服务器端的消息
        BufferedReader br = new BufferedReader(new InputStreamReader(
                socket.getInputStream()));
        while (!TCPClientActivity.this.isFinishing()) {
            String msg = br.readLine();
            System.out.println("receive :" + msg);
            if (msg != null) {
                String time = formatDateTime(System.currentTimeMillis());
                final String showedMsg = "server " + time + ":" + msg
                        + "\n";
                mHandler.obtainMessage(MESSAGE_RECEIVE_NEW_MSG, showed-
                Msg)
                        .sendToTarget();
            }
        }
        System.out.println("quit...");
        MyUtils.close(mPrintWriter);
        MyUtils.close(br);
        socket.close();
    } catch (IOException e) {
        e.printStackTrace();
    }
}
```

上述就是通过 Socket 来进行进程间通信的实例，除了采用 TCP 套接字，还可以采用 UDP 套接字。另外，上面的例子仅仅是一个示例，实际上通过 Socket 不仅仅能实现进程间的通信，还可以实现设备间的通信，当然前提是这些设备之间的 IP 地址互相可见，这其中又涉及许多复杂的概念，这里就不一一介绍了。下面看一下上述例子的运行效果，如图 2-9 所示。

图 2-9　Socket 通信示例

2.5　Binder 连接池

上面我们介绍了不同的 IPC 方式，我们知道，不同的 IPC 方式有不同的特点和适用场景，当然这个问题会在 2.6 节进行介绍，在本节中要再次介绍一下 ADIL，原因是 AIDL 是一种最常用的进程间通信方式，是日常开发中涉及进程间通信时的首选，所以我们需要额外强调一下它。

如何使用 AIDL 在上面的一节中已经进行了介绍，这里在回顾一下大致流程：首先创建一个 Service 和一个 AIDL 接口，接着创建一个类继承自 AIDL 接口中的 Stub 类并实现 Stub 中的抽象方法，在 Service 的 onBind 方法中返回这个类的对象，然后客户端就可以绑定服务端 Service，建立连接后就可以访问远程服务端的方法了。

上述过程就是典型的 AIDL 的使用流程。这本来也没什么问题，但是现在考虑一种情况：公司的项目越来越庞大了，现在有 10 个不同的业务模块都需要使用 AIDL 来进行进程间通信，那我们该怎么处理呢？也许你会说："就按照 AIDL 的实现方式一个个来吧"，这是可以的，如果用这种方法，首先我们需要创建 10 个 Service，这好像有点多啊！如果有 100 个地方需要用到 AIDL 呢，先创建 100 个 Service？到这里，读者应该明白问题所在了。随着 AIDL 数量的增加，我们不能无限制地增加 Service，Service 是四大组件之一，本身就是一种系统资源。而且太多的 Service 会使得我们的应用看起来很重量级，因为正在运行的

Service 可以在应用详情页看到，当我们的应用详情显示有 10 个服务正在运行时，这看起来并不是什么好事。针对上述问题，我们需要减少 Service 的数量，将所有的 AIDL 放在同一个 Service 中去管理。

在这种模式下，整个工作机制是这样的：每个业务模块创建自己的 AIDL 接口并实现此接口，这个时候不同业务模块之间是不能有耦合的，所有实现细节我们要单独开来，然后向服务端提供自己的唯一标识和其对应的 Binder 对象；对于服务端来说，只需要一个 Service 就可以了，服务端提供一个 queryBinder 接口，这个接口能够根据业务模块的特征来返回相应的 Binder 对象给它们，不同的业务模块拿到所需的 Binder 对象后就可以进行远程方法调用了。由此可见，Binder 连接池的主要作用就是将每个业务模块的 Binder 请求统一转发到远程 Service 中去执行，从而避免了重复创建 Service 的过程，它的工作原理如图 2-10 所示。

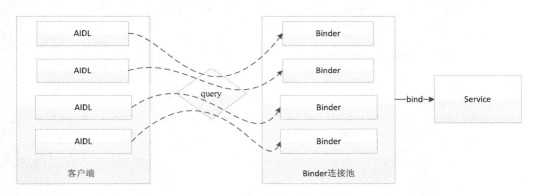

图 2-10　Binder 连接池的工作原理

通过上面的理论介绍，也许还有点不好理解，下面对 Binder 连接池的代码实现做一下说明。首先，为了说明问题，我们提供了两个 AIDL 接口（ISecurityCenter 和 ICompute）来模拟上面提到的多个业务模块都要使用 AIDL 的情况，其中 ISecurityCenter 接口提供加解密功能，声明如下：

```
interface ISecurityCenter {
    String encrypt(String content);
    String decrypt(String password);
}
```

而 ICompute 接口提供计算加法的功能，声明如下：

```
interface ICompute {
```

```
    int add(int a, int b);
}
```

虽然说上面两个接口的功能都比较简单，但是用于分析 Binder 连接池的工作原理已经足够了，读者可以写出更复杂的例子。接着看一下上面两个 AIDL 接口的实现，也比较简单，代码如下：

```
public class SecurityCenterImpl extends ISecurityCenter.Stub {

    private static final char SECRET_CODE = '^';

    @Override
    public String encrypt(String content) throws RemoteException {
        char[] chars = content.toCharArray();
        for (int i = 0; i < chars.length; i++) {
            chars[i] ^= SECRET_CODE;
        }
        return new String(chars);
    }

    @Override
    public String decrypt(String password) throws RemoteException {
        return encrypt(password);
    }

}

public class ComputeImpl extends ICompute.Stub {

    @Override
    public int add(int a, int b) throws RemoteException {
        return a + b;
    }

}
```

现在业务模块的 AIDL 接口定义和实现都已经完成了，注意这里并没有为每个模块的 AIDL 单独创建 Service，接下来就是服务端和 Binder 连接池的工作了。

首先，为 Binder 连接池创建 AIDL 接口 IBinderPool.aidl，代码如下所示。

```
interface IBinderPool {
```

```
/**
 * @param binderCode, the unique token of specific Binder<br/>
 * @return specific Binder who's token is binderCode.
 */
IBinder queryBinder(int binderCode);
}
```

接着，为 Binder 连接池创建远程 Service 并实现 IBinderPool，下面是 queryBinder 的具体实现，可以看到请求转发的实现方法，当 Binder 连接池连接上远程服务时，会根据不同模块的标识即 binderCode 返回不同的 Binder 对象，通过这个 Binder 对象所执行的操作全部发生在远程服务端。

```
@Override
public IBinder queryBinder(int binderCode) throws RemoteException {
    IBinder binder = null;
    switch (binderCode) {
    case BINDER_SECURITY_CENTER: {
        binder = new SecurityCenterImpl();
        break;
    }
    case BINDER_COMPUTE: {
        binder = new ComputeImpl();
        break;
    }
    default:
        break;
    }

    return binder;
}
```

远程 Service 的实现就比较简单了，代码如下所示。

```
public class BinderPoolService extends Service {

    private static final String TAG = "BinderPoolService";

    private Binder mBinderPool = new BinderPool.BinderPoolImpl();
```

```java
@Override
public void onCreate() {
    super.onCreate();
}

@Override
public IBinder onBind(Intent intent) {
    Log.d(TAG, "onBind");
    return mBinderPool;
}

@Override
public void onDestroy() {
    super.onDestroy();
}
```

下面还剩下 Binder 连接池的具体实现，在它的内部首先它要去绑定远程服务，绑定成功后，客户端就可以通过它的 queryBinder 方法去获取各自对应的 Binder，拿到所需的 Binder 以后，不同业务模块就可以进行各自的操作了，Binder 连接池的代码如下所示。

```java
public class BinderPool {
    private static final String TAG = "BinderPool";
    public static final int BINDER_NONE = -1;
    public static final int BINDER_COMPUTE = 0;
    public static final int BINDER_SECURITY_CENTER = 1;

    private Context mContext;
    private IBinderPool mBinderPool;
    private static volatile BinderPool sInstance;
    private CountDownLatch mConnectBinderPoolCountDownLatch;

    private BinderPool(Context context) {
        mContext = context.getApplicationContext();
        connectBinderPoolService();
    }

    public static BinderPool getInsance(Context context) {
        if (sInstance == null) {
```

```java
        synchronized (BinderPool.class) {
            if (sInstance == null) {
                sInstance = new BinderPool(context);
            }
        }
    }
    return sInstance;
}

private synchronized void connectBinderPoolService() {
    mConnectBinderPoolCountDownLatch = new CountDownLatch(1);
    Intent service = new Intent(mContext, BinderPoolService.class);
    mContext.bindService(service, mBinderPoolConnection,
            Context.BIND_AUTO_CREATE);
    try {
        mConnectBinderPoolCountDownLatch.await();
    } catch (InterruptedException e) {
        e.printStackTrace();
    }
}

/**
 * query binder by binderCode from binder pool
 *
 * @param binderCode
 *          the unique token of binder
 * @return binder who's token is binderCode<br>
 *          return null when not found or BinderPoolService died.
 */
public IBinder queryBinder(int binderCode) {
    IBinder binder = null;
    try {
        if (mBinderPool != null) {
            binder = mBinderPool.queryBinder(binderCode);
        }
    } catch (RemoteException e) {
        e.printStackTrace();
    }
    return binder;
}
```

```java
private ServiceConnection mBinderPoolConnection = new ServiceConnection() {

    @Override
    public void onServiceDisconnected(ComponentName name) {
        // ignored.
    }

    @Override
    public void onServiceConnected(ComponentName name, IBinder service) {
        mBinderPool = IBinderPool.Stub.asInterface(service);
        try {
            mBinderPool.asBinder().linkToDeath(mBinderPoolDeathRecipient, 0);
        } catch (RemoteException e) {
            e.printStackTrace();
        }
        mConnectBinderPoolCountDownLatch.countDown();
    }
};

private IBinder.DeathRecipient mBinderPoolDeathRecipient = new IBinder.DeathRecipient() {
    @Override
    public void binderDied() {
        Log.w(TAG, "binder died.");
        mBinderPool.asBinder().unlinkToDeath(mBinderPoolDeathRecipient, 0);
        mBinderPool = null;
        connectBinderPoolService();
    }
};

public static class BinderPoolImpl extends IBinderPool.Stub {

    public BinderPoolImpl() {
        super();
    }
```

```
        @Override
        public IBinder queryBinder(int binderCode) throws RemoteException {
            IBinder binder = null;
            switch (binderCode) {
            case BINDER_SECURITY_CENTER: {
                binder = new SecurityCenterImpl();
                break;
            }
            case BINDER_COMPUTE: {
                binder = new ComputeImpl();
                break;
            }
            default:
                break;
            }

            return binder;
        }
    }
}
```

Binder 连接池的具体实现就分析完了，它的好处是显然易见的，针对上面的例子，我们只需要创建一个 Service 即可完成多个 AIDL 接口的工作，下面我们来验证一下效果。新创建一个 Activity，在线程中执行如下操作：

```
private void doWork() {
    BinderPool binderPool = BinderPool.getInsance(BinderPoolActivity.
    this);
    IBinder securityBinder = binderPool
            .queryBinder(BinderPool.BINDER_SECURITY_CENTER);
    ;
    mSecurityCenter = (ISecurityCenter) SecurityCenterImpl
            .asInterface(securityBinder);
    Log.d(TAG, "visit ISecurityCenter");
    String msg = "helloworld-安卓";
    System.out.println("content:" + msg);
    try {
        String password = mSecurityCenter.encrypt(msg);
        System.out.println("encrypt:" + password);
```

```
        System.out.println("decrypt:" + mSecurityCenter.decrypt(password));
    } catch (RemoteException e) {
        e.printStackTrace();
    }

    Log.d(TAG, "visit ICompute");
    IBinder computeBinder = binderPool
            .queryBinder(BinderPool.BINDER_COMPUTE);
    mCompute = ComputeImpl.asInterface(computeBinder);
    try {
        System.out.println("3+5=" + mCompute.add(3, 5));
    } catch (RemoteException e) {
        e.printStackTrace();
    }
}
```

在上述代码中，我们先后调用了 ISecurityCenter 和 ICompute 这两个 AIDL 接口中的方法，看一下 log，很显然，工作正常。

```
D/BinderPoolActivity(20270): visit ISecurityCenter
I/System.out(20270): content:helloworld-安卓
I/System.out(20270): encrypt:6;221)1,2:s 骞匍
I/System.out(20270): decrypt:helloworld-安卓
D/BinderPoolActivity(20270): visit ICompute
I/System.out(20270): 3+5=8
```

这里需要额外说明一下，为什么要在线程中去执行呢？这是因为在 Binder 连接池的实现中，我们通过 CountDownLatch 将 bindService 这一异步操作转换成了同步操作，这就意味着它有可能是耗时的，然后就是 Binder 方法的调用过程也可能是耗时的，因此不建议放在主线程去执行。注意到 BinderPool 是一个单例实现，因此在同一个进程中只会初始化一次，所以如果我们提前初始化 BinderPool，那么可以优化程序的体验，比如我们可以放在 Application 中提前对 BinderPool 进行初始化，虽然这不能保证当我们调用 BinderPool 时它一定是初始化好的，但是在大多数情况下，这种初始化工作（绑定远程服务）的时间开销（如果 BinderPool 没有提前初始化完成的话）是可以接受的。另外，BinderPool 中有断线重连的机制，当远程服务意外终止时，BinderPool 会重新建立连接，这个时候如果业务模块中的 Binder 调用出现了异常，也需要手动去重新获取最新的 Binder 对象，这个是需要注意的。

有了 BinderPool 可以大大方便日常的开发工作，比如如果有一个新的业务模块需要添

加新的 AIDL，那么在它实现了自己的 AIDL 接口后，只需要修改 BinderPoolImpl 中的 queryBinder 方法，给自己添加一个新的 binderCode 并返回对应的 Binder 对象即可，不需要做其他修改，也不需要创建新的 Service。由此可见，BinderPool 能够极大地提高 AIDL 的开发效率，并且可以避免大量的 Service 创建，因此，建议在 AIDL 开发工作中引入 BinderPool 机制。

2.6 选用合适的 IPC 方式

在上面的一节中，我们介绍了各种各样的 IPC 方式，那么到底它们有什么不同呢？我们到底该使用哪一种呢？本节就为读者解答这些问题，具体内容如表 2-2 所示。通过表 2-2，可以明确地看出不同 IPC 方式的优缺点和适用场景，那么在实际的开发中，只要我们选择合适的 IPC 方式就可以轻松完成多进程的开发场景。

表 2-2　IPC 方式的优缺点和适用场景

名称	优点	缺点	适用场景
Bundle	简单易用	只能传输 Bundle 支持的数据类型	四大组件间的进程间通信
文件共享	简单易用	不适合高并发场景，并且无法做到进程间的即时通信	无并发访问情形，交换简单的数据实时性不高的场景
AIDL	功能强大，支持一对多并发通信，支持实时通信	使用稍复杂，需要处理好线程同步	一对多通信且有 RPC 需求
Messenger	功能一般，支持一对多串行通信，支持实时通信	不能很好处理高并发情形，不支持 RPC，数据通过 Message 进行传输，因此只能传输 Bundle 支持的数据类型	低并发的一对多即时通信，无 RPC 需求，或者无须要返回结果的 RPC 需求
ContentProvider	在数据源访问方面功能强大，支持一对多并发数据共享，可通过 Call 方法扩展其他操作	可以理解为受约束的 AIDL，主要提供数据源的 CRUD 操作	一对多的进程间的数据共享
Socket	功能强大，可以通过网络传输字节流，支持一对多并发实时通信	实现细节稍微有点烦琐，不支持直接的 RPC	网络数据交换

第 3 章　View 的事件体系

本章将介绍 Android 中十分重要的一个概念：View，虽然说 View 不属于四大组件，但是它的作用堪比四大组件，甚至比 Receiver 和 Provider 的重要性都要大。在 Android 开发中，Activity 承担这可视化的功能，同时 Android 系统提供了很多基础控件，常见的有 Button、TextView、CheckBox 等。很多时候仅仅使用系统提供的控件是不能满足需求的，因此我们就需要能够根据需求进行新控件的定义，而控件的自定义就需要对 Android 的 View 体系有深入的理解，只有这样才能写出完美的自定义控件。同时 Android 手机属于移动设备，移动设备的一个特点就是用户可以直接通过屏幕来进行一系列操作，一个典型的场景就是屏幕的滑动，用户可以通过滑动来切换到不同的界面。很多情况下我们的应用　都需要支持滑动操作，当处于不同层级的 View 都可以响应用户的滑动操作时，就会带来　一个问题，那就是滑动冲突。如何解决滑动冲突呢？这对于初学者来说的确是个头疼的问题，其实解决滑动冲突本不难，它需要读者对 View 的事件分发机制有一定的了解，在这个基础上，我们就可以利于这个特性从而得出滑动冲突的解决方法。上述这些内容就是本章所要介绍的内容，同时，View 的内部工作原理和自定义 View 相关的知识会在第 4 章进行介绍。

3.1　View 基础知识

本节主要介绍 View 的一些基础知识，从而为更好地介绍后续的内容做铺垫。主要介绍的内容有：View 的位置参数、MotionEvent 和 TouchSlop 对象、VelocityTracker、GestureDetector 和 Scroller 对象，通过对这些基础知识的介绍，可以方便读者理解更复杂的内容。类似的基础概念还有不少，但是本节所介绍的都是一些比较常用的，其他不常用的基础概念读者可以自行了解。

3.1.1 什么是 View

在介绍 View 的基础知识之前，我们首先要知道到底什么是 View。View 是 Android 中所有控件的基类，不管是简单的 Button 和 TextView 还是复杂的 RelativeLayout 和 ListView，它们的共同基类都是 View。所以说，View 是一种界面层的控件的一种抽象，它代表了一个控件。除了 View，还有 ViewGroup，从名字来看，它可以被翻译为控件组，言外之意是 ViewGroup 内部包含了许多个控件，即一组 View。在 Android 的设计中，ViewGroup 也继承了 View，这就意味着 View 本身就可以是单个控件也可以是由多个控件组成的一组控件，通过这种关系就形成了 View 树的结构，这和 Web 前端中的 DOM 树的概念是相似的。根据这个概念，我们知道，Button 显然是个 View，而 LinearLayout 不但是一个 View 而且还是一个 ViewGroup，而 ViewGroup 内部是可以有子 View 的，这个子 View 同样还可以是 ViewGroup，依此类推。

明白 View 的这种层级关系有助于理解 View 的工作机制。如图 3-1 所示，可以看到自定义的 TestButton 是一个 View，它继承了 TextView，而 TextView 则直接继承了 View，因此不管怎么说，TestButton 都是一个 View，同理我们也可以构造出一个继承自 ViewGroup 的控件。

图 3-1　TestButton 的层次结构

3.1.2 View 的位置参数

View 的位置主要由它的四个顶点来决定，分别对应于 View 的四个属性：top、left、right、bottom，其中 top 是左上角纵坐标，left 是左上角横坐标，right 是右下角横坐标，bottom 是右下角纵坐标。需要注意的是，这些坐标都是相对于 View 的父容器来说的，因此它是一种相对坐标，View 的坐标和父容器的关系如图 3-2 所示。在 Android 中，x 轴和 y 轴的正方向分别为右和下，这点不难理解，不仅仅是 Android，大部分显示系统都是按照这个标准来定义坐标系的。

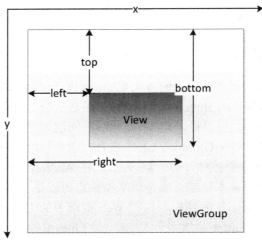

图 3-2　View 的位置坐标和父容器的关系

根据图 3-2，我们很容易得出 View 的宽高和坐标的关系：

```
width = right - left
height = bottom - top
```

那么如何得到 View 的这四个参数呢？也很简单，在 View 的源码中它们对应于 mLeft、mRight、mTop 和 mBottom 这四个成员变量，获取方式如下所示。

- Left = getLeft();

- Right = getRight();

- Top = getTop；

- Bottom = getBottom()。

从 Android3.0 开始，View 增加了额外的几个参数：x、y、translationX 和 translationY，其中 x 和 y 是 View 左上角的坐标，而 translationX 和 translationY 是 View 左上角相对于父容器的偏移量。这几个参数也是相对于父容器的坐标，并且 translationX 和 translationY 的默认值是 0，和 View 的四个基本的位置参数一样，View 也为它们提供了 get/set 方法，这几个参数的换算关系如下所示。

$$x = left + translationX$$

```
y = top + translationY
```

需要注意的是，View 在平移的过程中，top 和 left 表示的是原始左上角的位置信息，其值并不会发生改变，此时发生改变的是 x、y、translationX 和 translationY 这四个参数。

3.1.3 MotionEvent 和 TouchSlop

1. MotionEvent

在手指接触屏幕后所产生的一系列事件中，典型的事件类型有如下几种：

- ACTION_DOWN——手指刚接触屏幕；
- ACTION_MOVE——手指在屏幕上移动；
- ACTION_UP——手指从屏幕上松开的一瞬间。

正常情况下，一次手指触摸屏幕的行为会触发一系列点击事件，考虑如下几种情况：

- 点击屏幕后离开松开，事件序列为 DOWN -> UP；
- 点击屏幕滑动一会再松开，事件序列为 DOWN -> MOVE -> … > MOVE -> UP。

上述三种情况是典型的事件序列，同时通过 MotionEvent 对象我们可以得到点击事件发生的 x 和 y 坐标。为此，系统提供了两组方法：getX/getY 和 getRawX/getRawY。它们的区别其实很简单，getX/getY 返回的是相对于当前 View 左上角的 x 和 y 坐标，而 getRawX/getRawY 返回的是相对于手机屏幕左上角的 x 和 y 坐标。

2. TouchSlop

TouchSlop 是系统所能识别出的被认为是滑动的最小距离，换句话说，当手指在屏幕上滑动时，如果两次滑动之间的距离小于这个常量，那么系统就不认为你是在进行滑动操作。原因很简单：滑动的距离太短，系统不认为它是滑动。这是一个常量，和设备有关，在不同设备上这个值可能是不同的，通过如下方式即可获取这个常量：ViewConfiguration.get(getContext()).getScaledTouchSlop()。这个常量有什么意义呢？当我们在处理滑动时，可以利用这个常量来做一些过滤，比如当两次滑动事件的滑动距离小于这个值，我们就可以认为未达到滑动距离的临界值，因此就可以认为它们不是滑动，这样做可以有更好的用户体验。其实如果细心的话，可以在源码中找到这个常量的定义，在 frameworks/base/core/res/res/values/config.xml 文件中，如下所示。这个"config_viewConfigurationTouchSlop"对应的就是这个常量的定义。

```
<!-- Base "touch slop" value used by ViewConfiguration as a movement threshold
where scrolling should begin. -->
<dimen name="config_viewConfigurationTouchSlop">8dp</dimen>
```

3.1.4 VelocityTracker、GestureDetector 和 Scroller

1. VelocityTracker

速度追踪，用于追踪手指在滑动过程中的速度，包括水平和竖直方向的速度。它的使用过程很简单，首先，在 View 的 onTouchEvent 方法中追踪当前单击事件的速度：

```
VelocityTracker velocityTracker = VelocityTracker.obtain();
velocityTracker.addMovement(event);
```

接着，当我们想知道当前的滑动速度时，这个时候可以采用如下方式来获得当前的速度：

```
velocityTracker.computeCurrentVelocity(1000);
int xVelocity = (int) velocityTracker.getXVelocity();
int yVelocity = (int) velocityTracker.getYVelocity();
```

在这一步中有两点需要注意，第一点，获取速度之前必须先计算速度，即 getXVelocity 和 getYVelocity 这两个方法的前面必须要调用 computeCurrentVelocity 方法；第二点，这里的速度是指一段时间内手指所滑过的像素数，比如将时间间隔设为 1000ms 时，在 1s 内，手指在水平方向从左向右滑过 100 像素，那么水平速度就是 100。注意速度可以为负数，当手指从右往左滑动时，水平方向速度即为负值，这个需要理解一下。速度的计算可以用如下公式来表示：

速度 =（终点位置 – 起点位置）/ 时间段

根据上面的公式再加上 Android 系统的坐标系，可以知道，手指逆着坐标系的正方向滑动，所产生的速度就为负值。另外，computeCurrentVelocity 这个方法的参数表示的是一个时间单元或者说时间间隔，它的单位是毫秒（ms），计算速度时得到的速度就是在这个时间间隔内手指在水平或竖直方向上所滑动的像素数。针对上面的例子，如果我们通过 velocityTracker.computeCurrentVelocity(100) 来获取速度，那么得到的速度就是手指在 100ms 内所滑过的像素数，因此水平速度就成了 10 像素/每 100ms（这里假设滑动过程是匀速的），即水平速度为 10，这点需要好好理解一下。

最后，当不需要使用它的时候，需要调用 clear 方法来重置并回收内存：

```
velocityTracker.clear();
velocityTracker.recycle();
```

上面就是如何使用 VelocityTracker 对象的全过程，看起来并不复杂。

2．GestureDetector

手势检测，用于辅助检测用户的单击、滑动、长按、双击等行为。要使用 GestureDetector 也不复杂，参考如下过程。

首先，需要创建一个 GestureDetector 对象并实现 OnGestureListener 接口，根据需要我们还可以实现 OnDoubleTapListener 从而能够监听双击行为：

```
GestureDetector mGestureDetector = new GestureDetector(this);
//解决长按屏幕后无法拖动的现象
mGestureDetector.setIsLongpressEnabled(false);
```

接着，接管目标 View 的 onTouchEvent 方法，在待监听 View 的 onTouchEvent 方法中添加如下实现：

```
boolean consume = mGestureDetector.onTouchEvent(event);
return consume;
```

做完了上面两步，我们就可以有选择地实现 OnGestureListener 和 OnDoubleTapListener 中的方法了，这两个接口中的方法介绍如表 3-1 所示。

表 3-1　OnGestureListener 和 OnDoubleTapListener 中的方法介绍

方 法 名	描　　述	所 属 接 口
onDown	手指轻轻触摸屏幕的一瞬间，由 1 个 ACTION_DOWN 触发	OnGestureListener
onShowPress	手指轻轻触摸屏幕，尚未松开或拖动，由 1 个 ACTION_DOWN 触发 * 注意和 onDown() 的区别，它强调的是没有松开或者拖动的状态	OnGestureListener
onSingleTapUp	手指（轻轻触摸屏幕后）松开，伴随着 1 个 MotionEvent ACTION_UP 而触发，这是单击行为	OnGestureListener
onScroll	手指按下屏幕并拖动，由 1 个 ACTION_DOWN，多个 ACTION_MOVE 触发，这是拖动行为	OnGestureListener

续表

方 法 名	描 述	所 属 接 口
onLongPress	用户长久地按着屏幕不放，即长按	OnGestureListener
onFling	用户按下触摸屏，快速滑动后松开，由 1 个 ACTION_DOWN、多个 ACTION_MOVE 和 1 个 ACTION_UP 触发，这是快速滑动行为	OnGestureListener
onDoubleTap	双击，由 2 次连续的单击组成，它不可能和 onSingleTapConfirmed 共存	OnDoubleTapListener
onSingleTapConfirmed	严格的单击行为 *注意它和 onSingleTapUp 的区别，如果触发了 onSingleTapConfirmed，那么后面不可能再紧跟着另一个单击行为，即这只可能是单击，而不可能是双击中的一次单击	OnDoubleTapListener
onDoubleTapEvent	表示发生了双击行为，在双击的期间，ACTION_DOWN、ACTION_MOVE 和 ACTION_UP 都会触发此回调	OnDoubleTapListener

表 3-1 里面的方法很多，但是并不是所有的方法都会被时常用到，在日常开发中，比较常用的有：onSingleTapUp（单击）、onFling（快速滑动）、onScroll（拖动）、onLongPress（长按）和 onDoubleTap（双击）。另外这里要说明的是，实际开发中，可以不使用 GestureDetector，完全可以自己在 View 的 onTouchEvent 方法中实现所需的监听，这个就看个人的喜好了。这里有一个建议供读者参考：如果只是监听滑动相关的，建议自己在 onTouchEvent 中实现，如果要监听双击这种行为的话，那么就使用 GestureDetector。

3．Scroller

弹性滑动对象，用于实现 View 的弹性滑动。我们知道，当使用 View 的 scrollTo/scrollBy 方法来进行滑动时，其过程是瞬间完成的，这个没有过渡效果的滑动用户体验不好。这个时候就可以使用 Scroller 来实现有过渡效果的滑动，其过程不是瞬间完成的，而是在一定的时间间隔内完成的。Scroller 本身无法让 View 弹性滑动，它需要和 View 的 computeScroll 方法配合使用才能共同完成这个功能。那么如何使用 Scroller 呢？它的典型代码是固定的，如下所示。至于它为什么能实现弹性滑动，这个在 3.2 节中会进行详细介绍。

```
Scroller scroller = new Scroller(mContext);

// 缓慢滚动到指定位置
private void smoothScrollTo(int destX, int destY) {
    int scrollX = getScrollX();
    int delta = destX - scrollX;
    // 1000ms 内滑向 destX，效果就是慢慢滑动
    mScroller.startScroll(scrollX, 0, delta, 0, 1000);
```

```
        invalidate();
    }

    @Override
    public void computeScroll() {
        if (mScroller.computeScrollOffset()) {
            scrollTo(mScroller.getCurrX(), mScroller.getCurrY());
            postInvalidate();
        }
    }
```

3.2 View 的滑动

3.1 节介绍了 View 的一些基础知识和概念，本节开始介绍很重要的一个内容：View 的滑动。在 Android 设备上，滑动几乎是应用的标配，不管是下拉刷新还是 SlidingMenu，它们的基础都是滑动。从另外一方面来说，Android 手机由于屏幕比较小，为了给用户呈现更多的内容，就需要使用滑动来隐藏和显示一些内容。基于上述两点，可以知道，滑动在 Android 开发中具有很重要的作用，不管一些滑动效果多么绚丽，归根结底，它们都是由不同的滑动外加一些特效所组成的。因此，掌握滑动的方法是实现绚丽的自定义控件的基础。通过三种方式可以实现 View 的滑动：第一种是通过 View 本身提供的 scrollTo/scrollBy 方法来实现滑动；第二种是通过动画给 View 施加平移效果来实现滑动；第三种是通过改变 View 的 LayoutParams 使得 View 重新布局从而实现滑动。从目前来看，常见的滑动方式就这么三种，下面一一进行分析。

3.2.1 使用 scrollTo/scrollBy

为了实现 View 的滑动，View 提供了专门的方法来实现这个功能，那就是 scrollTo 和 scrollBy，我们先来看看这两个方法的实现，如下所示。

```
/**
 * Set the scrolled position of your view. This will cause a call to
 * {@link #onScrollChanged(int, int, int, int)} and the view will be
 * invalidated.
 * @param x the x position to scroll to
 * @param y the y position to scroll to
```

```java
 */
public void scrollTo(int x, int y) {
    if (mScrollX != x || mScrollY != y) {
        int oldX = mScrollX;
        int oldY = mScrollY;
        mScrollX = x;
        mScrollY = y;
        invalidateParentCaches();
        onScrollChanged(mScrollX, mScrollY, oldX, oldY);
        if (!awakenScrollBars()) {
            postInvalidateOnAnimation();
        }
    }
}

/**
 * Move the scrolled position of your view. This will cause a call to
 * {@link #onScrollChanged(int, int, int, int)} and the view will be
 * invalidated.
 * @param x the amount of pixels to scroll by horizontally
 * @param y the amount of pixels to scroll by vertically
 */
public void scrollBy(int x, int y) {
    scrollTo(mScrollX + x, mScrollY + y);
}
```

从上面的源码可以看出，scrollBy 实际上也是调用了 scrollTo 方法，它实现了基于当前位置的相对滑动，而 scrollTo 则实现了基于所传递参数的绝对滑动，这个不难理解。利用 scrollTo 和 scrollBy 来实现 View 的滑动，这不是一件困难的事，但是我们要明白滑动过程中 View 内部的两个属性 mScrollX 和 mScrollY 的改变规则，这两个属性可以通过 getScrollX 和 getScrollY 方法分别得到。这里先简要概况一下：在滑动过程中，mScrollX 的值总是等于 View 左边缘和 View 内容左边缘在水平方向的距离，而 mScrollY 的值总是等于 View 上边缘和 View 内容上边缘在竖直方向的距离。View 边缘是指 View 的位置，由四个顶点组成，而 View 内容边缘是指 View 中的内容的边缘，scrollTo 和 scrollBy 只能改变 View 内容的位置而不能改变 View 在布局中的位置。mScrollX 和 mScrollY 的单位为像素，并且当 View 左边缘在 View 内容左边缘的右边时，mScrollX 为正值，反之为负值；当 View 上边缘在 View 内容上边缘的下边时，mScrollY 为正值，反之为负值。换句话说，如果从左向右滑动，那么 mScrollX 为负值，反之为正值；如果从上往下滑动，那么 mScrollY 为负值，反之为正值。

为了更好地理解这个问题，下面举个例子，如图 3-3 所示。在图中假设水平和竖直方向的滑动距离都为 100 像素，针对图中各种滑动情况，都给出了对应的 mScrollX 和 mScrollY 的值。根据上面的分析，可以知道，使用 scrollTo 和 scrollBy 来实现 View 的滑动，只能将 View 的内容进行移动，并不能将 View 本身进行移动，也就是说，不管怎么滑动，也不可能将当前 View 滑动到附近 View 所在的区域，这个需要仔细体会一下。

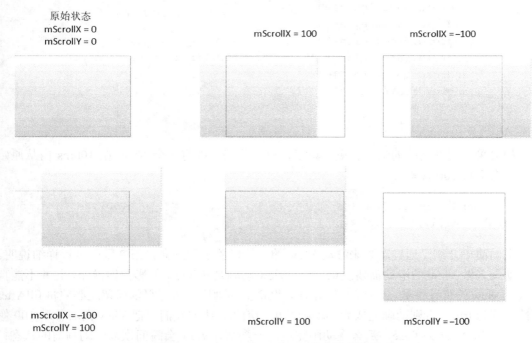

图 3-3　mScrollX 和 mScrollY 的变换规律示意

3.2.2　使用动画

上一节介绍了采用 scrollTo/scrollBy 来实现 View 的滑动，本节介绍另外一种滑动方式，即使用动画，通过动画我们能够让一个 View 进行平移，而平移就是一种滑动。使用动画来移动 View，主要是操作 View 的 translationX 和 translationY 属性，既可以采用传统的 View 动画，也可以采用属性动画，如果采用属性动画的话，为了能够兼容 3.0 以下的版本，需要采用开源动画库 nineoldandroids（http://nineoldandroids.com/）。

采用 View 动画的代码，如下所示。此动画可以在 100ms 内将一个 View 从原始位置向右下角移动 100 个像素。

```xml
<?xml version="1.0" encoding="utf-8"?>
<set xmlns:android="http://schemas.android.com/apk/res/android"
  android:fillAfter="true"
  android:zAdjustment="normal" >

  <translate
    android:duration="100"
    android:fromXDelta="0"
    android:fromYDelta="0"
    android:interpolator="@android:anim/linear_interpolator"
    android:toXDelta="100"
    android:toYDelta="100" />

</set>
```

如果采用属性动画的话，就更简单了，以下代码可以将一个 View 在 100ms 内从原始位置向右平移 100 像素。

```
ObjectAnimator.ofFloat(targetView, "translationX", 0, 100).setDuration
(100).start();
```

上面简单介绍了通过动画来移动 View 的方法，关于动画会在第 7 章中进行详细说明。使用动画来做 View 的滑动需要注意一点，View 动画是对 View 的影像做操作，它并不能真正改变 View 的位置参数，包括宽/高，并且如果希望动画后的状态得以保留还必须将 fillAfter 属性设置为 true，否则动画完成后其动画结果会消失。比如我们要把 View 向右移动 100 像素，如果 fillAfter 为 false，那么在动画完成的一刹那，View 会瞬间恢复到动画前的状态；如果 fillAfter 为 true，在动画完成后，View 会停留在距原始位置 100 像素的右边。使用属性动画并不会存在上述问题，但是在 Android 3.0 以下无法使用属性动画，这个时候我们可以使用动画兼容库 nineoldandroids 来实现属性动画，尽管如此，在 Android 3.0 以下的手机上通过 nineoldandroids 来实现的属性动画本质上仍然是 View 动画。

上面提到 View 动画并不能真正改变 View 的位置，这会带来一个很严重的问题。试想一下，比如我们通过 View 动画将一个 Button 向右移动 100px，并且这个 View 设置的有单击事件，然后你会惊奇地发现，单击新位置无法触发 onClick 事件，而单击原始位置仍然可以触发 onClick 事件，尽管 Button 已经不在原始位置了。这个问题带来的影响是致命的，但是它却又是可以理解的，因为不管 Button 怎么做变换，但是它的位置信息（四个顶点和宽/高）并不会随着动画而改变，因此在系统眼里，这个 Button 并没有发生任何改变，它的真身仍然在原始位置。在这种情况下，单击新位置当然不会触发 onClick 事件了，因为 Button 的真身并没有发生改变，在新位置上只是 View 的影像而已。基于这一点，我们不能简单地给一个 View 做平移动画并且还希望它在新位置继续触发一些单击事件。

从 Android 3.0 开始，使用属性动画可以解决上面的问题，但是大多数应用都需要兼容到 Android 2.2，在 Android 2.2 上无法使用属性动画，因此这里还是会有问题。那么这种问题难道就无法解决了吗？也不是的，虽然不能直接解决这个问题，但是还可以间接解决这个问题，这里给出一个简单的解决方法。针对上面 View 动画的问题，我们可以在新位置预先创建一个和目标 Button 一模一样的 Button，它们不但外观一样连 onClick 事件也一样。当目标 Button 完成平移动画后，就把目标 Button 隐藏，同时把预先创建的 Button 显示出来，通过这种间接的方式我们解决了上面的问题。这仅仅是个参考，面对这种问题时读者可以灵活应对。

3.2.3 改变布局参数

本节将介绍第三种实现 View 滑动的方法，那就是改变布局参数，即改变 LayoutParams。这个比较好理解了，比如我们想把一个 Button 向右平移 100px，我们只需要将这个 Button 的 LayoutParams 里的 marginLeft 参数的值增加 100px 即可，是不是很简单呢？还有一种情形，为了达到移动 Button 的目的，我们可以在 Button 的左边放置一个空的 View，这个空 View 的默认宽度为 0，当我们需要向右移动 Button 时，只需要重新设置空 View 的宽度即可，当空 View 的宽度增大时（假设 Button 的父容器是水平方向的 LinearLayout），Button 就自动被挤向右边，即实现了向右平移的效果。如何重新设置一个 View 的 LayoutParams 呢？很简单，如下所示。

```
MarginLayoutParams params = (MarginLayoutParams)mButton1.getLayoutParams();
params.width += 100;
params.leftMargin += 100;
mButton1.requestLayout();
//或者 mButton1.setLayoutParams(params);
```

通过改变 LayoutParams 的方式去实现 View 的滑动同样是一种很灵活的方法，需要根据不同情况去做不同的处理。

3.2.4 各种滑动方式的对比

上面分别介绍了三种不同的滑动方式，它们都能实现 View 的滑动，那么它们之间的差别是什么呢？

先看 scrollTo/scrollBy 这种方式，它是 View 提供的原生方法，其作用是专门用于 View 的滑动，它可以比较方便地实现滑动效果并且不影响内部元素的单击事件。但是它的缺点也是很显然的：它只能滑动 View 的内容，并不能滑动 View 本身。

再看动画，通过动画来实现 View 的滑动，这要分情况。如果是 Android 3.0 以上并采用属性动画，那么采用这种方式没有明显的缺点；如果是使用 View 动画或者在 Android 3.0 以下使用属性动画，均不能改变 View 本身的属性。在实际使用中，如果动画元素不需要响应用户的交互，那么使用动画来做滑动是比较合适的，否则就不太适合。但是动画有一个很明显的优点，那就是一些复杂的效果必须要通过动画才能实现。

最后再看一下改变布局这种方式，它除了使用起来麻烦点以外，也没有明显的缺点，它的主要适用对象是一些具有交互性的 View，因为这些 View 需要和用户交互，直接通过动画去实现会有问题，这在 3.2.2 节中已经有所介绍，所以这个时候我们可以使用直接改变布局参数的方式去实现。

针对上面的分析做一下总结，如下所示。

- scrollTo/scrollBy：操作简单，适合对 View 内容的滑动；
- 动画：操作简单，主要适用于没有交互的 View 和实现复杂的动画效果；
- 改变布局参数：操作稍微复杂，适用于有交互的 View。

下面我们实现一个跟手滑动的效果，这是一个自定义 View，拖动它可以让它在整个屏幕上随意滑动。这个 View 实现起来很简单，我们只要重写它的 onTouchEvent 方法并处理 ACTION_MOVE 事件，根据两次滑动之间的距离就可以实现它的滑动了。为了实现全屏滑动，我们采用动画的方式来实现。原因很简单，这个效果无法采用 scrollTo 来实现。另外，它还可以采用改变布局的方式来实现，这里仅仅是为了演示，所以就选择了动画的方式，核心代码如下所示。

```
public boolean onTouchEvent(MotionEvent event) {
    int x = (int) event.getRawX();
    int y = (int) event.getRawY();
    switch (event.getAction()) {
    case MotionEvent.ACTION_DOWN: {
        break;
    }
    case MotionEvent.ACTION_MOVE: {
```

```
            int deltaX = x - mLastX;
            int deltaY = y - mLastY;
            Log.d(TAG, "move, deltaX:" + deltaX + " deltaY:" + deltaY);
            int translationX = (int)ViewHelper.getTranslationX(this) + deltaX;
            int translationY = (int)ViewHelper.getTranslationY(this) + deltaY;
            ViewHelper.setTranslationX(this, translationX);
            ViewHelper.setTranslationY(this, translationY);
            break;
        }
        case MotionEvent.ACTION_UP: {
            break;
        }
        default:
            break;
        }

        mLastX = x;
        mLastY = y;
        return true;
    }
```

通过上述代码可以看出，这一全屏滑动的效果实现起来相当简单。首先我们通过 getRawX 和 getRawY 方法来获取手指当前的坐标，注意不能使用 getX 和 getY 方法，因为这个是要全屏滑动的，所以需要获取当前点击事件在屏幕中的坐标而不是相对于 View 本身的坐标；其次，我们要得到两次滑动之间的位移，有了这个位移就可以移动当前的 View，移动方法采用的是动画兼容库 nineoldandroids 中的 ViewHelper 类所提供的 setTranslationX 和 setTranslationY 方法。实际上，ViewHelper 类提供了一系列 get/set 方法，因为 View 的 setTranslationX 和 setTranslationY 只能在 Android 3.0 及其以上版本才能使用，但是 ViewHelper 所提供的方法是没有版本要求的，与此类似的还有 setX、setScaleX、setAlpha 等方法，这一系列方法实际上是为属性动画服务的，更详细的内容会在第 7 章进行进一步的介绍。这个自定义 View 可以在 2.x 及其以上版本工作，但是由于动画的性质，如果给它加上 onClick 事件，那么在 3.0 以下版本它将无法在新位置响应用户的点击，这个问题在前面已经提到过。

3.3 弹性滑动

知道了 View 的滑动，我们还要知道如何实现 View 的弹性滑动，比较生硬地滑动过去，这种方式的用户体验实在太差了,因此我们要实现渐进式滑动。那么如何实现弹性滑动呢?其实实现方法有很多,但是它们都有一个共同思想:将一次大的滑动分成若干次小的滑动并在一个时间段内完成,弹性滑动的具体实现方式有很多,比如通过 Scroller、Handler#postDelayed 以及 Thread#sleep 等,下面一一进行介绍。

3.3.1 使用 Scroller

Scroller 的使用方法在 3.1.4 节中已经进行了介绍,下面我们来分析一下它的源码,从而探究为什么它能实现 View 的弹性滑动。

```
Scroller scroller = new Scroller(mContext);

// 缓慢滚动到指定位置
private void smoothScrollTo(int destX, int destY) {
    int scrollX = getScrollX();
    int deltaX = destX - scrollX;
    // 1000ms 内滑向 destX,效果就是慢慢滑动
    mScroller.startScroll(scrollX, 0, deltaX, 0, 1000);
    invalidate();
}

@Override
public void computeScroll() {
    if (mScroller.computeScrollOffset()) {
        scrollTo(mScroller.getCurrX(), mScroller.getCurrY());
        postInvalidate();
    }
}
```

上面是 Scroller 的典型的使用方法,这里先描述它的工作原理:当我们构造一个 Scroller 对象并且调用它的 startScroll 方法时,Scroller 内部其实什么也没做,它只是保存了我们传递的几个参数,这几个参数从 startScroll 的原型上就可以看出来,如下所示。

```
public void startScroll(int startX, int startY, int dx, int dy, int duration) {
    mMode = SCROLL_MODE;
    mFinished = false;
    mDuration = duration;
    mStartTime = AnimationUtils.currentAnimationTimeMillis();
    mStartX = startX;
```

```
    mStartY = startY;
    mFinalX = startX + dx;
    mFinalY = startY + dy;
    mDeltaX = dx;
    mDeltaY = dy;
    mDurationReciprocal = 1.0f / (float) mDuration;
}
```

这个方法的参数含义很清楚，startX 和 startY 表示的是滑动的起点，dx 和 dy 表示的是要滑动的距离，而 duration 表示的是滑动时间，即整个滑动过程完成所需要的时间，注意这里的滑动是指 View 内容的滑动而非 View 本身位置的改变。可以看到，仅仅调用 startScroll 方法是无法让 View 滑动的，因为它内部并没有做滑动相关的事，那么 Scroller 到底是如何让 View 弹性滑动的呢？答案就是 startScroll 方法下面的 invalidate 方法，虽然有点不可思议，但是的确是这样的。invalidate 方法会导致 View 重绘，在 View 的 draw 方法中又会去调用 computeScroll 方法，computeScroll 方法在 View 中是一个空实现，因此需要我们自己去实现，上面的代码已经实现了 computeScroll 方法。正是因为这个 computeScroll 方法，View 才能实现弹性滑动。这看起来还是很抽象，其实这样的：当 View 重绘后会在 draw 方法中调用 computeScroll，而 computeScroll 又会去向 Scroller 获取当前的 scrollX 和 scrollY；然后通过 scrollTo 方法实现滑动；接着又调用 postInvalidate 方法来进行第二次重绘，这一次重绘的过程和第一次重绘一样，还是会导致 computeScroll 方法被调用；然后继续向 Scroller 获取当前的 scrollX 和 scrollY，并通过 scrollTo 方法滑动到新的位置，如此反复，直到整个滑动过程结束。

我们再看一下 Scroller 的 computeScrollOffset 方法的实现，如下所示。

```
/**
 * Call this when you want to know the new location.  If it returns true,
 * the animation is not yet finished.
 */
public boolean computeScrollOffset() {
    ...
    int timePassed = (int)(AnimationUtils.currentAnimationTimeMillis() -
    mStartTime);

    if (timePassed < mDuration) {
        switch (mMode) {
        case SCROLL_MODE:
            final float x = mInterpolator.getInterpolation(timePassed *
```

```
                mDurationReciprocal);
            mCurrX = mStartX + Math.round(x * mDeltaX);
            mCurrY = mStartY + Math.round(x * mDeltaY);
            break;
        ...
        }
    }
    return true;
}
```

是不是突然就明白了？这个方法会根据时间的流逝来计算出当前的 scrollX 和 scrollY 的值。计算方法也很简单，大意就是根据时间流逝的百分比来算出 scrollX 和 scrollY 改变的百分比并计算出当前的值，这个过程类似于动画中的插值器的概念，这里我们先不去深究这个具体过程。这个方法的返回值也很重要，它返回 true 表示滑动还未结束，false 则表示滑动已经结束，因此当这个方法返回 true 时，我们要继续进行 View 的滑动。

通过上面的分析，我们应该明白 Scroller 的工作原理了，这里做一下概括：Scroller 本身并不能实现 View 的滑动，它需要配合 View 的 computeScroll 方法才能完成弹性滑动的效果，它不断地让 View 重绘，而每一次重绘距滑动起始时间会有一个时间间隔，通过这个时间间隔 Scroller 就可以得出 View 当前的滑动位置，知道了滑动位置就可以通过 scrollTo 方法来完成 View 的滑动。就这样，View 的每一次重绘都会导致 View 进行小幅度的滑动，而多次的小幅度滑动就组成了弹性滑动，这就是 Scroller 的工作机制。由此可见，Scroller 的设计思想是多么值得称赞，整个过程中它对 View 没有丝毫的引用，甚至在它内部连计时器都没有。

3.3.2 通过动画

动画本身就是一种渐近的过程，因此通过它来实现的滑动天然就具有弹性效果，比如以下代码可以让一个 View 在 100ms 内向右移动 100 像素。

```
ObjectAnimator.ofFloat(targetView, "translationX", 0, 100).setDuration
(100).start();
```

不过这里想说的并不是这个问题，我们可以利用动画的特性来实现一些动画不能实现的效果。还拿 scrollTo 来说，我们也想模仿 Scroller 来实现 View 的弹性滑动，那么利用动画的特性，我们可以采用如下方式来实现：

```
final int startX = 0;
final int deltaX = 100;
```

```
ValueAnimator animator = ValueAnimator.ofInt(0, 1).setDuration(1000);
animator.addUpdateListener(new AnimatorUpdateListener() {
    @Override
    public void onAnimationUpdate(ValueAnimator animator) {
        float fraction = animator.getAnimatedFraction();
        mButton1.scrollTo(startX + (int) (deltaX * fraction), 0);
    }
});
animator.start();
```

在上述代码中,我们的动画本质上没有作用于任何对象上,它只是在 1000ms 内完成了整个动画过程。利用这个特性,我们就可以在动画的每一帧到来时获取动画完成的比例,然后再根据这个比例计算出当前 View 所要滑动的距离。注意,这里的滑动针对的是 View 的内容而非 View 本身。可以发现,这个方法的思想其实和 Scroller 比较类似,都是通过改变一个百分比配合 scrollTo 方法来完成 View 的滑动。需要说明一点,采用这种方法除了能够完成弹性滑动以外,还可以实现其他动画效果,我们完全可以在 onAnimationUpdate 方法中加上我们想要的其他操作。

3.3.3 使用延时策略

本节介绍另外一种实现弹性滑动的方法,那就是延时策略。它的核心思想是通过发送一系列延时消息从而达到一种渐近式的效果,具体来说可以使用 Handler 或 View 的 postDelayed 方法,也可以使用线程的 sleep 方法。对于 postDelayed 方法来说,我们可以通过它来延时发送一个消息,然后在消息中来进行 View 的滑动,如果接连不断地发送这种延时消息,那么就可以实现弹性滑动的效果。对于 sleep 方法来说,通过在 while 循环中不断地滑动 View 和 sleep,就可以实现弹性滑动的效果。

下面采用 Handler 来做个示例,其他方法请读者自行去尝试,思想都是类似的。下面的代码在大约 1000ms 内将 View 的内容向左移动了 100 像素,代码比较简单,就不再详细介绍了。之所以说大约 1000ms,是因为采用这种方式无法精确地定时,原因是系统的消息调度也是需要时间的,并且所需时间不定。

```
private static final int MESSAGE_SCROLL_TO = 1;
private static final int FRAME_COUNT = 30;
private static final int DELAYED_TIME = 33;

private int mCount = 0;

@SuppressLint("HandlerLeak")
```

```java
private Handler mHandler = new Handler() {
    public void handleMessage(Message msg) {
        switch (msg.what) {
        case MESSAGE_SCROLL_TO: {
            mCount++;
            if (mCount <= FRAME_COUNT) {
                float fraction = mCount / (float) FRAME_COUNT;
                int scrollX = (int) (fraction * 100);
                mButton1.scrollTo(scrollX, 0);
                mHandler.sendEmptyMessageDelayed(MESSAGE_SCROLL_TO,
                DELAYED_TIME);
            }
            break;
        }

        default:
            break;
        }
    };
};
```

上面几种弹性滑动的实现方法，在介绍中侧重更多的是实现思想，在实际使用中可以对其灵活地进行扩展从而实现更多复杂的效果。

3.4 View 的事件分发机制

上面几节介绍了 View 的基础知识以及 View 的滑动，本节将介绍 View 的一个核心知识点：事件分发机制。事件分发机制不仅仅是核心知识点更是难点，不少初学者甚至中级开发者面对这个问题时都会觉得困惑。另外，View 的另一大难题滑动冲突，它的解决方法的理论基础就是事件分发机制，因此掌握好 View 的事件分发机制是十分重要的。本节将深入介绍 View 的事件分发机制，在 3.4.1 节会对事件分发机制进行概括性地介绍，而在 3.4.2 节将结合系统源码去进一步分析事件分发机制。

3.4.1 点击事件的传递规则

在介绍点击事件的传递规则之前，首先我们要明白这里要分析的对象就是 MotionEvent，即点击事件，关于 MotionEvent 在 3.1 节中已经进行了介绍。所谓点击事件的事件分发，其实就是对 MotionEvent 事件的分发过程，即当一个 MotionEvent 产生了以后，系统需要把这个事件传递给一个具体的 View，而这个传递的过程就是分发过程。点击事件的分发过程由三个很重要的方法来共同完成：dispatchTouchEvent、onInterceptTouchEvent 和 onTouchEvent，下面我们先介绍一下这几个方法。

public boolean dispatchTouchEvent(MotionEvent ev)

用来进行事件的分发。如果事件能够传递给当前 View，那么此方法一定会被调用，返回结果受当前 View 的 onTouchEvent 和下级 View 的 dispatchTouchEvent 方法的影响，表示是否消耗当前事件。

public boolean onInterceptTouchEvent(MotionEvent event)

在上述方法内部调用，用来判断是否拦截某个事件，如果当前 View 拦截了某个事件，那么在同一个事件序列当中，此方法不会被再次调用，返回结果表示是否拦截当前事件。

public boolean onTouchEvent(MotionEvent event)

在 dispatchTouchEvent 方法中调用，用来处理点击事件，返回结果表示是否消耗当前事件，如果不消耗，则在同一个事件序列中，当前 View 无法再次接收到事件。

上述三个方法到底有什么区别呢？它们是什么关系呢？其实它们的关系可以用如下伪代码表示：

```java
public boolean dispatchTouchEvent(MotionEvent ev) {
    boolean consume = false;
    if (onInterceptTouchEvent(ev)) {
        consume = onTouchEvent(ev);
    } else {
        consume = child.dispatchTouchEvent(ev);
    }

    return consume;
}
```

上述伪代码已经将三者的关系表现得淋漓尽致。通过上面的伪代码，我们也可以大致了解点击事件的传递规则：对于一个根 ViewGroup 来说，点击事件产生后，首先会传递给它，这时它的 dispatchTouchEvent 就会被调用，如果这个 ViewGroup 的 onInterceptTouchEvent 方法返回 true 就表示它要拦截当前事件，接着事件就会交给这个 ViewGroup 处理，即它的 onTouchEvent 方法就会被调用；如果这个 ViewGroup 的 onInterceptTouchEvent 方法返回 false 就表示它不拦截当前事件，这时当前事件就会继续传递给它的子元素，接着子元素的 dispatchTouchEvent 方法就会被调用，如此反复直到事件被最终处理。

当一个 View 需要处理事件时，如果它设置了 OnTouchListener，那么 OnTouchListener 中的 onTouch 方法会被回调。这时事件如何处理还要看 onTouch 的返回值，如果返回 false，则当前 View 的 onTouchEvent 方法会被调用；如果返回 true，那么 onTouchEvent 方法将不会被调用。由此可见，给 View 设置的 OnTouchListener，其优先级比 onTouchEvent 要高。在 onTouchEvent 方法中，如果当前设置的有 OnClickListener，那么它的 onClick 方法会被调用。可以看出，平时我们常用的 OnClickListener，其优先级最低，即处于事件传递的尾端。

当一个点击事件产生后，它的传递过程遵循如下顺序：Activity -> Window -> View，即事件总是先传递给 Activity，Activity 再传递给 Window，最后 Window 再传递给顶级 View。顶级 View 接收到事件后，就会按照事件分发机制去分发事件。考虑一种情况，如果一个 View 的 onTouchEvent 返回 false，那么它的父容器的 onTouchEvent 将会被调用，依此类推。如果所有的元素都不处理这个事件，那么这个事件将会最终传递给 Activity 处理，即 Activity 的 onTouchEvent 方法会被调用。这个过程其实也很好理解，我们可以换一种思路，假如点击事件是一个难题，这个难题最终被上级领导分给了一个程序员去处理（这是事件分发过程），结果这个程序员搞不定（onTouchEvent 返回了 false），现在该怎么办呢？难题必须要解决，那只能交给水平更高的上级解决（上级的 onTouchEvent 被调用），如果上级再搞不定，那只能交给上级的上级去解决，就这样将难题一层层地向上抛，这是公司内部一种很常见的处理问题的过程。从这个角度来看，View 的事件传递过程还是很贴近现实的，毕竟程序员也生活在现实中。

关于事件传递的机制，这里给出一些结论，根据这些结论可以更好地理解整个传递机制，如下所示。

（1）同一个事件序列是指从手指接触屏幕的那一刻起，到手指离开屏幕的那一刻结束，在这个过程中所产生的一系列事件，这个事件序列以 down 事件开始，中间含有数量不定的 move 事件，最终以 up 事件结束。

（2）正常情况下，一个事件序列只能被一个 View 拦截且消耗。这一条的原因可以参考

（3），因为一旦一个元素拦截了某此事件，那么同一个事件序列内的所有事件都会直接交给它处理，因此同一个事件序列中的事件不能分别由两个 View 同时处理，但是通过特殊手段可以做到，比如一个 View 将本该自己处理的事件通过 onTouchEvent 强行传递给其他 View 处理。

（3）某个 View 一旦决定拦截，那么这一个事件序列都只能由它来处理（如果事件序列能够传递给它的话），并且它的 onInterceptTouchEvent 不会再被调用。这条也很好理解，就是说当一个 View 决定拦截一个事件后，那么系统会把同一个事件序列内的其他方法都直接交给它来处理，因此就不用再调用这个 View 的 onInterceptTouchEvent 去询问它是否要拦截了。

（4）某个 View 一旦开始处理事件，如果它不消耗 ACTION_DOWN 事件（onTouchEvent 返回了 false），那么同一事件序列中的其他事件都不会再交给它来处理，并且事件将重新交由它的父元素去处理，即父元素的 onTouchEvent 会被调用。意思就是事件一旦交给一个 View 处理，那么它就必须消耗掉，否则同一事件序列中剩下的事件就不再交给它来处理了，这就好比上级交给程序员一件事，如果这件事没有处理好，短期内上级就不敢再把事情交给这个程序员做了，二者是类似的道理。

（5）如果 View 不消耗除 ACTION_DOWN 以外的其他事件，那么这个点击事件会消失，此时父元素的 onTouchEvent 并不会被调用，并且当前 View 可以持续收到后续的事件，最终这些消失的点击事件会传递给 Activity 处理。

（6）ViewGroup 默认不拦截任何事件。Android 源码中 ViewGroup 的 onInterceptTouch-Event 方法默认返回 false。

（7）View 没有 onInterceptTouchEvent 方法，一旦有点击事件传递给它，那么它的 onTouchEvent 方法就会被调用。

（8）View 的 onTouchEvent 默认都会消耗事件（返回 true），除非它是不可点击的（clickable 和 longClickable 同时为 false）。View 的 longClickable 属性默认都为 false，clickable 属性要分情况，比如 Button 的 clickable 属性默认为 true，而 TextView 的 clickable 属性默认为 false。

（9）View 的 enable 属性不影响 onTouchEvent 的默认返回值。哪怕一个 View 是 disable 状态的，只要它的 clickable 或者 longClickable 有一个为 true，那么它的 onTouchEvent 就返回 true。

（10） onClick 会发生的前提是当前 View 是可点击的，并且它收到了 down 和 up 的事件。

（11）事件传递过程是由外向内的，即事件总是先传递给父元素，然后再由父元素分发给子 View，通过 requestDisallowInterceptTouchEvent 方法可以在子元素中干预父元素的事件分发过程，但是 ACTION_DOWN 事件除外。

3.4.2 事件分发的源码解析

上一节分析了 View 的事件分发机制，本节将会从源码的角度去进一步分析、证实上面的结论。

1. Activity 对点击事件的分发过程

点击事件用 MotionEvent 来表示，当一个点击操作发生时，事件最先传递给当前 Activity，由 Activity 的 dispatchTouchEvent 来进行事件派发，具体的工作是由 Activity 内部的 Window 来完成的。Window 会将事件传递给 decor view，decor view 一般就是当前界面的底层容器（即 setContentView 所设置的 View 的父容器），通过 Activity.getWindow.getDecorView()可以获得。我们先从 Activity 的 dispatchTouchEvent 开始分析。

源码：Activity#dispatchTouchEvent

```java
public boolean dispatchTouchEvent(MotionEvent ev) {
    if (ev.getAction() == MotionEvent.ACTION_DOWN) {
        onUserInteraction();
    }
    if (getWindow().superDispatchTouchEvent(ev)) {
        return true;
    }
    return onTouchEvent(ev);
}
```

现在分析上面的代码。首先事件开始交给 Activity 所附属的 Window 进行分发，如果返回 true，整个事件循环就结束了，返回 false 意味着事件没人处理，所有 View 的 onTouchEvent 都返回了 false，那么 Activity 的 onTouchEvent 就会被调用。

接下来看 Window 是如何将事件传递给 ViewGroup 的。通过源码我们知道，Window 是个抽象类，而 Window 的 superDispatchTouchEvent 方法也是个抽象方法，因此我们必须找到 Window 的实现类才行。

源码：Window#superDispatchTouchEvent

```
public abstract boolean superDispatchTouchEvent(MotionEvent event);
```

那么到底 Window 的实现类是什么呢？其实是 PhoneWindow，这一点从 Window 的源码中也可以看出来，在 Window 的说明中，有这么一段话：

Abstract base class for a top-level window look and behavior policy. An instance of this class should be used as the top-level view added to the window manager. It provides standard UI policies such as a background, title area, default key processing, etc.

The only existing implementation of this abstract class is android. policy.PhoneWindow, which you should instantiate when needing a Window. Eventually that class will be refactored and a factory method added for creating Window instances without knowing about a particular implementation.

上面这段话的大概意思是：Window 类可以控制顶级 View 的外观和行为策略，它的唯一实现位于 android.policy.PhoneWindow 中，当你要实例化这个 Window 类的时候，你并不知道它的细节，因为这个类会被重构，只有一个工厂方法可以使用。尽管这看起来有点模糊，不过我们可以看一下 android.policy.PhoneWindow 这个类，尽管实例化的时候此类会被重构，仅是重构而已，功能是类似的。

由于 Window 的唯一实现是 PhoneWindow，因此接下来看一下 PhoneWindow 是如何处理点击事件的，如下所示。

源码：PhoneWindow#superDispatchTouchEvent

```
public boolean superDispatchTouchEvent(MotionEvent event) {
    return mDecor.superDispatchTouchEvent(event);
}
```

到这里逻辑就很清晰了，PhoneWindow 将事件直接传递给了 DecorView，这个 DecorView 是什么呢？请看下面：

```
private final class DecorView extends FrameLayout implements RootViewSur-
faceTaker

// This is the top-level view of the window, containing the window decor.
```

```
private DecorView mDecor;

@Override
public final View getDecorView() {
    if (mDecor == null) {
        installDecor();
    }
    return mDecor;
}
```

我们知道，通过((ViewGroup)getWindow().getDecorView().findViewById(android.R.id.content)).getChildAt(0)这种方式就可以获取 Activity 所设置的 View，这个 mDecor 显然就是 getWindow().getDecorView()返回的 View，而我们通过 setContentView 设置的 View 是它的一个子 View。目前事件传递到了 DecorView 这里，由于 DecorView 继承自 FrameLayout 且是父 View，所以最终事件会传递给 View。换句话来说，事件肯定会传递到 View，不然应用如何响应点击事件呢？不过这不是我们的重点，重点是事件到了 View 以后应该如何传递，这对我们更有用。从这里开始，事件已经传递到顶级 View 了，即在 Activity 中通过 setContentView 所设置的 View，另外顶级 View 也叫根 View，顶级 View 一般来说都是 ViewGroup。

3．顶级 View 对点击事件的分发过程

关于点击事件如何在 View 中进行分发，上一节已经做了详细的介绍，这里再大致回顾一下。点击事件达到顶级 View（一般是一个 ViewGroup）以后，会调用 ViewGroup 的 dispatchTouchEvent 方法，然后的逻辑是这样的：如果顶级 ViewGroup 拦截事件即 onInterceptTouchEvent 返回 true，则事件由 ViewGroup 处理，这时如果 ViewGroup 的 mOnTouchListener 被设置，则 onTouch 会被调用，否则 onTouchEvent 会被调用。也就是说，如果都提供的话，onTouch 会屏蔽掉 onTouchEvent。在 onTouchEvent 中，如果设置了 mOnClickListener，则 onClick 会被调用。如果顶级 ViewGroup 不拦截事件，则事件会传递给它所在的点击事件链上的子 View，这时子 View 的 dispatchTouchEvent 会被调用。到此为止，事件已经从顶级 View 传递给了下一层 View，接下来的传递过程和顶级 View 是一致的，如此循环，完成整个事件的分发。

首先看 ViewGroup 对点击事件的分发过程，其主要实现在 ViewGroup 的 dispatchTouchEvent 方法中，这个方法比较长，这里分段说明。先看下面一段，很显然，它描述的是当前 View 是否拦截点击事情这个逻辑。

```
        // Check for interception.
        final boolean intercepted;
        if (actionMasked == MotionEvent.ACTION_DOWN
                || mFirstTouchTarget != null) {
            final boolean disallowIntercept = (mGroupFlags & FLAG_DISALLOW_
            INTERCEPT) != 0;
            if (!disallowIntercept) {
                intercepted = onInterceptTouchEvent(ev);
                ev.setAction(action); // restore action in case it was changed
            } else {
                intercepted = false;
            }
        } else {
            // There are no touch targets and this action is not an initial down
            // so this view group continues to intercept touches.
            intercepted = true;
        }
```

从上面代码我们可以看出，ViewGroup 在如下两种情况下会判断是否要拦截当前事件：事件类型为 ACTION_DOWN 或者 mFirstTouchTarget != null。ACTION_DOWN 事件好理解，那么 mFirstTouchTarget != null 是什么意思呢？这个从后面的代码逻辑可以看出来，当事件由 ViewGroup 的子元素成功处理时，mFirstTouchTarget 会被赋值并指向子元素，换种方式来说，当 ViewGroup 不拦截事件并将事件交由子元素处理时 mFirstTouchTarget != null。反过来，一旦事件由当前 ViewGroup 拦截时，mFirstTouchTarget != null 就不成立。那么当 ACTION_MOVE 和 ACTION_UP 事件到来时，由于(actionMasked == MotionEvent. ACTION_DOWN || mFirstTouchTarget != null)这个条件为 false，将导致 ViewGroup 的 onInterceptTouchEvent 不会再被调用，并且同一序列中的其他事件都会默认交给它处理。

当然，这里有一种特殊情况，那就是 FLAG_DISALLOW_INTERCEPT 标记位，这个标记位是通过 requestDisallowInterceptTouchEvent 方法来设置的，一般用于子 View 中。FLAG_DISALLOW_INTERCEPT 一旦设置后，ViewGroup 将无法拦截除了 ACTION_DOWN 以外的其他点击事件。为什么说是除了 ACTION_DOWN 以外的其他事件呢？这是因为 ViewGroup 在分发事件时，如果是 ACTION_DOWN 就会重置 FLAG_DISALLOW_INTERCEPT 这个标记位，将导致子 View 中设置的这个标记位无效。因此，当面对 ACTION_DOWN 事件时，ViewGroup 总是会调用自己的 onInterceptTouchEvent 方法来询问自己是否要拦截事件，这一点从源码中也可以看出来。

在下面的代码中，ViewGroup 会在 ACTION_DOWN 事件到来时做重置状态的操作，而在 resetTouchState 方法中会对 FLAG_DISALLOW_INTERCEPT 进行重置，因此子 View 调用 request- DisallowInterceptTouchEvent 方法并不能影响 ViewGroup 对 ACTION_DOWN 事件的处理。

```
// Handle an initial down.
if (actionMasked == MotionEvent.ACTION_DOWN) {
    // Throw away all previous state when starting a new touch gesture.
    // The framework may have dropped the up or cancel event for the previous
      gesture
    // due to an app switch, ANR, or some other state change.
    cancelAndClearTouchTargets(ev);
    resetTouchState();
}
```

从上面的源码分析，我们可以得出结论：当 ViewGroup 决定拦截事件后，那么后续的点击事件将会默认交给它处理并且不再调用它的 onInterceptTouchEvent 方法，这证实了 3.4.1 节末尾处的第 3 条结论。FLAG_DISALLOW_INTERCEPT 这个标志的作用是让 ViewGroup 不再拦截事件，当然前提是 ViewGroup 不拦截 ACTION_DOWN 事件，这证实了 3.4.1 节末尾处的第 11 条结论。那么这段分析对我们有什么价值呢？总结起来有两点：第一点，onInterceptTouchEvent 不是每次事件都会被调用的，如果我们想提前处理所有的点击事件，要选择 dispatchTouchEvent 方法，只有这个方法能确保每次都会调用，当然前提是事件能够传递到当前的 ViewGroup；另外一点，FLAG_DISALLOW_INTERCEPT 标记位的作用给我们提供了一个思路，当面对滑动冲突时，我们可以是不是考虑用这种方法去解决问题？关于滑动冲突，将在 3.5 节进行详细分析。

接着再看当 ViewGroup 不拦截事件的时候，事件会向下分发交由它的子 View 进行处理，这段源码如下所示。

```
final View[] children = mChildren;
for (int i = childrenCount - 1; i >= 0; i--) {
    final int childIndex = customOrder
            ? getChildDrawingOrder(childrenCount, i) : i;
    final View child = (preorderedList == null)
            ? children[childIndex] : preorderedList.get(childIndex);
    if (!canViewReceivePointerEvents(child)
            || !isTransformedTouchPointInView(x, y, child, null)) {
        continue;
```

```
        }

        newTouchTarget = getTouchTarget(child);
        if (newTouchTarget != null) {
            // Child is already receiving touch within its bounds.
            // Give it the new pointer in addition to the ones it is handling.
            newTouchTarget.pointerIdBits |= idBitsToAssign;
            break;
        }

        resetCancelNextUpFlag(child);
        if (dispatchTransformedTouchEvent(ev, false, child, idBitsToAssign)) {
            // Child wants to receive touch within its bounds.
            mLastTouchDownTime = ev.getDownTime();
            if (preorderedList != null) {
                // childIndex points into presorted list, find original index
                for (int j = 0; j < childrenCount; j++) {
                    if (children[childIndex] == mChildren[j]) {
                        mLastTouchDownIndex = j;
                        break;
                    }
                }
            } else {
                mLastTouchDownIndex = childIndex;
            }
            mLastTouchDownX = ev.getX();
            mLastTouchDownY = ev.getY();
            newTouchTarget = addTouchTarget(child, idBitsToAssign);
            alreadyDispatchedToNewTouchTarget = true;
            break;
        }
    }
}
```

上面这段代码逻辑也很清晰，首先遍历 ViewGroup 的所有子元素，然后判断子元素是否能够接收到点击事件。是否能够接收点击事件主要由两点来衡量：子元素是否在播动画和点击事件的坐标是否落在子元素的区域内。如果某个子元素满足这两个条件，那么事件就会传递给它来处理。可以看到，dispatchTransformedTouchEvent 实际上调用的就是子元素的 dispatchTouchEvent 方法，在它的内部有如下一段内容，而在上面的代码中 child 传递的不是 null，因此它会直接调用子元素的 dispatchTouchEvent 方法，这样事件就交由子元素处理了，从而完成了一轮事件分发。

```
if (child == null) {
    handled = super.dispatchTouchEvent(event);
} else {
    handled = child.dispatchTouchEvent(event);
}
```

如果子元素的 dispatchTouchEvent 返回 true，这时我们暂时不用考虑事件在子元素内部是怎么分发的，那么 mFirstTouchTarget 就会被赋值同时跳出 for 循环，如下所示。

```
newTouchTarget = addTouchTarget(child, idBitsToAssign);
alreadyDispatchedToNewTouchTarget = true;
break;
```

这几行代码完成了 mFirstTouchTarget 的赋值并终止对子元素的遍历。如果子元素的 dispatchTouchEvent 返回 false，ViewGroup 就会把事件分发给下一个子元素（如果还有下一个子元素的话）。

其实 mFirstTouchTarget 真正的赋值过程是在 addTouchTarget 内部完成的，从下面的 addTouchTarget 方法的内部结构可以看出，mFirstTouchTarget 其实是一种单链表结构。mFirstTouchTarget 是否被赋值，将直接影响到 ViewGroup 对事件的拦截策略，如果 mFirstTouchTarget 为 null，那么 ViewGroup 就默认拦截接下来同一序列中所有的点击事件，这一点在前面已经做了分析。

```
private TouchTarget addTouchTarget(View child, int pointerIdBits) {
    TouchTarget target = TouchTarget.obtain(child, pointerIdBits);
    target.next = mFirstTouchTarget;
    mFirstTouchTarget = target;
    return target;
}
```

如果遍历所有的子元素后事件都没有被合适地处理，这包含两种情况：第一种是 ViewGroup 没有子元素；第二种是子元素处理了点击事件，但是在 dispatchTouchEvent 中返回了 false，这一般是因为子元素在 onTouchEvent 中返回了 false。在这两种情况下，ViewGroup 会自己处理点击事件，这里就证实了 3.4.1 节中的第 4 条结论，代码如下所示。

```
// Dispatch to touch targets.
if (mFirstTouchTarget == null) {
```

```
        // No touch targets so treat this as an ordinary view.
        handled = dispatchTransformedTouchEvent(ev, canceled, null,
                TouchTarget.ALL_POINTER_IDS);
}
```

注意上面这段代码,这里第三个参数 child 为 null,从前面的分析可以知道,它会调用 super.dispatchTouchEvent(event),很显然,这里就转到了 View 的 dispatchTouchEvent 方法,即点击事件开始交由 View 来处理,请看下面的分析。

4. View 对点击事件的处理过程

View 对点击事件的处理过程稍微简单一些,注意这里的 View 不包含 ViewGroup。先看它的 dispatchTouchEvent 方法,如下所示。

```
public boolean dispatchTouchEvent(MotionEvent event) {
    boolean result = false;
    ...

    if (onFilterTouchEventForSecurity(event)) {
        //noinspection SimplifiableIfStatement
        ListenerInfo li = mListenerInfo;
        if (li != null && li.mOnTouchListener != null
                && (mViewFlags & ENABLED_MASK) == ENABLED
                && li.mOnTouchListener.onTouch(this, event)) {
            result = true;
        }

        if (!result && onTouchEvent(event)) {
            result = true;
        }
    }
    ...

    return result;
}
```

View 对点击事件的处理过程就比较简单了,因为 View(这里不包含 ViewGroup)是一个单独的元素,它没有子元素因此无法向下传递事件,所以它只能自己处理事件。从上面的源码可以看出 View 对点击事件的处理过程,首先会判断有没有设置 OnTouchListener,

如果 OnTouchListener 中的 onTouch 方法返回 true，那么 onTouchEvent 就不会被调用，可见 OnTouchListener 的优先级高于 onTouchEvent，这样做的好处是方便在外界处理点击事件。

接着再分析 onTouchEvent 的实现。先看当 View 处于不可用状态下点击事件的处理过程，如下所示。很显然，不可用状态下的 View 照样会消耗点击事件，尽管它看起来不可用。

```
if ((viewFlags & ENABLED_MASK) == DISABLED) {
    if (event.getAction() == MotionEvent.ACTION_UP && (mPrivateFlags &
    PFLAG_PRESSED) != 0) {
        setPressed(false);
    }
    // A disabled view that is clickable still consumes the touch
    // events, it just doesn't respond to them.
    return (((viewFlags & CLICKABLE) == CLICKABLE ||
            (viewFlags & LONG_CLICKABLE) == LONG_CLICKABLE));
}
```

接着，如果 View 设置有代理，那么还会执行 TouchDelegate 的 onTouchEvent 方法，这个 onTouchEvent 的工作机制看起来和 OnTouchListener 类似，这里不深入研究了。

```
if (mTouchDelegate != null) {
    if (mTouchDelegate.onTouchEvent(event)) {
        return true;
    }
}
```

下面再看一下 onTouchEvent 中对点击事件的具体处理，如下所示。

```
if (((viewFlags & CLICKABLE) == CLICKABLE ||
        (viewFlags & LONG_CLICKABLE) == LONG_CLICKABLE)) {
    switch (event.getAction()) {
        case MotionEvent.ACTION_UP:
            boolean prepressed = (mPrivateFlags & PFLAG_PREPRESSED) != 0;
            if ((mPrivateFlags & PFLAG_PRESSED) != 0 || prepressed) {
                ...
                if (!mHasPerformedLongPress) {
                    // This is a tap, so remove the longpress check
                    removeLongPressCallback();

                    // Only perform take click actions if we were in the
```

```
pressed state
                    if (!focusTaken) {
                        // Use a Runnable and post this rather than calling
                        // performClick directly. This lets other visual
                           state
                        // of the view update before click actions start.
                        if (mPerformClick == null) {
                            mPerformClick = new PerformClick();
                        }
                        if (!post(mPerformClick)) {
                            performClick();
                        }
                    }
                    ...
                }
                break;
        }
        ...
        return true;
    }
```

从上面的代码来看，只要 View 的 CLICKABLE 和 LONG_CLICKABLE 有一个为 true，那么它就会消耗这个事件，即 onTouchEvent 方法返回 true，不管它是不是 DISABLE 状态，这就证实了 3.4.1 节末尾处的第 8、第 9 和第 10 条结论。然后就是当 ACTION_UP 事件发生时，会触发 performClick 方法，如果 View 设置了 OnClickListener，那么 performClick 方法内部会调用它的 onClick 方法，如下所示。

```
public boolean performClick() {
    final boolean result;
    final ListenerInfo li = mListenerInfo;
    if (li != null && li.mOnClickListener != null) {
        playSoundEffect(SoundEffectConstants.CLICK);
        li.mOnClickListener.onClick(this);
        result = true;
    } else {
        result = false;
    }
```

```
        sendAccessibilityEvent(AccessibilityEvent.TYPE_VIEW_CLICKED);
        return result;
}
```

View 的 LONG_CLICKABLE 属性默认为 false，而 CLICKABLE 属性是否为 false 和具体的 View 有关，确切来说是可点击的 View 其 CLICKABLE 为 true，不可点击的 View 其 CLICKABLE 为 false，比如 Button 是可点击的，TextView 是不可点击的。通过 setClickable 和 setLongClickable 可以分别改变 View 的 CLICKABLE 和 LONG_CLICKABLE 属性。另外，setOnClickListener 会自动将 View 的 CLICKABLE 设为 true，setOnLongClickListener 则会自动将 View 的 LONG_CLICKABLE 设为 true，这一点从源码中可以看出来，如下所示。

```
public void setOnClickListener(OnClickListener l) {
    if (!isClickable()) {
        setClickable(true);
    }
    getListenerInfo().mOnClickListener = l;
}

public void setOnLongClickListener(OnLongClickListener l) {
    if (!isLongClickable()) {
        setLongClickable(true);
    }
    getListenerInfo().mOnLongClickListener = l;
}
```

到这里，点击事件的分发机制的源码实现已经分析完了，结合 3.4.1 节中的理论分析和相关结论，读者就可以更好地理解事件分发了。在 3.5 节将介绍滑动冲突相关的知识，具体情况请看下面的分析。

3.5 View 的滑动冲突

本节开始介绍 View 体系中一个深入的话题：滑动冲突。相信开发 Android 的人都会有这种体会：滑动冲突实在是太坑人了，本来从网上下载的 demo 运行得好好的，但是只要出现滑动冲突，demo 就无法正常工作了。那么滑动冲突是如何产生的呢？其实在界面中只要内外两层同时可以滑动，这个时候就会产生滑动冲突。如何解决滑动冲突呢？这既是一件困难的事又是一件简单的事，说困难是因为许多开发者面对滑动冲突都会显得束手无策，说简单是因为滑动冲突的解决有固定的套路，只要知道了这个固定套路问题就好解决了。本节是 View 体系的核心章节，前面 4 节均是为本节服务的，通过本节的学习，滑动冲突将不再是个问题。

3.5.1 常见的滑动冲突场景

常见的滑动冲突场景可以简单分为如下三种（详情请参看图 3-4）：

- 场景 1——外部滑动方向和内部滑动方向不一致；
- 场景 2——外部滑动方向和内部滑动方向一致；
- 场景 3——上面两种情况的嵌套。

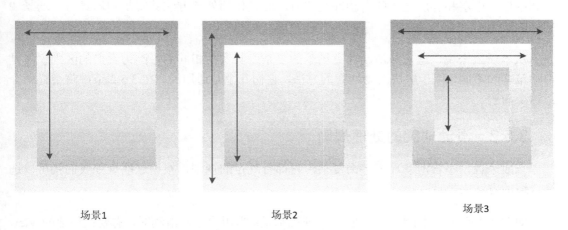

图 3-4　滑动冲突的场景

先说场景 1，主要是将 ViewPager 和 Fragment 配合使用所组成的页面滑动效果，主流应用几乎都会使用这个效果。在这种效果中，可以通过左右滑动来切换页面，而每个页面内部往往又是一个 ListView。本来这种情况下是有滑动冲突的，但是 ViewPager 内部处理了这种滑动冲突，因此采用 ViewPager 时我们无须关注这个问题，如果我们采用的不是 ViewPager 而是 ScrollView 等，那就必须手动处理滑动冲突了，否则造成的后果就是内外两层只能有一层能够滑动，这是因为两者之间的滑动事件有冲突。除了这种典型情况外，还存在其他情况，比如外部上下滑动、内部左右滑动等，但是它们属于同一类滑动冲突。

再说场景 2，这种情况就稍微复杂一些，当内外两层都在同一个方向可以滑动的时候，显然存在逻辑问题。因为当手指开始滑动的时候，系统无法知道用户到底是想让哪一层滑动，所以当手指滑动的时候就会出现问题，要么只有一层能滑动，要么就是内外两层都滑动得很卡顿。在实际的开发中，这种场景主要是指内外两层同时能上下滑动或者内外两层同时能左右滑动。

最后说下场景 3，场景 3 是场景 1 和场景 2 两种情况的嵌套，因此场景 3 的滑动冲突看起来就更加复杂了。比如在许多应用中会有这么一个效果：内层有一个场景 1 中的滑动效果，然后外层又有一个场景 2 中的滑动效果。具体说就是，外部有一个 SlideMenu 效果，然后内部有一个 ViewPager，ViewPager 的每一个页面中又是一个 ListView。虽然说场景 3 的滑动冲突看起来更复杂，但是它是几个单一的滑动冲突的叠加，因此只需要分别处理内层和中层、中层和外层之间的滑动冲突即可，而具体的处理方法其实是和场景 1、场景 2 相同的。

从本质上来说，这三种滑动冲突场景的复杂度其实是相同的，因为它们的区别仅仅是滑动策略的不同，至于解决滑动冲突的方法，它们几个是通用的，在 3.5.2 节中将会详细介绍这个问题。

3.5.2 滑动冲突的处理规则

一般来说，不管滑动冲突多么复杂，它都有既定的规则，根据这些规则我们就可以选择合适的方法去处理。

如图 3-4 所示，对于场景 1，它的处理规则是：当用户左右滑动时，需要让外部的 View 拦截点击事件，当用户上下滑动时，需要让内部 View 拦截点击事件。这个时候我们就可以根据它们的特征来解决滑动冲突，具体来说是：根据滑动是水平滑动还是竖直滑动来判断到底由谁来拦截事件，如图 3-5 所示，根据滑动过程中两个点之间的坐标就可以得出到底是水平滑动还是竖直滑动。如何根据坐标来得到滑动的方向呢？这个很简单，有很多可以参考，比如可以依据滑动路径和水平方向所形成的夹角，也可以依据水平方向和竖直方向上的距离差来判断，某些特殊时候还可以依据水平和竖直方向的速度差来做判断。这里我们可以通过水平和竖直方向的距离差来判断，比如竖直方向滑动的距离大就判断为竖直滑动，否则判断为水平滑动。根据这个规则就可以进行下一步的解决方法制定了。

对于场景 2 来说，比较特殊，它无法根据滑动的角度、距离差以及速度差来做判断，但是这个时候一般都能在业务上找到突破点，比如业务上有规定：当处于某种状态时需要外部 View 响应用户的滑动，而处于另外一种状态时则需要内部 View 来响应 View 的滑动，根据这种业务上的需求我们也能得出相应的处理规则，有了处理规则同样可以进行下一步处理。这种场景通过文字描述可能比较抽象，在下一节会通过实际的例子来演示这种情况的解决方案，那时就容易理解了，这里先有这个概念即可。

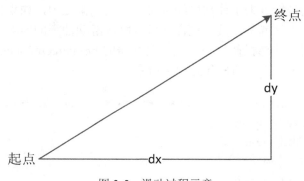

图 3-5　滑动过程示意

对于场景 3 来说，它的滑动规则就更复杂了，和场景 2 一样，它也无法直接根据滑动的角度、距离差以及速度差来做判断，同样还是只能从业务上找到突破点，具体方法和场景 2 一样，都是从业务的需求上得出相应的处理规则，在下一节将会通过实际的例子来演示这种情况的解决方案。

3.5.3　滑动冲突的解决方式

在 3.5.1 节中描述了三种典型的滑动冲突场景，在本节将会一一分析各种场景并给出具体的解决方法。首先我们要分析第一种滑动冲突场景，这也是最简单、最典型的一种滑动冲突，因为它的滑动规则比较简单，不管多复杂的滑动冲突，它们之间的区别仅仅是滑动规则不同而已。抛开滑动规则不说，我们需要找到一种不依赖具体的滑动规则的通用的解决方法，在这里，我们就根据场景 1 的情况来得出通用的解决方案，然后场景 2 和场景 3 我们只需要修改有关滑动规则的逻辑即可。

上面说过，针对场景 1 中的滑动，我们可以根据滑动的距离差来进行判断，这个距离差就是所谓的滑动规则。如果用 ViewPager 去实现场景 1 中的效果，我们不需要手动处理滑动冲突，因为 ViewPager 已经帮我们做了，但是这里为了更好地演示滑动冲突的解决思想，没有采用 ViewPager。其实在滑动过程中得到滑动的角度这个是相当简单的，但是到底要怎么做才能将点击事件交给合适的 View 去处理呢？这时就要用到 3.4 节所讲述的事件分发机制了。针对滑动冲突，这里给出两种解决滑动冲突的方式：外部拦截法和内部拦截法。

1. 外部拦截法

所谓外部拦截法是指点击事件都先经过父容器的拦截处理，如果父容器需要此事件就拦截，如果不需要此事件就不拦截，这样就可以解决滑动冲突的问题，这种方法比较符合点击事件的分发机制。外部拦截法需要重写父容器的 onInterceptTouchEvent 方法，在内部做相应的拦截即可，这种方法的伪代码如下所示。

```java
public boolean onInterceptTouchEvent(MotionEvent event) {
    boolean intercepted = false;
    int x = (int) event.getX();
    int y = (int) event.getY();
    switch (event.getAction()) {
    case MotionEvent.ACTION_DOWN: {
        intercepted = false;
        break;
    }
    case MotionEvent.ACTION_MOVE: {
        if (父容器需要当前点击事件) {
            intercepted = true;
        } else {
            intercepted = false;
        }
        break;
    }
    case MotionEvent.ACTION_UP: {
        intercepted = false;
        break;
    }
    default:
        break;
```

```
    }
    mLastXIntercept = x;
    mLastYIntercept = y;
    return intercepted;
}
```

上述代码是外部拦截法的典型逻辑，针对不同的滑动冲突，只需要修改父容器需要当前点击事件这个条件即可，其他均不需做修改并且也不能修改。这里对上述代码再描述一下，在 onInterceptTouchEvent 方法中，首先是 ACTION_DOWN 这个事件，父容器必须返回 false，即不拦截 ACTION_DOWN 事件，这是因为一旦父容器拦截了 ACTION_DOWN，那么后续的 ACTION_MOVE 和 ACTION_UP 事件都会直接交由父容器处理，这个时候事件没法再传递给子元素了；其次是 ACTION_MOVE 事件，这个事件可以根据需要来决定是否拦截，如果父容器需要拦截就返回 true，否则返回 false；最后是 ACTION_UP 事件，这里必须要返回 false，因为 ACTION_UP 事件本身没有太多意义。

考虑一种情况，假设事件交由子元素处理，如果父容器在 ACTION_UP 时返回了 true，就会导致子元素无法接收到 ACTION_UP 事件，这个时候子元素中的 onClick 事件就无法触发，但是父容器比较特殊，一旦它开始拦截任何一个事件，那么后续的事件都会交给它来处理，而 ACTION_UP 作为最后一个事件也必定可以传递给父容器，即便父容器的 onInterceptTouchEvent 方法在 ACTION_UP 时返回了 false。

2．内部拦截法

内部拦截法是指父容器不拦截任何事件，所有的事件都传递给子元素，如果子元素需要此事件就直接消耗掉，否则就交由父容器进行处理，这种方法和 Android 中的事件分发机制不一致，需要配合 requestDisallowInterceptTouchEvent 方法才能正常工作，使用起来较外部拦截法稍显复杂。它的伪代码如下，我们需要重写子元素的 dispatchTouchEvent 方法：

```
public boolean dispatchTouchEvent(MotionEvent event) {
    int x = (int) event.getX();
    int y = (int) event.getY();

    switch (event.getAction()) {
    case MotionEvent.ACTION_DOWN: {
        parent.requestDisallowInterceptTouchEvent(true);
```

```
        break;
    }
    case MotionEvent.ACTION_MOVE: {
        int deltaX = x - mLastX;
        int deltaY = y - mLastY;
        if (父容器需要此类点击事件)) {
            parent.requestDisallowInterceptTouchEvent(false);
        }
        break;
    }
    case MotionEvent.ACTION_UP: {
        break;
    }
    default:
        break;
    }

    mLastX = x;
    mLastY = y;
    return super.dispatchTouchEvent(event);
}
```

上述代码是内部拦截法的典型代码，当面对不同的滑动策略时只需要修改里面的条件即可，其他不需要做改动而且也不能有改动。除了子元素需要做处理以外，父元素也要默认拦截除了 ACTION_DOWN 以外的其他事件，这样当子元素调用 parent.requestDisallowInterceptTouchEvent(false)方法时，父元素才能继续拦截所需的事件。

为什么父容器不能拦截 ACTION_DOWN 事件呢？那是因为 ACTION_DOWN 事件并不受 FLAG_DISALLOW_INTERCEPT 这个标记位的控制，所以一旦父容器拦截 ACTION_DOWN 事件，那么所有的事件都无法传递到子元素中去，这样内部拦截就无法起作用了。父元素所做的修改如下所示。

```
public boolean onInterceptTouchEvent(MotionEvent event) {
    int action = event.getAction();
    if (action == MotionEvent.ACTION_DOWN) {
        return false;
```

```
        } else {
            return true;
        }
    }
```

下面通过一个实例来分别介绍这两种方法。我们来实现一个类似于 ViewPager 中嵌套 ListView 的效果，为了制造滑动冲突，我们写一个类似于 ViewPager 的控件即可，名字就叫 HorizontalScrollViewEx，这个控件的具体实现思想会在第 4 章进行详细介绍，这里只讲述滑动冲突的部分。

为了实现 ViewPager 的效果，我们定义了一个类似于水平的 LinearLayout 的东西，只不过它可以水平滑动，初始化时我们在它的内部添加若干个 ListView，这样一来，由于它内部的 Listview 可以竖直滑动。而它本身又可以水平滑动，因此一个典型的滑动冲突场景就出现了，并且这种冲突属于类型 1 的冲突。根据滑动策略，我们可以选择水平和竖直的滑动距离差来解决滑动冲突。

首先来看一下 Activity 中的初始化代码，如下所示。

```
public class DemoActivity_1 extends Activity {
    private static final String TAG = "SecondActivity";
    private HorizontalScrollViewEx mListContainer;

    @Override
    protected void onCreate(Bundle savedInstanceState) {
        super.onCreate(savedInstanceState);
        setContentView(R.layout.demo_1);
        Log.d(TAG, "onCreate");
        initView();
    }

    private void initView() {
        LayoutInflater inflater = getLayoutInflater();
        mListContainer = (HorizontalScrollViewEx) findViewById(R.id.
        container);
        final int screenWidth = MyUtils.getScreenMetrics(this).widthPixels;
        final int screenHeight = MyUtils.getScreenMetrics(this).height-
        Pixels;
```

```java
        for (int i = 0; i < 3; i++) {
            ViewGroup layout = (ViewGroup) inflater.inflate(
                    R.layout.content_layout, mListContainer, false);
            layout.getLayoutParams().width = screenWidth;
            TextView textView = (TextView) layout.findViewById(R.id.title);
            textView.setText("page " + (i + 1));
            layout.setBackgroundColor(Color.rgb(255/(i+1),255 / (i + 1), 0));
            createList(layout);
            mListContainer.addView(layout);
        }
    }

    private void createList(ViewGroup layout) {
        ListView listView = (ListView) layout.findViewById(R.id.list);
        ArrayList<String> datas = new ArrayList<String>();
        for (int i = 0; i < 50; i++) {
            datas.add("name " + i);
        }
        ArrayAdapter<String> adapter = new ArrayAdapter<String>(this,
                R.layout.content_list_item, R.id.name, datas);
        listView.setAdapter(adapter);
    }
}
```

上述初始化代码很简单，就是创建了 3 个 ListView 并且把 ListView 加入到我们自定义的 HorizontalScrollViewEx 中，这里 HorizontalScrollViewEx 是父容器，而 ListView 则是子元素，这里就不再多介绍了。

首先采用外部拦截法来解决这个问题，按照前面的分析，我们只需要修改父容器需要拦截事件的条件即可。对于本例来说，父容器的拦截条件就是滑动过程中水平距离差比竖直距离差大，在这种情况下，父容器就拦截当前点击事件，根据这一条件进行相应修改，修改后的 HorizontalScrollViewEx 的 onInterceptTouchEvent 方法如下所示。

```java
public boolean onInterceptTouchEvent(MotionEvent event) {
    boolean intercepted = false;
    int x = (int) event.getX();
    int y = (int) event.getY();
```

```
switch (event.getAction()) {
case MotionEvent.ACTION_DOWN: {
    intercepted = false;
    if (!mScroller.isFinished()) {
        mScroller.abortAnimation();
        intercepted = true;
    }
    break;
}
case MotionEvent.ACTION_MOVE: {
    int deltaX = x - mLastXIntercept;
    int deltaY = y - mLastYIntercept;
    if (Math.abs(deltaX) > Math.abs(deltaY)) {
        intercepted = true;
    } else {
        intercepted = false;
    }
    break;
}
case MotionEvent.ACTION_UP: {
    intercepted = false;
    break;
}
default:
    break;
}

Log.d(TAG, "intercepted=" + intercepted);
mLastXIntercept = x;
mLastYIntercept = y;

return intercepted;
}
```

从上面的代码来看，它和外部拦截法的伪代码的差别很小，只是把父容器的拦截条件换成了具体的逻辑。在滑动过程中，当水平方向的距离大时就判断为水平滑动，为了能够

水平滑动所以让父容器拦截事件；而竖直距离大时父容器就不拦截事件，于是事件就传递给了 ListView，所以 ListView 也能上下滑动，如此滑动冲突就解决了。至于 mScroller.abortAnimation()这一句话主要是为了优化滑动体验而加入的。

考虑一种情况，如果此时用户正在水平滑动，但是在水平滑动停止之前如果用户再迅速进行竖直滑动，就会导致界面在水平方向无法滑动到终点从而处于一种中间状态。为了避免这种不好的体验，当水平方向正在滑动时，下一个序列的点击事件仍然交给父容器处理，这样水平方向就不会停留在中间状态了。

下面是 HorizontalScrollViewEx 的具体实现，只展示了和滑动冲突相关的代码：

```java
public class HorizontalScrollViewEx extends ViewGroup {
    private static final String TAG = "HorizontalScrollViewEx";

    private int mChildrenSize;
    private int mChildWidth;
    private int mChildIndex;
    // 分别记录上次滑动的坐标
    private int mLastX = 0;
    private int mLastY = 0;
    // 分别记录上次滑动的坐标(onInterceptTouchEvent)
    private int mLastXIntercept = 0;
    private int mLastYIntercept = 0;

    private Scroller mScroller;
    private VelocityTracker mVelocityTracker;
    …
    private void init() {
        mScroller = new Scroller(getContext());
        mVelocityTracker = VelocityTracker.obtain();
    }

    @Override
    public boolean onInterceptTouchEvent(MotionEvent event) {
        boolean intercepted = false;
        int x = (int) event.getX();
        int y = (int) event.getY();

        switch (event.getAction()) {
```

```
        case MotionEvent.ACTION_DOWN: {
            intercepted = false;
            if (!mScroller.isFinished()) {
                mScroller.abortAnimation();
                intercepted = true;
            }
            break;
        }
        case MotionEvent.ACTION_MOVE: {
            int deltaX = x - mLastXIntercept;
            int deltaY = y - mLastYIntercept;
            if (Math.abs(deltaX) > Math.abs(deltaY)) {
                intercepted = true;
            } else {
                intercepted = false;
            }
            break;
        }
        case MotionEvent.ACTION_UP: {
            intercepted = false;
            break;
        }
        default:
            break;
        }
        
        Log.d(TAG, "intercepted=" + intercepted);
        mLastX = x;
        mLastY = y;
        mLastXIntercept = x;
        mLastYIntercept = y;
        
        return intercepted;
    }
    
    @Override
    public boolean onTouchEvent(MotionEvent event) {
        mVelocityTracker.addMovement(event);
        int x = (int) event.getX();
        int y = (int) event.getY();
```

```java
switch (event.getAction()) {
case MotionEvent.ACTION_DOWN: {
    if (!mScroller.isFinished()) {
        mScroller.abortAnimation();
    }
    break;
}
case MotionEvent.ACTION_MOVE: {
    int deltaX = x - mLastX;
    int deltaY = y - mLastY;
    scrollBy(-deltaX, 0);
    break;
}
case MotionEvent.ACTION_UP: {
    int scrollX = getScrollX();
    int scrollToChildIndex = scrollX / mChildWidth;
    mVelocityTracker.computeCurrentVelocity(1000);
    float xVelocity = mVelocityTracker.getXVelocity();
    if (Math.abs(xVelocity) >= 50) {
        mChildIndex = xVelocity>0? mChildIndex - 1 : mChildIndex + 1;
    } else {
        mChildIndex = (scrollX + mChildWidth / 2) / mChildWidth;
    }
    mChildIndex = Math.max(0, Math.min(mChildIndex, mChildrenSize - 1));
    int dx = mChildIndex * mChildWidth - scrollX;
    smoothScrollBy(dx, 0);
    mVelocityTracker.clear();
    break;
}
default:
    break;
}

mLastX = x;
mLastY = y;
return true;
}
```

```
    private void smoothScrollBy(int dx, int dy) {
        mScroller.startScroll(getScrollX(), 0, dx, 0, 500);
        invalidate();
    }

    @Override
    public void computeScroll() {
        if (mScroller.computeScrollOffset()) {
            scrollTo(mScroller.getCurrX(), mScroller.getCurrY());
            postInvalidate();
        }
    }
...
}
```

如果采用内部拦截法也是可以的，按照前面对内部拦截法的分析，我们只需要修改 ListView 的 dispatchTouchEvent 方法中的父容器的拦截逻辑，同时让父容器拦截 ACTION_MOVE 和 ACTION_UP 事件即可。为了重写 ListView 的 dispatchTouchEvent 方法，我们必须自定义一个 ListView，称为 ListViewEx，然后对内部拦截法的模板代码进行修改，根据需要，ListViewEx 的实现如下所示。

```
public class ListViewEx extends ListView {
    private static final String TAG = "ListViewEx";

    private HorizontalScrollViewEx2 mHorizontalScrollViewEx2;
    // 分别记录上次滑动的坐标
    private int mLastX = 0;
    private int mLastY = 0;
    ...
    @Override
    public boolean dispatchTouchEvent(MotionEvent event) {
        int x = (int) event.getX();
        int y = (int) event.getY();

        switch (event.getAction()) {
        case MotionEvent.ACTION_DOWN: {
            mHorizontalScrollViewEx2.requestDisallowInterceptTouchEvent
                (true);
            break;
```

```
        }
        case MotionEvent.ACTION_MOVE: {
            int deltaX = x - mLastX;
            int deltaY = y - mLastY;
            if (Math.abs(deltaX) > Math.abs(deltaY)) {
                mHorizontalScrollViewEx2.requestDisallowInterceptTouch-
                    Event(false);
            }
            break;
        }
        case MotionEvent.ACTION_UP: {
            break;
        }
        default:
            break;
        }

        mLastX = x;
        mLastY = y;
        return super.dispatchTouchEvent(event);
    }

}
```

除了上面对 ListView 所做的修改，我们还需要修改 HorizontalScrollViewEx 的 onInterceptTouchEvent 方法，修改后的类暂且叫 HorizontalScrollViewEx2，其 onInterceptTouchEvent 方法如下所示。

```
public boolean onInterceptTouchEvent(MotionEvent event) {
    int x = (int) event.getX();
    int y = (int) event.getY();
    int action = event.getAction();
    if (action == MotionEvent.ACTION_DOWN) {
        mLastX = x;
        mLastY = y;
        if (!mScroller.isFinished()) {
            mScroller.abortAnimation();
            return true;
        }
        return false;
```

```
    } else {
        return true;
    }
}
```

上面的代码就是内部拦截法的示例，其中 mScroller.abortAnimation()这一句不是必须的，在当前这种情形下主要是为了优化滑动体验。从实现上来看，内部拦截法的操作要稍微复杂一些，因此推荐采用外部拦截法来解决常见的滑动冲突。

前面说过，只要我们根据场景 1 的情况来得出通用的解决方案，那么对于场景 2 和场景 3 来说我们只需要修改相关滑动规则的逻辑即可，下面我们就来演示如何利用场景 1 得出的通用的解决方案来解决更复杂的滑动冲突。这里只详细分析场景 2 中的滑动冲突，对于场景 3 中的叠加型滑动冲突，由于它可以拆解为单一的滑动冲突，所以其滑动冲突的解决思想和场景 1、场景 2 中的单一滑动冲突的解决思想一致，只需要分别解决每层之间的滑动冲突即可，再加上本书的篇幅有限，这里就不对场景 3 进行详细分析了。

对于场景 2 来说，它的解决方法和场景 1 一样，只是滑动规则不同而已，在前面我们已经得出了通用的解决方案，因此这里我们只需要替换父容器的拦截规则即可。注意，这里不再演示如何通过内部拦截法来解决场景 2 中的滑动冲突，因为内部拦截法没有外部拦截法简单易用，所以推荐采用外部拦截法来解决常见的滑动冲突。

下面通过一个实际的例子来分析场景 2，首先我们提供一个可以上下滑动的父容器，这里就叫 StickyLayout，它看起来就像是可以上下滑动的竖直的 LinearLayout，然后在它的内部分别放一个 Header 和一个 ListView，这样内外两层都能上下滑动，于是就形成了场景 2 中的滑动冲突了。当然这个 StickyLayout 是有滑动规则的：当 Header 显示时或者 ListView 滑动到顶部时，由 StickyLayout 拦截事件；当 Header 隐藏时，这要分情况，如果 ListView 已经滑动到顶部并且当前手势是向下滑动的话，这个时候还是 StickyLayout 拦截事件，其他情况则由 ListView 拦截事件。这种滑动规则看起来有点复杂，为了解决它们之间的滑动冲突，我们还是需要重写父容器 StickyLayout 的 onInterceptTouchEvent 方法，至于 ListView 则不用做任何修改，我们来看一下 StickyLayout 的具体实现，滑动冲突相关的主要代码如下所示。

```
public class StickyLayout extends LinearLayout {
    private int mTouchSlop;
    // 分别记录上次滑动的坐标
    private int mLastX = 0;
```

```java
private int mLastY = 0;
// 分别记录上次滑动的坐标(onInterceptTouchEvent)
private int mLastXIntercept = 0;
private int mLastYIntercept = 0;
...
@Override
public boolean onInterceptTouchEvent(MotionEvent event) {
    int intercepted = 0;
    int x = (int) event.getX();
    int y = (int) event.getY();

    switch (event.getAction()) {
    case MotionEvent.ACTION_DOWN: {
        mLastXIntercept = x;
        mLastYIntercept = y;
        mLastX = x;
        mLastY = y;
        intercepted = 0;
        break;
    }
    case MotionEvent.ACTION_MOVE: {
        int deltaX = x - mLastXIntercept;
        int deltaY = y - mLastYIntercept;
        if (mDisallowInterceptTouchEventOnHeader && y <= getHeader-
        Height()) {
            intercepted = 0;
        } else if (Math.abs(deltaY) <= Math.abs(deltaX)) {
            intercepted = 0;
        }else if (mStatus == STATUS_EXPANDED && deltaY <= -mTouchSlop) {
            intercepted = 1;
        } else if (mGiveUpTouchEventListener != null) {
            if (mGiveUpTouchEventListener.giveUpTouchEvent(event) &&
            deltaY >= mTouchSlop) {
                intercepted = 1;
            }
        }
        break;
    }
    case MotionEvent.ACTION_UP: {
        intercepted = 0;
```

```java
            mLastXIntercept = mLastYIntercept = 0;
            break;
        }
        default:
            break;
        }

        if (DEBUG) {
            Log.d(TAG, "intercepted=" + intercepted);
        }
        return intercepted != 0 && mIsSticky;
    }

    @Override
    public boolean onTouchEvent(MotionEvent event) {
        if (!mIsSticky) {
            return true;
        }
        int x = (int) event.getX();
        int y = (int) event.getY();
        switch (event.getAction()) {
        case MotionEvent.ACTION_DOWN: {
            break;
        }
        case MotionEvent.ACTION_MOVE: {
            int deltaX = x - mLastX;
            int deltaY = y - mLastY;
            if (DEBUG) {
                Log.d(TAG, "mHeaderHeight=" + mHeaderHeight + " deltaY=" +
                    deltaY + " mlastY=" + mLastY);
            }
            mHeaderHeight += deltaY;
            setHeaderHeight(mHeaderHeight);
            break;
        }
        case MotionEvent.ACTION_UP: {
            // 这里做了一下判断，当松开手的时候，会自动向两边滑动，具体向哪边滑，要看当
            前所处的位置
            int destHeight = 0;
            if (mHeaderHeight <= mOriginalHeaderHeight * 0.5) {
```

```
            destHeight = 0;
            mStatus = STATUS_COLLAPSED;
        } else {
            destHeight = mOriginalHeaderHeight;
            mStatus = STATUS_EXPANDED;
        }
        // 慢慢滑向终点
        this.smoothSetHeaderHeight(mHeaderHeight, destHeight, 500);
        break;
    }
    default:
        break;
    }
    mLastX = x;
    mLastY = y;
    return true;
}
...
}
```

从上面的代码来看，这个 StickyLayout 的实现有点复杂，在第 4 章会详细介绍这个自定义 View 的实现思想，这里先有大概的印象即可。下面我们主要看它的 onInterceptTouchEvent 方法中对 ACTION_MOVE 的处理，如下所示。

```
case MotionEvent.ACTION_MOVE: {
    int deltaX = x - mLastXIntercept;
    int deltaY = y - mLastYIntercept;
    if (mDisallowInterceptTouchEventOnHeader && y <= getHeaderHeight()) {
        intercepted = 0;
    } else if (Math.abs(deltaY) <= Math.abs(deltaX)) {
        intercepted = 0;
    } else if (mStatus == STATUS_EXPANDED && deltaY <= -mTouchSlop) {
        intercepted = 1;
    } else if (mGiveUpTouchEventListener != null) {
        if (mGiveUpTouchEventListener.giveUpTouchEvent(event) && deltaY >=
        mTouchSlop) {
            intercepted = 1;
        }
    }
    break;
```

}

　　我们来分析上面这段代码的逻辑，这里的父容器是 StickyLayout，子元素是 ListView。首先，当事件落在 Header 上面时父容器不会拦截事件；接着，如果竖直距离差小于水平距离差，那么父容器也不会拦截事件；然后，当 Header 是展开状态并且向上滑动时父容器拦截事件。另外一种情况，当 ListView 滑动到顶部了并且向下滑动时，父容器也会拦截事件，经过这些层层判断就可以达到我们想要的效果了。另外，giveUpTouchEvent 是一个接口方法，由外部实现，在本例中主要是用来判断 ListView 是否滑动到顶部，它的具体实现如下：

```
public boolean giveUpTouchEvent(MotionEvent event) {
    if (expandableListView.getFirstVisiblePosition() == 0) {
        View view = expandableListView.getChildAt(0);
        if (view != null && view.getTop() >= 0) {
            return true;
        }
    }
    return false;
}
```

　　上面这个例子比较复杂，需要读者多多体会其中的写法和思想。到这里滑动冲突的解决方法就介绍完毕了，至于场景 3 中的滑动冲突，利用本节所给出的通用的方法是可以轻松解决的，读者可以自己练习一下。在第 4 章会介绍 View 的底层工作原理，并且会介绍如何写出一个好的自定义 View。同时，在本节中所提到的两个自定义 View：Horizontal-ScrollViewEx 和 StickyLayout 将会在第 4 章中进行详细的介绍，它们的完整源码请查看本书所提供的示例代码。

第 4 章　View 的工作原理

在本章中主要介绍两方面的内容，首先介绍 View 的工作原理，接着介绍自定义 View 的实现方式。在 Android 的知识体系中，View 扮演着很重要的角色，简单来理解，View 是 Android 在视觉上的呈现。在界面上 Android 提供了一套 GUI 库，里面有很多控件，但是很多时候我们并不满足于系统提供的控件，因为这样就意味这应用界面的同类化比较严重。那么怎么才能做出与众不同的效果呢？答案是自定义 View，也可以叫自定义控件，通过自定义 View 我们可以实现各种五花八门的效果。但是自定义 View 是有一定难度的，尤其是复杂的自定义 View，大部分时候我们仅仅了解基本控件的使用方法是无法做出复杂的自定义控件的。为了更好地自定义 View，还需要掌握 View 的底层工作原理，比如 View 的测量流程、布局流程以及绘制流程，掌握这几个基本流程后，我们就对 View 的底层更加了解，这样我们就可以做出一个比较完善的自定义 View。

除了 View 的三大流程以外，View 常见的回调方法也是需要熟练掌握的，比如构造方法、onAttach、onVisibilityChanged、onDetach 等。另外对于一些具有滑动效果的自定义 View，我们还需要处理 View 的滑动，如果遇到滑动冲突就还需要解决相应的滑动冲突，关于滑动和滑动冲突这一块内容已经在第 3 章中进行了全面介绍。自定义 View 的实现看起来很复杂，实际上说简单也简单。总结来说，自定义 View 是有几种固定类型的，有的直接继承自 View 和 ViewGroup，而有的则选择继承现有的系统控件，这些都可以，关键是要选择最适合当前需要的方式，选对自定义 View 的实现方式可以起到事半功倍的效果，下面就围绕着这些话题一一展开。

4.1　初识 ViewRoot 和 DecorView

在正式介绍 View 的三大流程之前，我们必须先介绍一些基本概念，这样才能更好地理解 View 的 measure、layout 和 draw 过程，本节主要介绍 ViewRoot 和 DecorView 的概念。

ViewRoot 对应于 ViewRootImpl 类，它是连接 WindowManager 和 DecorView 的纽带，View 的三大流程均是通过 ViewRoot 来完成的。在 ActivityThread 中，当 Activity 对象被创建完毕后，会将 DecorView 添加到 Window 中，同时会创建 ViewRootImpl 对象，并将 ViewRootImpl 对象和 DecorView 建立关联，这个过程可参看如下源码：

```
root = new ViewRootImpl(view.getContext(), display);
root.setView(view, wparams, panelParentView);
```

View 的绘制流程是从 ViewRoot 的 performTraversals 方法开始的，它经过 measure、layout 和 draw 三个过程才能最终将一个 View 绘制出来，其中 measure 用来测量 View 的宽和高，layout 用来确定 View 在父容器中的放置位置，而 draw 则负责将 View 绘制在屏幕上。针对 performTraversals 的大致流程，可用流程图 4-1 来表示。

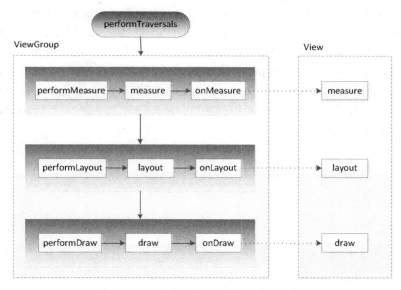

图 4-1　performTraversals 的工作流程图

如图 4-1 所示，performTraversals 会依次调用 performMeasure、performLayout 和 performDraw 三个方法，这三个方法分别完成顶级 View 的 measure、layout 和 draw 这三大流程，其中在 performMeasure 中会调用 measure 方法，在 measure 方法中又会调用 onMeasure 方法，在 onMeasure 方法中则会对所有的子元素进行 measure 过程，这个时候 measure 流程就从父容器传递到子元素中了，这样就完成了一次 measure 过程。接着子元素会重复父容器的 measure 过程，如此反复就完成了整个 View 树的遍历。同理，performLayout 和 performDraw 的传递流程和 performMeasure 是类似的，唯一不同的是，performDraw 的传递过程是在 draw 方法中通过 dispatchDraw 来实现的，不过这并没有本质区别。

measure 过程决定了 View 的宽/高，Measure 完成以后，可以通过 getMeasuredWidth 和 getMeasuredHeight 方法来获取到 View 测量后的宽/高，在几乎所有的情况下它都等同于 View 最终的宽/高，但是特殊情况除外，这点在本章后面会进行说明。Layout 过程决定了 View 的四个顶点的坐标和实际的 View 的宽/高，完成以后，可以通过 getTop、getBottom、getLeft 和 getRight 来拿到 View 的四个顶点的位置，并可以通过 getWidth 和 getHeight 方法来拿到 View 的最终宽/高。Draw 过程则决定了 View 的显示，只有 draw 方法完成以后 View 的内容才能呈现在屏幕上。

如图 4-2 所示，DecorView 作为顶级 View，一般情况下它内部会包含一个竖直方向的 LinearLayout，在这个 LinearLayout 里面有上下两个部分（具体情况和 Android 版本及主题有关），上面是标题栏，下面是内容栏。在 Activity 中我们通过 setContentView 所设置的布局文件其实就是被加到内容栏之中的，而内容栏的 id 是 content，因此可以理解为 Activity 指定布局的方法不叫 setview 而叫 setContentView，因为我们的布局的确加到了 id 为 content 的 FrameLayout 中。如何得到 content 呢？可以这样：ViewGroup content = (ViewGroup) findViewById(android.R.id.content)。如何得到我们设置的 View 呢？可以这样：content.getChildAt(0)。同时，通过源码我们可以知道，DecorView 其实是一个 FrameLayout，View 层的事件都先经过 DecorView，然后才传递给我们的 View。

图 4-2　顶级 View：DecorView 的结构

4.2 理解 MeasureSpec

为了更好地理解 View 的测量过程，我们还需要理解 MeasureSpec。从名字上来看，MeasureSpec 看起来像"测量规格"或者"测量说明书"，不管怎么翻译，它看起来都好像是或多或少地决定了 View 的测量过程。通过源码可以发现，MeasureSpec 的确参与了 View 的 measure 过程。读者可能有疑问，MeasureSpec 是干什么的呢？确切来说，MeasureSpec 在很大程度上决定了一个 View 的尺寸规格，之所以说是很大程度上是因为这个过程还受父容器的影响，因为父容器影响 View 的 MeasureSpec 的创建过程。在测量过程中，系统会将 View 的 LayoutParams 根据父容器所施加的规则转换成对应的 MeasureSpec，然后再根据这个 measureSpec 来测量出 View 的宽/高。上面提到过，这里的宽/高是测量宽/高，不一定等于 View 的最终宽/高。MeasureSpec 看起来有点复杂，其实它的实现是很简单的，下面会详细地分析 MeasureSpec。

4.2.1 MeasureSpec

MeasureSpec 代表一个 32 位 int 值，高 2 位代表 SpecMode，低 30 位代表 SpecSize，SpecMode 是指测量模式，而 SpecSize 是指在某种测量模式下的规格大小。下面先看一下 MeasureSpec 内部的一些常量的定义，通过下面的代码，应该不难理解 MeasureSpec 的工作原理：

```
private static final int MODE_SHIFT = 30;
private static final int MODE_MASK = 0x3 << MODE_SHIFT;
public static final int UNSPECIFIED = 0 << MODE_SHIFT;
public static final int EXACTLY = 1 << MODE_SHIFT;
public static final int AT_MOST = 2 << MODE_SHIFT;

public static int makeMeasureSpec(int size, int mode) {
    if (sUseBrokenMakeMeasureSpec) {
        return size + mode;
    } else {
        return (size & ~MODE_MASK) | (mode & MODE_MASK);
    }
}
```

```
public static int getMode(int measureSpec) {
    return (measureSpec & MODE_MASK);
}

public static int getSize(int measureSpec) {
    return (measureSpec & ~MODE_MASK);
}
```

MeasureSpec 通过将 SpecMode 和 SpecSize 打包成一个 int 值来避免过多的对象内存分配，为了方便操作，其提供了打包和解包方法。SpecMode 和 SpecSize 也是一个 int 值，一组 SpecMode 和 SpecSize 可以打包为一个 MeasureSpec，而一个 MeasureSpec 可以通过解包的形式来得出其原始的 SpecMode 和 SpecSize，需要注意的是这里提到的 MeasureSpec 是指 MeasureSpec 所代表的 int 值，而并非 MeasureSpec 本身。

SpecMode 有三类，每一类都表示特殊的含义，如下所示。

UNSPECIFIED

父容器不对 View 有任何限制，要多大给多大，这种情况一般用于系统内部，表示一种测量的状态。

EXACTLY

父容器已经检测出 View 所需要的精确大小，这个时候 View 的最终大小就是 SpecSize 所指定的值。它对应于 LayoutParams 中的 match_parent 和具体的数值这两种模式。

AT_MOST

父容器指定了一个可用大小即 SpecSize， View 的大小不能大于这个值，具体是什么值要看不同 View 的具体实现。它对应于 LayoutParams 中的 wrap_content。

4.2.2　MeasureSpec 和 LayoutParams 的对应关系

上面提到，系统内部是通过 MeasureSpec 来进行 View 的测量，但是正常情况下我们使用 View 指定 MeasureSpec，尽管如此，但是我们可以给 View 设置 LayoutParams。在 View 测量的时候，系统会将 LayoutParams 在父容器的约束下转换成对应的 MeasureSpec，然后再根据这个 MeasureSpec 来确定 View 测量后的宽/高。需要注意的是，MeasureSpec 不是唯一由 LayoutParams 决定的，LayoutParams 需要和父容器一起才能决定 View 的 MeasureSpec，从而进一步决定 View 的宽/高。另外，对于顶级 View（即 DecorView）和普通 View 来说，MeasureSpec 的转换过程略有不同。对于 DecorView，其 MeasureSpec 由窗口的尺寸和其自身的 LayoutParams 来共同确定；对于普通 View，其 MeasureSpec 由父容器的 MeasureSpec 和自身的 LayoutParams 来共同决定，MeasureSpec 一旦确定后，onMeasure 中就可以确定 View 的测量宽/高。

对于 DecorView 来说，在 ViewRootImpl 中的 measureHierarchy 方法中有如下一段代码，它展示了 DecorView 的 MeasureSpec 的创建过程，其中 desiredWindowWidth 和 desiredWindowHeight 是屏幕的尺寸：

```
childWidthMeasureSpec = getRootMeasureSpec(desiredWindowWidth, lp.width);
childHeightMeasureSpec = getRootMeasureSpec(desiredWindowHeight, lp.height);
performMeasure(childWidthMeasureSpec, childHeightMeasureSpec);
```

接着再看一下 getRootMeasureSpec 方法的实现：

```
private static int getRootMeasureSpec(int windowSize, int rootDimension) {
    int measureSpec;
    switch (rootDimension) {
    case ViewGroup.LayoutParams.MATCH_PARENT:
        // Window can't resize. Force root view to be windowSize.
        measureSpec = MeasureSpec.makeMeasureSpec(windowSize, MeasureSpec.EXACTLY);
        break;
    case ViewGroup.LayoutParams.WRAP_CONTENT:
        // Window can resize. Set max size for root view.
        measureSpec = MeasureSpec.makeMeasureSpec(windowSize, MeasureSpec.AT_MOST);
        break;
    default:
        // Window wants to be an exact size. Force root view to be that size.
        measureSpec = MeasureSpec.makeMeasureSpec(rootDimension, MeasureSpec.EXACTLY);
        break;
    }
    return measureSpec;
}
```

通过上述代码，DecorView 的 MeasureSpec 的产生过程就很明确了，具体来说其遵守

如下规则，根据它的 LayoutParams 中的宽/高的参数来划分。

- LayoutParams.MATCH_PARENT：精确模式，大小就是窗口的大小；
- LayoutParams.WRAP_CONTENT：最大模式，大小不定，但是不能超过窗口的大小；
- 固定大小（比如 100dp）：精确模式，大小为 LayoutParams 中指定的大小。

对于普通 View 来说，这里是指我们布局中的 View，View 的 measure 过程由 ViewGroup 传递而来，先看一下 ViewGroup 的 measureChildWithMargins 方法：

```
protected void measureChildWithMargins(View child,
        int parentWidthMeasureSpec, int widthUsed,
        int parentHeightMeasureSpec, int heightUsed) {
    final MarginLayoutParams lp = (MarginLayoutParams) child.getLayout-
Params();

    final int childWidthMeasureSpec = getChildMeasureSpec(parentWidth
MeasureSpec,mPaddingLeft + mPaddingRight + lp.leftMargin + lp.rightMargin
            + widthUsed, lp.width);
    final int childHeightMeasureSpec = getChildMeasureSpec(parentHeight-
MeasureSpec,mPaddingTop + mPaddingBottom + lp.topMargin + lp.bottomMargin
            + heightUsed, lp.height);

    child.measure(childWidthMeasureSpec, childHeightMeasureSpec);
}
```

上述方法会对子元素进行 measure，在调用子元素的 measure 方法之前会先通过 getChildMeasureSpec 方法来得到子元素的 MeasureSpec。从代码来看，很显然，子元素的 MeasureSpec 的创建与父容器的 MeasureSpec 和子元素本身的 LayoutParams 有关，此外还和 View 的 margin 及 padding 有关，具体情况可以看一下 ViewGroup 的 getChildMeasureSpec 方法，如下所示。

```
public static int getChildMeasureSpec(int spec, int padding, int child-
Dimension) {
    int specMode = MeasureSpec.getMode(spec);
    int specSize = MeasureSpec.getSize(spec);

    int size = Math.max(0, specSize - padding);

    int resultSize = 0;
```

```
        int resultMode = 0;

        switch (specMode) {
        // Parent has imposed an exact size on us
        case MeasureSpec.EXACTLY:
            if (childDimension >= 0) {
                resultSize = childDimension;
                resultMode = MeasureSpec.EXACTLY;
            } else if (childDimension == LayoutParams.MATCH_PARENT) {
                // Child wants to be our size. So be it.
                resultSize = size;
                resultMode = MeasureSpec.EXACTLY;
            } else if (childDimension == LayoutParams.WRAP_CONTENT) {
                // Child wants to determine its own size. It can't be
                // bigger than us.
                resultSize = size;
                resultMode = MeasureSpec.AT_MOST;
            }
            break;

        // Parent has imposed a maximum size on us
        case MeasureSpec.AT_MOST:
            if (childDimension >= 0) {
                // Child wants a specific size... so be it
                resultSize = childDimension;
                resultMode = MeasureSpec.EXACTLY;
            } else if (childDimension == LayoutParams.MATCH_PARENT) {
                // Child wants to be our size, but our size is not fixed.
                // Constrain child to not be bigger than us.
                resultSize = size;
                resultMode = MeasureSpec.AT_MOST;
            } else if (childDimension == LayoutParams.WRAP_CONTENT) {
                // Child wants to determine its own size. It can't be
                // bigger than us.
                resultSize = size;
                resultMode = MeasureSpec.AT_MOST;
            }
            break;

        // Parent asked to see how big we want to be
        case MeasureSpec.UNSPECIFIED:
```

```
            if (childDimension >= 0) {
                // Child wants a specific size... let him have it
                resultSize = childDimension;
                resultMode = MeasureSpec.EXACTLY;
            } else if (childDimension == LayoutParams.MATCH_PARENT) {
                // Child wants to be our size... find out how big it should
                // be
                resultSize = 0;
                resultMode = MeasureSpec.UNSPECIFIED;
            } else if (childDimension == LayoutParams.WRAP_CONTENT) {
                // Child wants to determine its own size.... find out how
                // big it should be
                resultSize = 0;
                resultMode = MeasureSpec.UNSPECIFIED;
            }
        break;
    }
    return MeasureSpec.makeMeasureSpec(resultSize, resultMode);
}
```

上述方法不难理解，它的主要作用是根据父容器的 MeasureSpec 同时结合 View 本身的 LayoutParams 来确定子元素的 MeasureSpec，参数中的 padding 是指父容器中已占用的空间大小，因此子元素可用的大小为父容器的尺寸减去 padding，具体代码如下所示。

```
int specSize = MeasureSpec.getSize(spec);
int size = Math.max(0, specSize - padding);
```

getChildMeasureSpec 清楚展示了普通 View 的 MeasureSpec 的创建规则，为了更清晰地理解 getChildMeasureSpec 的逻辑，这里提供一个表，表中对 getChildMeasureSpec 的工作原理进行了梳理，请看表 4-1。注意，表中的 parentSize 是指父容器中目前可使用的大小。

表 4-1 普通 View 的 MeasureSpec 的创建规则

parentSpecMode chlidLayoutParams	EXACTLY	AT_MOST	UNSPECIFIED
dp/px	EXACTLY childSize	EXACTLY childSize	EXACTLY childSize
match_parent	EXACTLY parentSize	AT_MOST parentSize	UNSPECIFIED 0
wrap_content	AT_MOST parentSize	AT_MOST parentSize	UNSPECIFIED 0

针对表 4-1，这里再做一下说明。前面已经提到，对于普通 View，其 MeasureSpec 由父容器的 MeasureSpec 和自身的 LayoutParams 来共同决定，那么针对不同的父容器和 View 本身不同的 LayoutParams，View 就可以有多种 MeasureSpec。这里简单说一下，当 View 采用固定宽/高的时候，不管父容器的 MeasureSpec 是什么，View 的 MeasureSpec 都是精确模式并且其大小遵循 Layoutparams 中的大小。当 View 的宽/高是 match_parent 时，如果父容器的模式是精准模式，那么 View 也是精准模式并且其大小是父容器的剩余空间；如果父容器是最大模式，那么 View 也是最大模式并且其大小不会超过父容器的剩余空间。当 View 的宽/高是 wrap_content 时，不管父容器的模式是精准还是最大化，View 的模式总是最大化并且大小不能超过父容器的剩余空间。可能读者会发现，在我们的分析中漏掉了 UNSPECIFIED 模式，那是因为这个模式主要用于系统内部多次 Measure 的情形，一般来说，我们不需要关注此模式。

通过表 4-1 可以看出，只要提供父容器的 MeasureSpec 和子元素的 LayoutParams，就可以快速地确定出子元素的 MeasureSpec 了，有了 MeasureSpec 就可以进一步确定出子元素测量后的大小了。需要说明的是，表 4-1 并非是什么经验总结，它只是 getChildMeasureSpec 这个方法以表格的方式呈现出来而已。

4.3 View 的工作流程

View 的工作流程主要是指 measure、layout、draw 这三大流程，即测量、布局和绘制，其中 measure 确定 View 的测量宽/高，layout 确定 View 的最终宽/高和四个顶点的位置，而 draw 则将 View 绘制到屏幕上。

4.3.1 measure 过程

measure 过程要分情况来看，如果只是一个原始的 View，那么通过 measure 方法就完成了其测量过程，如果是一个 ViewGroup，除了完成自己的测量过程外，还会遍历去调用所有子元素的 measure 方法，各个子元素再递归去执行这个流程，下面针对这两种情况分别讨论。

1. View 的 measure 过程

View 的 measure 过程由其 measure 方法来完成，measure 方法是一个 final 类型的方法，这意味着子类不能重写此方法，在 View 的 measure 方法中会去调用 View 的 onMeasure 方法，因此只需要看 onMeasure 的实现即可，View 的 onMeasure 方法如下所示。

```
protected void onMeasure(int widthMeasureSpec, int heightMeasureSpec) {
    setMeasuredDimension(getDefaultSize(getSuggestedMinimumWidth(),
        widthMeasureSpec),getDefaultSize(getSuggestedMinimumHeight(),
        heightMeasureSpec));
}
```

上述代码很简洁，但是简洁并不代表简单，setMeasuredDimension 方法会设置 View 宽/高的测量值，因此我们只需要看 getDefaultSize 这个方法即可：

```
public static int getDefaultSize(int size, int measureSpec) {
    int result = size;
    int specMode = MeasureSpec.getMode(measureSpec);
    int specSize = MeasureSpec.getSize(measureSpec);

    switch (specMode) {
    case MeasureSpec.UNSPECIFIED:
        result = size;
        break;
    case MeasureSpec.AT_MOST:
    case MeasureSpec.EXACTLY:
        result = specSize;
        break;
    }
    return result;
}
```

可以看出，getDefaultSize 这个方法的逻辑很简单，对于我们来说，我们只需要看 AT_MOST 和 EXACTLY 这两种情况。简单地理解，其实 getDefaultSize 返回的大小就是 measureSpec 中的 specSize，而这个 specSize 就是 View 测量后的大小，这里多次提到测量后的大小，是因为 View 最终的大小是在 layout 阶段确定的，所以这里必须要加以区分，但是几乎所有情况下 View 的测量大小和最终大小是相等的。

至于 UNSPECIFIED 这种情况，一般用于系统内部的测量过程，在这种情况下，View 的大小为 getDefaultSize 的第一个参数 size，即宽/高分别为 getSuggestedMinimumWidth 和 getSuggestedMinimumHeight 这两个方法的返回值，看一下它们的源码：

```java
protected int getSuggestedMinimumWidth() {
    return (mBackground == null) ? mMinWidth : max(mMinWidth, mBackground.
    getMinimumWidth());
}

protected int getSuggestedMinimumHeight() {
    return (mBackground == null) ? mMinHeight : max(mMinHeight, mBackground.
    getMinimumHeight());
}
```

这里只分析 getSuggestedMinimumWidth 方法的实现，getSuggestedMinimumHeight 和它的实现原理是一样的。从 getSuggestedMinimumWidth 的代码可以看出，如果 View 没有设置背景，那么 View 的宽度为 mMinWidth，而 mMinWidth 对应于 android:minWidth 这个属性所指定的值，因此 View 的宽度即为 android:minWidth 属性所指定的值。这个属性如果不指定，那么 mMinWidth 则默认为 0；如果 View 指定了背景，则 View 的宽度为 max(mMinWidth, mBackground.getMinimumWidth())。mMinWidth 的含义我们已经知道了，那么 mBackground.getMinimumWidth()是什么呢？我们看一下 Drawable 的 getMinimumWidth 方法，如下所示。

```java
public int getMinimumWidth() {
    final int intrinsicWidth = getIntrinsicWidth();
    return intrinsicWidth > 0 ? intrinsicWidth : 0;
}
```

可以看出，getMinimumWidth 返回的就是 Drawable 的原始宽度，前提是这个 Drawable 有原始宽度，否则就返回 0。那么 Drawable 在什么情况下有原始宽度呢？这里先举个例子说明一下，ShapeDrawable 无原始宽/高，而 BitmapDrawable 有原始宽/高（图片的尺寸），详细内容会在第 6 章进行介绍。

这里再总结一下 getSuggestedMinimumWidth 的逻辑：如果 View 没有设置背景，那么返回 android:minWidth 这个属性所指定的值，这个值可以为 0；如果 View 设置了背景，则返回 android:minWidth 和背景的最小宽度这两者中的最大值，getSuggestedMinimumWidth 和 getSuggestedMinimumHeight 的返回值就是 View 在 UNSPECIFIED 情况下的测量宽/高。

从 getDefaultSize 方法的实现来看，View 的宽/高由 specSize 决定，所以我们可以得出如下结论：直接继承 View 的自定义控件需要重写 onMeasure 方法并设置 wrap_content 时的自身大小，否则在布局中使用 wrap_content 就相当于使用 match_parent。为什么呢？这个原因需要结合上述代码和表 4-1 才能更好地理解。从上述代码中我们知道，如果 View 在布局中使用 wrap_content，那么它的 specMode 是 AT_MOST 模式，在这种模式下，它的宽/高等于 specSize；查表 4-1 可知，这种情况下 View 的 specSize 是 parentSize，而 parentSize 是父容器中目前可以使用的大小，也就是父容器当前剩余的空间大小。很显然，View 的宽/高就等于父容器当前剩余的空间大小，这种效果和在布局中使用 match_parent 完全一致。如何解决这个问题呢？也很简单，代码如下所示。

```
protected void onMeasure(int widthMeasureSpec, int heightMeasureSpec) {
    super.onMeasure(widthMeasureSpec, heightMeasureSpec);
    int widthSpecMode = MeasureSpec.getMode(widthMeasureSpec);
    int widthSpecSize = MeasureSpec.getSize(widthMeasureSpec);
    int heightSpecMode = MeasureSpec.getMode(heightMeasureSpec);
    int heightSpecSize = MeasureSpec.getSize(heightMeasureSpec);
    if (widthSpecMode == MeasureSpec.AT_MOST && heightSpecMode ==
MeasureSpec.AT_MOST) {
        setMeasuredDimension(mWidth, mHeight);
    } else if (widthSpecMode == MeasureSpec.AT_MOST) {
        setMeasuredDimension(mWidth, heightSpecSize);
    } else if (heightSpecMode == MeasureSpec.AT_MOST) {
        setMeasuredDimension(widthSpecSize, mHeight);
    }
}
```

在上面的代码中，我们只需要给 View 指定一个默认的内部宽/高（mWidth 和 mHeight），并在 wrap_content 时设置此宽/高即可。对于非 wrap_content 情形，我们沿用系统的测量值即可，至于这个默认的内部宽/高的大小如何指定，这个没有固定的依据，根据需要灵活指定即可。如果查看 TextView、ImageView 等的源码就可以知道，针对 wrap_content 情形，它们的 onMeasure 方法均做了特殊处理，读者可以自行查看它们的源码。

2. ViewGroup 的 measure 过程

对于 ViewGroup 来说，除了完成自己的 measure 过程以外，还会遍历去调用所有子元素的 measure 方法，各个子元素再递归去执行这个过程。和 View 不同的是，ViewGroup 是一个抽象类，因此它没有重写 View 的 onMeasure 方法，但是它提供了一个叫 measureChildren 的方法，如下所示。

```
protected void measureChildren(int widthMeasureSpec,int heightMeasureSpec) {
    final int size = mChildrenCount;
    final View[] children = mChildren;
```

```
        for (int i = 0; i < size; ++i) {
            final View child = children[i];
            if ((child.mViewFlags & VISIBILITY_MASK) != GONE) {
                measureChild(child, widthMeasureSpec, heightMeasureSpec);
            }
        }
    }
```

从上述代码来看，ViewGroup 在 measure 时，会对每一个子元素进行 measure，measureChild 这个方法的实现也很好理解，如下所示。

```
protected void measureChild(View child, int parentWidthMeasureSpec,
        int parentHeightMeasureSpec) {
    final LayoutParams lp = child.getLayoutParams();

    final int childWidthMeasureSpec = getChildMeasureSpec(parentWidth-
MeasureSpec,
            mPaddingLeft + mPaddingRight, lp.width);
    final int childHeightMeasureSpec = getChildMeasureSpec(parentHeight-
MeasureSpec,
            mPaddingTop + mPaddingBottom, lp.height);

    child.measure(childWidthMeasureSpec, childHeightMeasureSpec);
}
```

很显然，measureChild 的思想就是取出子元素的 LayoutParams，然后再通过 getChildMeasureSpec 来创建子元素的 MeasureSpec，接着将 MeasureSpec 直接传递给 View 的 measure 方法来进行测量。getChildMeasureSpec 的工作过程已经在上面进行了详细分析，通过表 4-1 可以更清楚地了解它的逻辑。

我们知道，ViewGroup 并没有定义其测量的具体过程，这是因为 ViewGroup 是一个抽象类，其测量过程的 onMeasure 方法需要各个子类去具体实现，比如 LinearLayout、RelativeLayout 等，为什么 ViewGroup 不像 View 一样对其 onMeasure 方法做统一的实现呢？那是因为不同的 ViewGroup 子类有不同的布局特性，这导致它们的测量细节各不相同，比如 LinearLayout 和 RelativeLayout 这两者的布局特性显然不同，因此 ViewGroup 无法做统一实现。下面就通过 LinearLayout 的 onMeasure 方法来分析 ViewGroup 的 measure 过程，其他 Layout 类型读者可以自行分析。

首先来看 LinearLayout 的 onMeasure 方法，如下所示。

```
protected void onMeasure(int widthMeasureSpec, int heightMeasureSpec) {
    if (mOrientation == VERTICAL) {
        measureVertical(widthMeasureSpec, heightMeasureSpec);
    } else {
        measureHorizontal(widthMeasureSpec, heightMeasureSpec);
    }
}
```

上述代码很简单，我们选择一个来看一下，比如选择查看竖直布局的 LinearLayout 的测量过程，即 measureVertical 方法，measureVertical 的源码比较长，下面只描述其大概逻辑，首先看一段代码：

```
// See how tall everyone is. Also remember max width.
for (int i = 0; i < count; ++i) {
    final View child = getVirtualChildAt(i);
    ...
    // Determine how big this child would like to be. If this or
    // previous children have given a weight, then we allow it to
    // use all available space (and we will shrink things later
    // if needed).
    measureChildBeforeLayout(
           child, i, widthMeasureSpec, 0, heightMeasureSpec,
           totalWeight == 0 ? mTotalLength : 0);

    if (oldHeight != Integer.MIN_VALUE) {
       lp.height = oldHeight;
    }

    final int childHeight = child.getMeasuredHeight();
    final int totalLength = mTotalLength;
    mTotalLength=Math.max(totalLength,totalLength+childHeight+lp.topMargin +
           lp.bottomMargin + getNextLocationOffset(child));
}
```

从上面这段代码可以看出，系统会遍历子元素并对每个子元素执行 measureChildBeforeLayout 方法，这个方法内部会调用子元素的 measure 方法，这样各个子元素就开始依次进入 measure 过程，并且系统会通过 mTotalLength 这个变量来存储 LinearLayout 在竖直方向的初步高度。每测量一个子元素，mTotalLength 就会增加，增加的部分主要包括了子元素的高度以及子元素在竖直方向上的 margin 等。当子元素测量完毕后，LinearLayout 会测量自己的大小，源码如下所示。

```
    // Add in our padding
mTotalLength += mPaddingTop + mPaddingBottom;
int heightSize = mTotalLength;
// Check against our minimum height
heightSize = Math.max(heightSize, getSuggestedMinimumHeight());
// Reconcile our calculated size with the heightMeasureSpec
int heightSizeAndState=resolveSizeAndState(heightSize, heightMeasureSpec, 0);
heightSize = heightSizeAndState & MEASURED_SIZE_MASK;
...
setMeasuredDimension(resolveSizeAndState(maxWidth, widthMeasureSpec,
childState),
heightSizeAndState);
```

这里对上述代码进行说明，当子元素测量完毕后，LinearLayout 会根据子元素的情况来测量自己的大小。针对竖直的 LinearLayout 而言，它在水平方向的测量过程遵循 View 的测量过程，在竖直方向的测量过程则和 View 有所不同。具体来说是指，如果它的布局中高度采用的是 match_parent 或者具体数值，那么它的测量过程和 View 一致，即高度为 specSize；如果它的布局中高度采用的是 wrap_content，那么它的高度是所有子元素所占用的高度总和，但是仍然不能超过它的父容器的剩余空间，当然它的最终高度还需要考虑其在竖直方向的 padding，这个过程可以进一步参看如下源码：

```
public static int resolveSizeAndState(int size, int measureSpec, int
childMeasuredState) {
    int result = size;
    int specMode = MeasureSpec.getMode(measureSpec);
    int specSize =  MeasureSpec.getSize(measureSpec);
    switch (specMode) {
    case MeasureSpec.UNSPECIFIED:
        result = size;
        break;
    case MeasureSpec.AT_MOST:
        if (specSize < size) {
            result = specSize | MEASURED_STATE_TOO_SMALL;
        } else {
```

```
                result = size;
            }
            break;
        case MeasureSpec.EXACTLY:
            result = specSize;
            break;
    }
    return result | (childMeasuredState&MEASURED_STATE_MASK);
}
```

View 的 measure 过程是三大流程中最复杂的一个，measure 完成以后，通过 getMeasured-Width/Height 方法就可以正确地获取到 View 的测量宽/高。需要注意的是，在某些极端情况下，系统可能需要多次 measure 才能确定最终的测量宽/高，在这种情形下，在 onMeasure 方法中拿到的测量宽/高很可能是不准确的。一个比较好的习惯是在 onLayout 方法中去获取 View 的测量宽/高或者最终宽/高。

上面已经对 View 的 measure 过程进行了详细的分析，现在考虑一种情况，比如我们想在 Activity 已启动的时候就做一件任务，但是这一件任务需要获取某个 View 的宽/高。读者可能会说，这很简单啊，在 onCreate 或者 onResume 里面去获取这个 View 的宽/高不就行了？读者可以自行试一下，实际上在 onCreate、onStart、onResume 中均无法正确得到某个 View 的宽/高信息，这是因为 View 的 measure 过程和 Activity 的生命周期方法不是同步执行的，因此无法保证 Activity 执行了 onCreate、onStart、onResume 时某个 View 已经测量完毕了，如果 View 还没有测量完毕，那么获得的宽/高就是 0。有没有什么方法能解决这个问题呢？答案是有的，这里给出四种方法来解决这个问题：

（1）Activity/View#onWindowFocusChanged。

onWindowFocusChanged 这个方法的含义是：View 已经初始化完毕了，宽/高已经准备好了，这个时候去获取宽/高是没问题的。需要注意的是，onWindowFocusChanged 会被调用多次，当 Activity 的窗口得到焦点和失去焦点时均会被调用一次。具体来说，当 Activity 继续执行和暂停执行时，onWindowFocusChanged 均会被调用，如果频繁地进行 onResume 和 onPause，那么 onWindowFocusChanged 也会被频繁地调用。典型代码如下：

```
public void onWindowFocusChanged(boolean hasFocus) {
    super.onWindowFocusChanged(hasFocus);
    if (hasFocus) {
        int width = view.getMeasuredWidth();
```

```
        int height = view.getMeasuredHeight();
    }
}
```

（2）view.post(runnable)。

通过 post 可以将一个 runnable 投递到消息队列的尾部，然后等待 Looper 调用此 runnable 的时候，View 也已经初始化好了。典型代码如下：

```
protected void onStart() {
    super.onStart();
    view.post(new Runnable() {

        @Override
        public void run() {
            int width = view.getMeasuredWidth();
            int height = view.getMeasuredHeight();
        }
    });
}
```

（3）ViewTreeObserver。

使用 ViewTreeObserver 的众多回调可以完成这个功能，比如使用 OnGlobalLayoutListener 这个接口，当 View 树的状态发生改变或者 View 树内部的 View 的可见性发现改变时，onGlobalLayout 方法将被回调，因此这是获取 View 的宽/高一个很好的时机。需要注意的是，伴随着 View 树的状态改变等，onGlobalLayout 会被调用多次。典型代码如下：

```
protected void onStart() {
    super.onStart();

    ViewTreeObserver observer = view.getViewTreeObserver();
    observer.addOnGlobalLayoutListener(new OnGlobalLayoutListener() {

        @SuppressWarnings("deprecation")
        @Override
        public void onGlobalLayout() {
```

```
            view.getViewTreeObserver().removeGlobalOnLayoutListener(this);
            int width = view.getMeasuredWidth();
            int height = view.getMeasuredHeight();
        }
    });
}
```

（4）view.measure(int widthMeasureSpec, int heightMeasureSpec)。

通过手动对 View 进行 measure 来得到 View 的宽/高。这种方法比较复杂，这里要分情况处理，根据 View 的 LayoutParams 来分：

match_parent

直接放弃，无法 measure 出具体的宽/高。原因很简单，根据 View 的 measure 过程，如表 4-1 所示，构造此种 MeasureSpec 需要知道 parentSize，即父容器的剩余空间，而这个时候我们无法知道 parentSize 的大小，所以理论上不可能测量出 View 的大小。

具体的数值（dp/px）

比如宽/高都是 100px，如下 measure：

```
int widthMeasureSpec = MeasureSpec.makeMeasureSpec(100, MeasureSpec.EXACTLY);
int heightMeasureSpec = MeasureSpec.makeMeasureSpec(100, MeasureSpec.EXACTLY);
    view.measure(widthMeasureSpec, heightMeasureSpec);
```

wrap_content

如下 measure：

```
int widthMeasureSpec = MeasureSpec.makeMeasureSpec( (1 << 30) - 1, MeasureSpec.AT_MOST);
int heightMeasureSpec = MeasureSpec.makeMeasureSpec( (1 << 30) - 1, MeasureSpec.AT_MOST);
    view.measure(widthMeasureSpec, heightMeasureSpec);
```

注意到(1 << 30)-1，通过分析 MeasureSpec 的实现可以知道，View 的尺寸使用 30 位二进制表示，也就是说最大是 30 个 1（即 $2^{30} - 1$），也就是(1 << 30) – 1，在最大化模式下，

我们用 View 理论上能支持的最大值去构造 MeasureSpec 是合理的。

关于 View 的 measure，网络上有两个错误的用法。为什么说是错误的，首先其违背了系统的内部实现规范（因为无法通过错误的 MeasureSpec 去得出合法的 SpecMode，从而导致 measure 过程出错），其次不能保证一定能 measure 出正确的结果。

第一种错误用法：

```
int widthMeasureSpec = MeasureSpec.makeMeasureSpec(-1, MeasureSpec.
UNSPECIFIED);
int heightMeasureSpec = MeasureSpec.makeMeasureSpec(-1, MeasureSpec.
UNSPECIFIED);
view.measure(widthMeasureSpec, heightMeasureSpec);
```

第二种错误用法：

```
view.measure(LayoutParams.WRAP_CONTENT, LayoutParams.WRAP_CONTENT)
```

4.3.2　layout 过程

Layout 的作用是 ViewGroup 用来确定子元素的位置，当 ViewGroup 的位置被确定后，它在 onLayout 中会遍历所有的子元素并调用其 layout 方法，在 layout 方法中 onLayout 方法又会被调用。Layout 过程和 measure 过程相比就简单多了，layout 方法确定 View 本身的位置，而 onLayout 方法则会确定所有子元素的位置，先看 View 的 layout 方法，如下所示。

```
public void layout(int l, int t, int r, int b) {
    if ((mPrivateFlags3 & PFLAG3_MEASURE_NEEDED_BEFORE_LAYOUT) != 0) {
        onMeasure(mOldWidthMeasureSpec, mOldHeightMeasureSpec);
        mPrivateFlags3 &= ~PFLAG3_MEASURE_NEEDED_BEFORE_LAYOUT;
    }

    int oldL = mLeft;
    int oldT = mTop;
    int oldB = mBottom;
    int oldR = mRight;

    boolean changed = isLayoutModeOptical(mParent) ?
            setOpticalFrame(l, t, r, b) : setFrame(l, t, r, b);
```

```
        if (changed || (mPrivateFlags & PFLAG_LAYOUT_REQUIRED) == PFLAG_
LAYOUT_REQUIRED) {
            onLayout(changed, l, t, r, b);
            mPrivateFlags &= ~PFLAG_LAYOUT_REQUIRED;

            ListenerInfo li = mListenerInfo;
            if (li != null && li.mOnLayoutChangeListeners != null) {
                ArrayList<OnLayoutChangeListener> listenersCopy =
                        (ArrayList<OnLayoutChangeListener>)li.mOnLayout-
                        ChangeListeners.clone();
                int numListeners = listenersCopy.size();
                for (int i = 0; i < numListeners; ++i) {
                    listenersCopy.get(i).onLayoutChange(this, l, t, r, b, oldL,
                    oldT, oldR, oldB);
                }
            }
        }

        mPrivateFlags &= ~PFLAG_FORCE_LAYOUT;
        mPrivateFlags3 |= PFLAG3_IS_LAID_OUT;
}
```

layout 方法的大致流程如下：首先会通过 setFrame 方法来设定 View 的四个顶点的位置，即初始化 mLeft、mRight、mTop 和 mBottom 这四个值，View 的四个顶点一旦确定，那么 View 在父容器中的位置也就确定了；接着会调用 onLayout 方法，这个方法的用途是父容器确定子元素的位置，和 onMeasure 方法类似，onLayout 的具体实现同样和具体的布局有关，所以 View 和 ViewGroup 均没有真正实现 onLayout 方法。接下来，我们可以看一下 LinearLayout 的 onLayout 方法，如下所示。

```
protected void onLayout(boolean changed, int l, int t, int r, int b) {
    if (mOrientation == VERTICAL) {
        layoutVertical(l, t, r, b);
    } else {
        layoutHorizontal(l, t, r, b);
    }
}
```

LinearLayout 中 onLayout 的实现逻辑和 onMeasure 的实现逻辑类似，这里选择 layoutVertical 继续讲解，为了更好地理解其逻辑，这里只给出了主要的代码：

```
void layoutVertical(int left, int top, int right, int bottom) {
    ...
    final int count = getVirtualChildCount();
    for (int i = 0; i < count; i++) {
        final View child = getVirtualChildAt(i);
        if (child == null) {
            childTop += measureNullChild(i);
        } else if (child.getVisibility() != GONE) {
            final int childWidth = child.getMeasuredWidth();
            final int childHeight = child.getMeasuredHeight();

            final LinearLayout.LayoutParams lp =
                    (LinearLayout.LayoutParams) child.getLayoutParams();
            ...
            if (hasDividerBeforeChildAt(i)) {
                childTop += mDividerHeight;
            }

            childTop += lp.topMargin;
            setChildFrame(child, childLeft, childTop + getLocationOffset
            (child),childWidth, childHeight);
            childTop += childHeight + lp.bottomMargin + getNextLocation-
            Offset(child);

            i += getChildrenSkipCount(child, i);
        }
    }
}
```

这里分析一下 layoutVertical 的代码逻辑，可以看到，此方法会遍历所有子元素并调用 setChildFrame 方法来为子元素指定对应的位置，其中 childTop 会逐渐增大，这就意味着后面的子元素会被放置在靠下的位置，这刚好符合竖直方向的 LinearLayout 的特性。至于 setChildFrame，它仅仅是调用子元素的 layout 方法而已，这样父元素在 layout 方法中完成自己的定位以后，就通过 onLayout 方法去调用子元素的 layout 方法，子元素又会通过自己的 layout 方法来确定自己的位置，这样一层一层地传递下去就完成了整个 View 树的 layout 过程。setChildFrame 方法的实现如下所示。

```
private void setChildFrame(View child, int left, int top, int width, int
height) {
```

```
    child.layout(left, top, left + width, top + height);
}
```

我们注意到，setChildFrame 中的 width 和 height 实际上就是子元素的测量宽/高，从下面的代码可以看出这一点：

```
final int childWidth = child.getMeasuredWidth();
final int childHeight = child.getMeasuredHeight();
setChildFrame(child, childLeft, childTop + getLocationOffset(child),
    childWidth, childHeight);
```

而在 layout 方法中会通过 setFrame 去设置子元素的四个顶点的位置，在 setFrame 中有如下几句赋值语句，这样一来子元素的位置就确定了：

```
mLeft = left;
mTop = top;
mRight = right;
mBottom = bottom;
```

下面我们来回答一个在 4.3.2 节中提到的问题：View 的测量宽/高和最终宽/高有什么区别？这个问题可以具体为：View 的 getMeasuredWidth 和 getWidth 这两个方法有什么区别，至于 getMeasuredHeight 和 getHeight 的区别和前两者完全一样。为了回答这个问题，首先，我们看一下 getwidth 和 getHeight 这两个方法的具体实现：

```
public final int getWidth() {
    return mRight - mLeft;
}

public final int getHeight() {
    return mBottom - mTop;
}
```

从 getWidth 和 getHeight 的源码再结合 mLeft、mRight、mTop 和 mBottom 这四个变量的赋值过程来看，getWidth 方法的返回值刚好就是 View 的测量宽度，而 getHeight 方法的返回值也刚好就是 View 的测量高度。经过上述分析，现在我们可以回答这个问题了：在 View 的默认实现中，View 的测量宽/高和最终宽/高是相等的，只不过测量宽/高形成于 View 的 measure 过程，而最终宽/高形成于 View 的 layout 过程，即两者的赋值时机不同，测量宽/高的赋值时机稍微早一些。因此，在日常开发中，我们可以认为 View 的测量宽/高就等于最终宽/高，但是的确存在某些特殊情况会导致两者不一致，下面举例说明。

如果重写 View 的 layout 方法，代码如下：

```
public void layout(int l, int t, int r, int b) {
    super.layout(l, t, r + 100, b + 100);
}
```

上述代码会导致在任何情况下 View 的最终宽/高总是比测量宽/高大 100px，虽然这样做会导致 View 显示不正常并且也没有实际意义，但是这证明了测量宽/高的确可以不等于最终宽/高。另外一种情况是在某些情况下，View 需要多次 measure 才能确定自己的测量宽/高，在前几次的测量过程中，其得出的测量宽/高有可能和最终宽/高不一致，但最终来说，测量宽/高还是和最终宽/高相同。

4.3.3　draw 过程

Draw 过程就比较简单了，它的作用是将 View 绘制到屏幕上面。View 的绘制过程遵循如下几步：

（1）绘制背景 background.draw(canvas)。

（2）绘制自己（onDraw）。

（3）绘制 children（dispatchDraw）。

（4）绘制装饰（onDrawScrollBars）。

这一点通过 draw 方法的源码可以明显看出来，如下所示。

```
public void draw(Canvas canvas) {
    final int privateFlags = mPrivateFlags;
    final boolean dirtyOpaque = (privateFlags & PFLAG_DIRTY_MASK) ==
    PFLAG_DIRTY_OPAQUE &&
            (mAttachInfo == null || !mAttachInfo.mIgnoreDirtyState);
    mPrivateFlags = (privateFlags & ~PFLAG_DIRTY_MASK) | PFLAG_DRAWN;

    /*
     * Draw traversal performs several drawing steps which must be executed
     * in the appropriate order:
```

```
 *      1. Draw the background
 *      2. If necessary, save the canvas' layers to prepare for fading
 *      3. Draw view's content
 *      4. Draw children
 *      5. If necessary, draw the fading edges and restore layers
 *      6. Draw decorations (scrollbars for instance)
 */

// Step 1, draw the background, if needed
int saveCount;

if (!dirtyOpaque) {
    drawBackground(canvas);
}

// skip step 2 & 5 if possible (common case)
final int viewFlags = mViewFlags;
boolean horizontalEdges = (viewFlags & FADING_EDGE_HORIZONTAL) != 0;
boolean verticalEdges = (viewFlags & FADING_EDGE_VERTICAL) != 0;
if (!verticalEdges && !horizontalEdges) {
    // Step 3, draw the content
    if (!dirtyOpaque) onDraw(canvas);

    // Step 4, draw the children
    dispatchDraw(canvas);

    // Step 6, draw decorations (scrollbars)
    onDrawScrollBars(canvas);

    if (mOverlay != null && !mOverlay.isEmpty()) {
        mOverlay.getOverlayView().dispatchDraw(canvas);
    }

    // we're done...
    return;
}
...
}
```

View 绘制过程的传递是通过 dispatchDraw 来实现的，dispatchDraw 会遍历调用所有子元素的 draw 方法，如此 draw 事件就一层层地传递了下去。View 有一个特殊的方法 setWillNotDraw，先看一下它的源码，如下所示。

```
/**
 * If this view doesn't do any drawing on its own, set this flag to
 * allow further optimizations. By default, this flag is not set on
 * View, but could be set on some View subclasses such as ViewGroup.
 *
 * Typically, if you override {@link #onDraw(android.graphics.Canvas)}
 * you should clear this flag.
 *
 * @param willNotDraw whether or not this View draw on its own
 */
public void setWillNotDraw(boolean willNotDraw) {
    setFlags(willNotDraw ? WILL_NOT_DRAW : 0, DRAW_MASK);
}
```

从 setWillNotDraw 这个方法的注释中可以看出，如果一个 View 不需要绘制任何内容，那么设置这个标记位为 true 以后，系统会进行相应的优化。默认情况下，View 没有启用这个优化标记位，但是 ViewGroup 会默认启用这个优化标记位。这个标记位对实际开发的意义是：当我们的自定义控件继承于 ViewGroup 并且本身不具备绘制功能时，就可以开启这个标记位从而便于系统进行后续的优化。当然，当明确知道一个 ViewGroup 需要通过 onDraw 来绘制内容时，我们需要显式地关闭 WILL_NOT_DRAW 这个标记位。

4.4 自定义 View

本节将详细介绍自定义 View 相关的知识。自定义 View 的作用不用多说，这个读者都应该清楚，如果想要做出绚丽的界面效果仅仅靠系统的控件是远远不够的，这个时候就必须通过自定义 View 来实现这些绚丽的效果。自定义 View 是一个综合的技术体系，它涉及 View 的层次结构、事件分发机制和 View 的工作原理等技术细节，而这些技术细节每一项又都是初学者难以掌握的，因此就不难理解为什么初学者都觉得自定义 View 很难这一现状了。考虑到这一点，本书在第 3 章和第 4 章的前半部分对自定义 View 的各种技术细节都做了详细的分析，目的就是为了让读者更好地掌握本节的内容。尽管自定义 View 很难，甚至面对各种复杂的效果时往往还会觉得有点无章可循。但是，本节将从一定高度来重新审视

自定义 View，并以综述的形式介绍自定义 View 的分类和须知，旨在帮助初学者能够透过现象看本质，避免陷入只见树木不见森林的状态之中。同时为了让读者更好地理解自定义 View，在本节最后还会针对自定义 View 的不同类别分别提供一个实际的例子，通过这些例子能够让读者更深入地掌握自定义 View。

4.4.1 自定义 View 的分类

自定义 View 的分类标准不唯一，而笔者则把自定义 View 分为 4 类。

1．继承 View 重写 onDraw 方法

这种方法主要用于实现一些不规则的效果，即这种效果不方便通过布局的组合方式来达到，往往需要静态或者动态地显示一些不规则的图形。很显然这需要通过绘制的方式来实现，即重写 onDraw 方法。采用这种方式需要自己支持 wrap_content，并且 padding 也需要自己处理。

2．继承 ViewGroup 派生特殊的 Layout

这种方法主要用于实现自定义的布局，即除了 LinearLayout、RelativeLayout、FrameLayout 这几种系统的布局之外，我们重新定义一种新布局，当某种效果看起来很像几种 View 组合在一起的时候，可以采用这种方法来实现。采用这种方式稍微复杂一些，需要合适地处理 ViewGroup 的测量、布局这两个过程，并同时处理子元素的测量和布局过程。

3．继承特定的 View（比如 TextView）

这种方法比较常见，一般是用于扩展某种已有的 View 的功能，比如 TextView，这种方法比较容易实现。这种方法不需要自己支持 wrap_content 和 padding 等。

4．继承特定的 ViewGroup（比如 LinearLayout）

这种方法也比较常见，当某种效果看起来很像几种 View 组合在一起的时候，可以采用这种方法来实现。采用这种方法不需要自己处理 ViewGroup 的测量和布局这两个过程。需要注意这种方法和方法 2 的区别，一般来说方法 2 能实现的效果方法 4 也都能实现，两者的主要差别在于方法 2 更接近 View 的底层。

上面介绍了自定义 View 的 4 种方式，读者可以仔细体会一下，是不是的确可以这么划分？但是这里要说的是，自定义 View 讲究的是灵活性，一种效果可能多种方法都可以实现，我们需要做的就是找到一种代价最小、最高效的方法去实现，在 4.4.2 节会列举一些自定义

View 过程中常见的注意事项。

4.4.2　自定义 View 须知

本节将介绍自定义 View 过程中的一些注意事项，这些问题如果处理不好，有些会影响 View 的正常使用，而有些则会导致内存泄露等，具体的注意事项如下所示。

1．让 View 支持 wrap_content

这是因为直接继承 View 或者 ViewGroup 的控件，如果不在 onMeasure 中对 wrap_content 做特殊处理，那么当外界在布局中使用 wrap_content 时就无法达到预期的效果，具体情形已经在 4.3.1 节中进行了详细的介绍，这里不再重复了。

2．如果有必要，让你的 View 支持 padding

这是因为直接继承 View 的控件，如果不在 draw 方法中处理 padding，那么 padding 属性是无法起作用的。另外，直接继承自 ViewGroup 的控件需要在 onMeasure 和 onLayout 中考虑 padding 和子元素的 margin 对其造成的影响，不然将导致 padding 和子元素的 margin 失效。

3．尽量不要在 View 中使用 Handler，没必要

这是因为 View 内部本身就提供了 post 系列的方法，完全可以替代 Handler 的作用，当然除非你很明确地要使用 Handler 来发送消息。

4．View 中如果有线程或者动画，需要及时停止，参考 View#onDetachedFromWindow

这一条也很好理解，如果有线程或者动画需要停止时，那么 onDetachedFromWindow 是一个很好的时机。当包含此 View 的 Activity 退出或者当前 View 被 remove 时，View 的 onDetachedFromWindow 方法会被调用，和此方法对应的是 onAttachedToWindow，当包含此 View 的 Activity 启动时，View 的 onAttachedToWindow 方法会被调用。同时，当 View 变得不可见时我们也需要停止线程和动画，如果不及时处理这种问题，有可能会造成内存泄漏。

5．View 带有滑动嵌套情形时，需要处理好滑动冲突

如果有滑动冲突的话，那么要合适地处理滑动冲突，否则将会严重影响 View 的效果，具体怎么解决滑动冲突请参看第 3 章。

4.4.3 自定义 View 示例

4.4.1 节和 4.4.2 节分别介绍了自定义 View 的类别和注意事项，本节将通过几个实际的例子来演示如何自定义一个规范的 View，通过本节的例子再结合上面两节的内容，可以让读者更好地掌握自定义 View。下面仍然按照自定义 View 的分类来介绍具体的实现细节。

1. 继承 View 重写 onDraw 方法

这种方法主要用于实现一些不规则的效果，一般需要重写 onDraw 方法。采用这种方式需要自己支持 wrap_content，并且 padding 也需要自己处理。下面通过一个具体的例子来演示如何实现这种自定义 View。

为了更好地展示一些平时不容易注意到的问题，这里选择实现一个很简单的自定义控件，简单到只是绘制一个圆，尽管如此，需要注意的细节还是很多的。为了实现一个规范的控件，在实现过程中必须考虑到 wrap_content 模式以及 padding，同时为了提高便捷性，还要对外提供自定义属性。我们先来看一下最简单的实现，代码如下所示。

```java
public class CircleView extends View {

    private int mColor = Color.RED;
    private Paint mPaint = new Paint(Paint.ANTI_ALIAS_FLAG);

    public CircleView(Context context) {
        super(context);
        init();
    }

    public CircleView(Context context, AttributeSet attrs) {
        super(context, attrs);
        init();
    }

    public CircleView(Context context, AttributeSet attrs, int defStyleAttr) {
        super(context, attrs, defStyleAttr);
        init();
    }
```

```
    private void init() {
        mPaint.setColor(mColor);
    }

    @Override
    protected void onDraw(Canvas canvas) {
        super.onDraw(canvas);
        int width = getWidth();
        int height = getHeight();
        int radius = Math.min(width, height) / 2;
        canvas.drawCircle(width / 2, height / 2, radius, mPaint);
    }
}
```

上面的代码实现了一个具有圆形效果的自定义 View，它会在自己的中心点以宽/高的最小值为直径绘制一个红色的实心圆，它的实现很简单，并且上面的代码相信大部分初学者都能写出来，但是不得不说，上面的代码只是一种初级的实现，并不是一个规范的自定义 View，为什么这么说呢？我们通过调整布局参数来对比一下。

请看下面的布局：

```
<LinearLayout xmlns:android="http://schemas.android.com/apk/res/android"
    xmlns:tools="http://schemas.android.com/tools"
    android:layout_width="match_parent"
    android:layout_height="match_parent"
    android:background="#ffffff"
    android:orientation="vertical" >

    <com.ryg.chapter_4.ui.CircleView
        android:id="@+id/circleView1"
        android:layout_width="match_parent"
        android:layout_height="100dp"
        android:background="#000000"/>

</LinearLayout>
```

再看一下运行的效果，如图 4-3 中的（1）所示，这是我们预期的效果。接着再调整 CircleView 的布局参数，为其设置 20dp 的 margin，调整后的布局如下所示。

```
<com.ryg.chapter_4.ui.CircleView
```

```
    android:id="@+id/circleView1"
    android:layout_width=" match_parent"
    android:layout_height="100dp"
    android:layout_margin="20dp"
    android:background="#000000"/>
```

运行后看一下效果，如图 4-3 中的（2）所示，这也是我们预期的效果，这说明 margin 属性是生效的。这是因为 margin 属性是由父容器控制的，因此不需要在 CircleView 中做特殊处理。再调整 CircleView 的布局参数，为其设置 20dp 的 padding，如下所示。

```
<com.ryg.chapter_4.ui.CircleView
    android:id="@+id/circleView1"
    android:layout_width="match_parent"
    android:layout_height="100dp"
    android:layout_margin="20dp"
    android:padding="20dp"
    android:background="#000000"/>
```

运行后看一下效果，如图 4-3 中的（3）所示。结果发现 padding 根本没有生效，这就是我们在前面提到的直接继承自 View 和 ViewGroup 的控件，padding 是默认无法生效的，需要自己处理。再调整一下 CircleView 的布局参数，将其宽度设置为 wrap_content，如下所示。

```
<com.ryg.chapter_4.ui.CircleView
    android:id="@+id/circleView1"
    android:layout_width="wrap_content"
    android:layout_height="100dp"
    android:layout_margin="20dp"
    android:padding="20dp"
    android:background="#000000"/>
```

运行后看一下效果，如图 4-3 中的（4）所示，结果发现 wrap_content 并没有达到预期的效果。对比下（3）和（4）的效果图，发现宽度使用 wrap_content 和使用 match_parent 没有任何区别。的确是这样的，这一点在前面也已经提到过：对于直接继承自 View 的控件，如果不对 wrap_content 做特殊处理，那么使用 wrap_content 就相当于使用 match_parent。

为了解决上面提到的几种问题，我们需要做如下处理：

首先，针对 wrap_content 的问题，其解决方法在 4.3.1 节中已经做了详细的介绍，这里只需要指定一个 wrap_content 模式的默认宽/高即可,比如选择 200px 作为默认的宽/高。

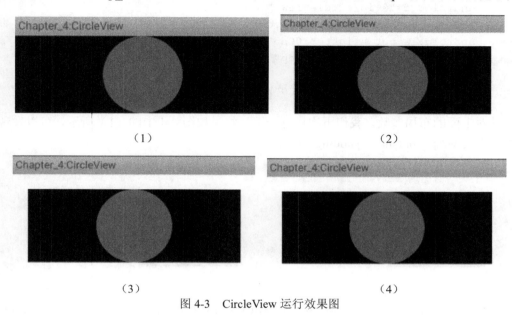

图 4-3　CircleView 运行效果图

其次,针对 padding 的问题,也很简单,只要在绘制的时候考虑一下 padding 即可,因此我们需要对 onDraw 稍微做一下修改,修改后的代码如下所示。

```
protected void onDraw(Canvas canvas) {
    super.onDraw(canvas);
    final int paddingLeft = getPaddingLeft();
    final int paddingRight = getPaddingRight();
    final int paddingTop = getPaddingTop();
    final int paddingBottom = getPaddingBottom();
    int width = getWidth() - paddingLeft - paddingRight;
    int height = getHeight() - paddingTop - paddingBottom;
    int radius = Math.min(width, height) / 2;
    canvas.drawCircle(paddingLeft + width / 2, paddingTop + height/2, radius,
    mPaint);
}
```

上面的代码很简单,中心思想就是在绘制的时候考虑到 View 四周的空白即可,其中圆心和半径都会考虑到 View 四周的 padding,从而做相应的调整。

```
<com.ryg.chapter_4.ui.CircleView
    android:id="@+id/circleView1"
    android:layout_width="wrap_content"
    android:layout_height="100dp"
    android:layout_margin="20dp"
    android:padding="20dp"
    android:background="#000000"/>
```

针对上面的布局参数，我们再次运行一下，结果如图 4-4 中的（1）所示，可以发现布局参数中的 wrap_content 和 padding 均生效了。

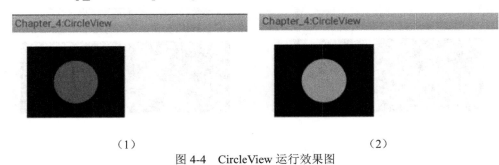

（1）　　　　　　　　　　　　　（2）

图 4-4　CircleView 运行效果图

最后，为了让我们的 View 更加容易使用，很多情况下我们还需要为其提供自定义属性，像 android:layout_width 和 android:padding 这种以 android 开头的属性是系统自带的属性，那么如何添加自定义属性呢？这也不是什么难事，遵循如下几步：

第一步，在 values 目录下面创建自定义属性的 XML，比如 attrs.xml，也可以选择类似于 attrs_circle_view.xml 等这种以 attrs_ 开头的文件名，当然这个文件名并没有什么限制，可以随便取名字。针对本例来说，选择创建 attrs.xml 文件，文件内容如下：

```
<?xml version="1.0" encoding="utf-8"?>
<resources>
    <declare-styleable name="CircleView">
        <attr name="circle_color" format="color" />
    </declare-styleable>
</resources>
```

在上面的 XML 中声明了一个自定义属性集合"CircleView"，在这个集合里面可以有很多自定义属性，这里只定义了一个格式为"color"的属性"circle_color"，这里的格式 color 指的是颜色。除了颜色格式，自定义属性还有其他格式，比如 reference 是指资源 id，dimension

是指尺寸，而像 string、integer 和 boolean 这种是指基本数据类型。除了列举的这些还有其他类型，这里就不一一描述了，读者查看一下文档即可，这并没有什么难度。

第二步，在 View 的构造方法中解析自定义属性的值并做相应处理。对于本例来说，我们需要解析 circle_color 这个属性的值，代码如下所示。

```
public CircleView(Context context, AttributeSet attrs, int defStyleAttr) {
    super(context, attrs, defStyleAttr);
    TypedArray a = context.obtainStyledAttributes(attrs, R.styleable.CircleView);
    mColor = a.getColor(R.styleable.CircleView_circle_color, Color.RED);
    a.recycle();
    init();
}
```

这看起来很简单，首先加载自定义属性集合 CircleView，接着解析 CircleView 属性集合中的 circle_color 属性，它的 id 为 R.styleable.CircleView_circle_color。在这一步骤中，如果在使用时没有指定 circle_color 这个属性，那么就会选择红色作为默认的颜色值，解析完自定义属性后，通过 recycle 方法来释放资源，这样 CircleView 中所做的工作就完成了。

第三步，在布局文件中使用自定义属性，如下所示。

```
<LinearLayout xmlns:android="http://schemas.android.com/apk/res/android"
    xmlns:app="http://schemas.android.com/apk/res-auto"
    android:layout_width="match_parent"
    android:layout_height="match_parent"
    android:background="#ffffff"
    android:orientation="vertical" >

    <com.ryg.chapter_4.ui.CircleView
        android:id="@+id/circleView1"
        android:layout_width="wrap_content"
        android:layout_height="100dp"
        android:layout_margin="20dp"
        app:circle_color="@color/light_green"
        android:padding="20dp"
        android:background="#000000"/>

</LinearLayout>
```

上面的布局文件中有一点需要注意，首先，为了使用自定义属性，必须在布局文件中添加 schemas 声明：xmlns:app=http://schemas.android.com/apk/res-auto。在这个声明中，app 是自定义属性的前缀，当然可以换其他名字，但是 CircleView 中的自定义属性的前缀必须和这里的一致，然后就可以在 CircleView 中使用自定义属性了，比如：app:circle_color= "@color/light_green"。另外，也有按照如下方式声明 schemas：xmlns:app=http:// schemas.android.com/apk/res/com.ryg.chapter_4，这种方式会在 apk/res/后面附加应用的包名。但是这两种方式并没有本质区别，笔者比较喜欢的是 xmlns:app=http://schemas.android.com/ apk/res-auto 这种声明方式。

到这里自定义属性的使用过程就完成了，运行一下程序，效果如图 4-4 中的（2）所示，很显然，CircleView 的自定义属性 circle_color 生效了。下面给出 CircleView 的完整代码，这时的 CircleView 已经是一个很规范的自定义 View 了，如下所示。

```java
public class CircleView extends View {

    private int mColor = Color.RED;
    private Paint mPaint = new Paint(Paint.ANTI_ALIAS_FLAG);

    public CircleView(Context context) {
        super(context);
        init();
    }

    public CircleView(Context context, AttributeSet attrs) {
        this(context, attrs, 0);
    }

    public CircleView(Context context, AttributeSet attrs, int defStyleAttr) {
        super(context, attrs, defStyleAttr);
        TypedArray a = context.obtainStyledAttributes(attrs, R.styleable.CircleView);
        mColor = a.getColor(R.styleable.CircleView_circle_color, Color.RED);
        a.recycle();
        init();
    }

    private void init() {
        mPaint.setColor(mColor);
```

```java
    }

    @Override
    protected void onMeasure(int widthMeasureSpec, int heightMeasureSpec) {
        super.onMeasure(widthMeasureSpec, heightMeasureSpec);
        int widthSpecMode = MeasureSpec.getMode(widthMeasureSpec);
        int widthSpecSize = MeasureSpec.getSize(widthMeasureSpec);
        int heightSpecMode = MeasureSpec.getMode(heightMeasureSpec);
        int heightSpecSize = MeasureSpec.getSize(heightMeasureSpec);
        if (widthSpecMode == MeasureSpec.AT_MOST && heightSpecMode ==
        MeasureSpec.AT_MOST) {
            setMeasuredDimension(200, 200);
        } else if (widthSpecMode == MeasureSpec.AT_MOST) {
            setMeasuredDimension(200, heightSpecSize);
        } else if (heightSpecMode == MeasureSpec.AT_MOST) {
            setMeasuredDimension(widthSpecSize, 200);
        }
    }

    @Override
    protected void onDraw(Canvas canvas) {
        super.onDraw(canvas);
        final int paddingLeft = getPaddingLeft();
        final int paddingRight = getPaddingRight();
        final int paddingTop = getPaddingTop();
        final int paddingBottom = getPaddingBottom();
        int width = getWidth() - paddingLeft - paddingRight;
        int height = getHeight() - paddingTop - paddingBottom;
        int radius = Math.min(width, height) / 2;
        canvas.drawCircle(paddingLeft + width / 2, paddingTop + height / 2,
        radius, mPaint);
    }
}
```

2．继承 ViewGroup 派生特殊的 Layout

这种方法主要用于实现自定义的布局，采用这种方式稍微复杂一些，需要合适地处理 ViewGroup 的测量、布局这两个过程，并同时处理子元素的测量和布局过程。在第 3 章的 3.5.3 节中，我们分析了滑动冲突的两种方式并实现了两个自定义 View：HorizontalScroll-

ViewEx 和 StickyLayout，其中 HorizontalScrollViewEx 就是通过继承 ViewGroup 来实现的自定义 View，这里会再次分析它的 measure 和 layout 过程。

需要说明的是，如果要采用此种方法实现一个很规范的自定义 View，是有一定的代价的，这点通过查看 LinearLayout 等的源码就知道，它们的实现都很复杂。对于 HorizontalScrollViewEx 来说，这里不打算实现它的方方面面，仅仅是完成主要功能，但是需要规范化的地方会给出说明。

这里再回顾一下 HorizontalScrollViewEx 的功能，它主要是一个类似于 ViewPager 的控件，也可以说是一个类似于水平方向的 LinearLayout 的控件，它内部的子元素可以进行水平滑动并且子元素的内部还可以进行竖直滑动，这显然是存在滑动冲突的，但是 HorizontalScrollViewEx 内部解决了水平和竖直方向的滑动冲突问题。关于 HorizontalScrollViewEx 是如何解决滑动冲突的，请参看第 3 章的相关内容。这里有一个假设，那就是所有子元素的宽/高都是一样的。下面主要看一下它的 onMeasure 和 onLayout 方法的实现，先看 onMeasure，如下所示。

```java
protected void onMeasure(int widthMeasureSpec, int heightMeasureSpec) {
    super.onMeasure(widthMeasureSpec, heightMeasureSpec);
    int measuredWidth = 0;
    int measuredHeight = 0;
    final int childCount = getChildCount();
    measureChildren(widthMeasureSpec, heightMeasureSpec);

    int widthSpaceSize = MeasureSpec.getSize(widthMeasureSpec);
    int widthSpecMode = MeasureSpec.getMode(widthMeasureSpec);
    int heightSpaceSize = MeasureSpec.getSize(heightMeasureSpec);
    int heightSpecMode = MeasureSpec.getMode(heightMeasureSpec);
    if (childCount == 0) {
        setMeasuredDimension(0, 0);
    } else if (widthSpecMode == MeasureSpec.AT_MOST && heightSpecMode ==
    MeasureSpec.AT_MOST) {
        final View childView = getChildAt(0);
        measuredWidth = childView.getMeasuredWidth() * childCount;
        measuredHeight = childView.getMeasuredHeight();
        setMeasuredDimension(measuredWidth, measuredHeight);
    } else if (heightSpecMode == MeasureSpec.AT_MOST) {
        final View childView = getChildAt(0);
        measuredHeight = childView.getMeasuredHeight();
```

```
            setMeasuredDimension(widthSpaceSize, childView.getMeasured-
            Height());
        } else if (widthSpecMode == MeasureSpec.AT_MOST) {
            final View childView = getChildAt(0);
            measuredWidth = childView.getMeasuredWidth() * childCount;
            setMeasuredDimension(measuredWidth, heightSpaceSize);
        }
    }
```

这里说明一下上述代码的逻辑，首先会判断是否有子元素，如果没有子元素就直接把自己的宽/高设为 0；然后就是判断宽和高是不是采用了 wrap_content，如果宽采用了 wrap_content，那么 HorizontalScrollViewEx 的宽度就是所有子元素的宽度之和；如果高度采用了 wrap_content，那么 HorizontalScrollViewEx 的高度就是第一个子元素的高度。

上述代码不太规范的地方有两点：第一点是没有子元素的时候不应该直接把宽/高设为 0，而应该根据 LayoutParams 中的宽/高来做相应处理；第二点是在测量 HorizontalScrollViewEx 的宽/高时没有考虑到它的 padding 以及子元素的 margin，因为它的 padding 以及子元素的 margin 会影响到 HorizontalScrollViewEx 的宽/高。这是很好理解的，因为不管是自己的 padding 还是子元素的 margin，占用的都是 HorizontalScrollViewEx 的空间。

接着再看一下 HorizontalScrollViewEx 的 onLayout 方法，如下所示。

```
protected void onLayout(boolean changed, int l, int t, int r, int b) {
    int childLeft = 0;
    final int childCount = getChildCount();
    mChildrenSize = childCount;

    for (int i = 0; i < childCount; i++) {
        final View childView = getChildAt(i);
        if (childView.getVisibility() != View.GONE) {
            final int childWidth = childView.getMeasuredWidth();
            mChildWidth = childWidth;
            childView.layout(childLeft, 0, childLeft + childWidth,
                    childView.getMeasuredHeight());
            childLeft += childWidth;
        }
    }
}
```

上述代码的逻辑并不复杂，其作用是完成子元素的定位。首先会遍历所有的子元素，如果这个子元素不是处于 GONE 这个状态，那么就通过 layout 方法将其放置在合适的位置上。从代码上来看，这个放置过程是由左向右的，这和水平方向的 LinearLayout 比较类似。上述代码的不完美之处仍然在于放置子元素的过程没有考虑到自身的 padding 以及子元素的 margin，而从一个规范的控件的角度来看，这些都是应该考虑的。下面给出 HorizontalScrollViewEx 的完整代码，如下所示。

```java
public class HorizontalScrollViewEx extends ViewGroup {
    private static final String TAG = "HorizontalScrollViewEx";

    private int mChildrenSize;
    private int mChildWidth;
    private int mChildIndex;

    // 分别记录上次滑动的坐标
    private int mLastX = 0;
    private int mLastY = 0;
    // 分别记录上次滑动的坐标(onInterceptTouchEvent)
    private int mLastXIntercept = 0;
    private int mLastYIntercept = 0;

    private Scroller mScroller;
    private VelocityTracker mVelocityTracker;

    public HorizontalScrollViewEx(Context context) {
        super(context);
        init();
    }

    public HorizontalScrollViewEx(Context context, AttributeSet attrs) {
        super(context, attrs);
        init();
    }

    public HorizontalScrollViewEx(Context context, AttributeSet attrs,
            int defStyle) {
        super(context, attrs, defStyle);
        init();
```

```java
    }

    private void init() {
        if (mScroller == null) {
            mScroller = new Scroller(getContext());
            mVelocityTracker = VelocityTracker.obtain();
        }
    }

    @Override
    public boolean onInterceptTouchEvent(MotionEvent event) {
        boolean intercepted = false;
        int x = (int) event.getX();
        int y = (int) event.getY();

        switch (event.getAction()) {
        case MotionEvent.ACTION_DOWN: {
            intercepted = false;
            if (!mScroller.isFinished()) {
                mScroller.abortAnimation();
                intercepted = true;
            }
            break;
        }
        case MotionEvent.ACTION_MOVE: {
            int deltaX = x - mLastXIntercept;
            int deltaY = y - mLastYIntercept;
            if (Math.abs(deltaX) > Math.abs(deltaY)) {
                intercepted = true;
            } else {
                intercepted = false;
            }
            break;
        }
        case MotionEvent.ACTION_UP: {
            intercepted = false;
            break;
        }
        default:
            break;
```

```java
        }
        Log.d(TAG, "intercepted=" + intercepted);
        mLastX = x;
        mLastY = y;
        mLastXIntercept = x;
        mLastYIntercept = y;

        return intercepted;
    }

    @Override
    public boolean onTouchEvent(MotionEvent event) {
        mVelocityTracker.addMovement(event);
        int x = (int) event.getX();
        int y = (int) event.getY();
        switch (event.getAction()) {
        case MotionEvent.ACTION_DOWN: {
            if (!mScroller.isFinished()) {
                mScroller.abortAnimation();
            }
            break;
        }
        case MotionEvent.ACTION_MOVE: {
            int deltaX = x - mLastX;
            int deltaY = y - mLastY;
            scrollBy(-deltaX, 0);
            break;
        }
        case MotionEvent.ACTION_UP: {
            int scrollX = getScrollX();
            mVelocityTracker.computeCurrentVelocity(1000);
            float xVelocity = mVelocityTracker.getXVelocity();
            if (Math.abs(xVelocity) >= 50) {
                mChildIndex = xVelocity > 0 ? mChildIndex - 1 : mChildIndex + 1;
            } else {
                mChildIndex = (scrollX + mChildWidth / 2) / mChildWidth;
            }
            mChildIndex = Math.max(0, Math.min(mChildIndex, mChildrenSize - 1));
            int dx = mChildIndex * mChildWidth - scrollX;
```

```
            smoothScrollBy(dx, 0);
            mVelocityTracker.clear();
            break;
        }
        default:
            break;
        }

        mLastX = x;
        mLastY = y;
        return true;
    }

    @Override
    protected void onMeasure(int widthMeasureSpec, int heightMeasureSpec) {
        super.onMeasure(widthMeasureSpec, heightMeasureSpec);
        int measuredWidth = 0;
        int measuredHeight = 0;
        final int childCount = getChildCount();
        measureChildren(widthMeasureSpec, heightMeasureSpec);

        int widthSpaceSize = MeasureSpec.getSize(widthMeasureSpec);
        int widthSpecMode = MeasureSpec.getMode(widthMeasureSpec);
        int heightSpaceSize = MeasureSpec.getSize(heightMeasureSpec);
        int heightSpecMode = MeasureSpec.getMode(heightMeasureSpec);
        if (childCount == 0) {
            setMeasuredDimension(0, 0);
        } else if (widthSpecMode == MeasureSpec.AT_MOST && heightSpecMode ==
MeasureSpec.AT_MOST) {
            final View childView = getChildAt(0);
            measuredWidth = childView.getMeasuredWidth() * childCount;
            measuredHeight = childView.getMeasuredHeight();
            setMeasuredDimension(measuredWidth, measuredHeight);
        } else if (heightSpecMode == MeasureSpec.AT_MOST) {
            final View childView = getChildAt(0);
            measuredHeight = childView.getMeasuredHeight();
            setMeasuredDimension(widthSpaceSize, childView.getMeasured-
            Height());
        } else if (widthSpecMode == MeasureSpec.AT_MOST) {
            final View childView = getChildAt(0);
```

```java
            measuredWidth = childView.getMeasuredWidth() * childCount;
            setMeasuredDimension(measuredWidth, heightSpaceSize);
        }
    }
}

@Override
protected void onLayout(boolean changed, int l, int t, int r, int b) {
    int childLeft = 0;
    final int childCount = getChildCount();
    mChildrenSize = childCount;

    for (int i = 0; i < childCount; i++) {
        final View childView = getChildAt(i);
        if (childView.getVisibility() != View.GONE) {
            final int childWidth = childView.getMeasuredWidth();
            mChildWidth = childWidth;
            childView.layout(childLeft, 0, childLeft + childWidth,
                    childView.getMeasuredHeight());
            childLeft += childWidth;
        }
    }
}

private void smoothScrollBy(int dx, int dy) {
    mScroller.startScroll(getScrollX(), 0, dx, 0, 500);
    invalidate();
}

@Override
public void computeScroll() {
    if (mScroller.computeScrollOffset()) {
        scrollTo(mScroller.getCurrX(), mScroller.getCurrY());
        postInvalidate();
    }
}

@Override
protected void onDetachedFromWindow() {
    mVelocityTracker.recycle();
    super.onDetachedFromWindow();
```

```
        }
    }
```

继承特定的 View（比如 TextView）和继承特定的 ViewGroup（比如 LinearLayout）这两种方式比较简单，这里就不再举例说明了，关于第 3 章中提到的 StickyLayout 的具体实现，大家可以参看笔者在 Github 上的开源项目：https://github.com/singwhatiwanna/PinnedHeaderExpandableListView。

4.4.4 自定义 View 的思想

到这里，自定义 View 相关的知识都已经介绍完了，可能读者还是觉得有点模糊。前面说过，自定义 View 是一个综合的技术体系，很多情况下需要灵活地分析从而找出最高效的方法，因此本章不可能去分析一个个具体的自定义 View 的实现，因为自定义 View 五花八门，是不可能全部分析一遍的。虽然我们不能把自定义 View 都分析一遍，但是我们能够提取出一种思想，在面对陌生的自定义 View 时，运用这个思想去快速地解决问题。这种思想的描述如下：首先要掌握基本功，比如 View 的弹性滑动、滑动冲突、绘制原理等，这些东西都是自定义 View 所必须的，尤其是那些看起来很炫的自定义 View，它们往往对这些技术点的要求更高；熟练掌握基本功以后，在面对新的自定义 View 时，要能够对其分类并选择合适的实现思路，自定义 View 的实现方法的分类在 4.4.1 节中已经介绍过了；另外平时还需要多积累一些自定义 View 相关的经验，并逐渐做到融会贯通，通过这种思想慢慢地就可以提高自定义 View 的水平了。

第 5 章　理解 RemoteViews

本章所讲述的主题是 RemoteViews，从名字可以看出，RemoteViews 应该是一种远程 View，那么什么是远程 View 呢？如果说远程服务可能比较好理解，但是远程 View 的确没听说过，其实它和远程 Service 是一样的，RemoteViews 表示的是一个 View 结构，它可以在其他进程中显示，由于它在其他进程中显示，为了能够更新它的界面，RemoteViews 提供了一组基础的操作用于跨进程更新它的界面。这听起来有点神奇，竟然能跨进程更新界面！但是 RemoteViews 的确能够实现这个效果。RemoteViews 在 Android 中的使用场景有两种：通知栏和桌面小部件，为了更好地分析 RemoteViews 的内部机制，本章先简单介绍 RemoteViews 在通知栏和桌面小部件上的应用，接着分析 RemoteViews 的内部机制，最后分析 RemoteViews 的意义并给出一个采用 RemoteViews 来跨进程更新界面的示例。

5.1　RemoteViews 的应用

RemoteViews 在实际开发中，主要用在通知栏和桌面小部件的开发过程中。通知栏每个人都不陌生，主要是通过 NotificationManager 的 notify 方法来实现的，它除了默认效果外，还可以另外定义布局。桌面小部件则是通过 AppWidgetProvider 来实现的，AppWidget-Provider 本质上是一个广播。通知栏和桌面小部件的开发过程中都会用到 RemoteViews，它们在更新界面时无法像在 Activity 里面那样去直接更新 View，这是因为二者的界面都运行在其他进程中，确切来说是系统的 SystemServer 进程。为了跨进程更新界面，RemoteViews 提供了一系列 set 方法，并且这些方法只是 View 全部方法的子集，另外 RemoteViews 中所支持的 View 类型也是有限的，这一点会在 5.2 节中进行详细说明。下面简单介绍一下

RemoteViews 在通知栏和桌面小部件中的使用方法，至于它们更详细的使用方法请读者阅读相关资料即可，本章的重点是分析 RemoteViews 的内部机制。

5.1.1　RemoteViews 在通知栏上的应用

首先我们看一下 RemoteViews 在通知栏上的应用，我们知道，通知栏除了默认的效果外还支持自定义布局，下面分别说明这两种情况。

使用系统默认的样式弹出一个通知是很简单的，代码如下：

```
Notification notification = new Notification();
notification.icon = R.drawable.ic_launcher;
notification.tickerText = "hello world";
notification.when = System.currentTimeMillis();
notification.flags = Notification.FLAG_AUTO_CANCEL;
Intent intent = new Intent(this, DemoActivity_1.class);
PendingIntent pendingIntent = PendingIntent.getActivity(this,
        0, intent, PendingIntent.FLAG_UPDATE_CURRENT);
notification.setLatestEventInfo(this,"chapter_5","this is notification.",
pendingIntent);
NotificationManager manager = (NotificationManager)getSystemService
(Context.NOTIFICATION_SERVICE);
manager.notify(1, notification);
```

上述代码会弹出一个系统默认样式的通知，单击通知后会打开 DemoActivity_1 同时会清除本身。为了满足个性化需求，我们还可能会用到自定义通知。自定义通知也很简单，首先我们要提供一个布局文件，然后通过 RemoteViews 来加载这个布局文件即可改变通知的样式，代码如下所示。

```
Notification notification = new Notification();
notification.icon = R.drawable.ic_launcher;
notification.tickerText = "hello world";
notification.when = System.currentTimeMillis();
notification.flags = Notification.FLAG_AUTO_CANCEL;
Intent intent = new Intent(this, DemoActivity_1.class);
PendingIntent pendingIntent = PendingIntent.getActivity(this,
        0, intent, PendingIntent.FLAG_UPDATE_CURRENT);
RemoteViews remoteViews = new RemoteViews(getPackageName(), R.layout.
layout_notification);
```

```
remoteViews.setTextViewText(R.id.msg, "chapter_5");
remoteViews.setImageViewResource(R.id.icon, R.drawable.icon1);
PendingIntent openActivity2PendingIntent = PendingIntent.getActivity(this,
        0, new Intent(this, DemoActivity_2.class), PendingIntent.FLAG_
        UPDATE_CURRENT);
remoteViews.setOnClickPendingIntent(R.id.open_activity2, openActivity2-
PendingIntent);
notification.contentView = remoteViews;
notification.contentIntent = pendingIntent;
NotificationManager manager = (NotificationManager)getSystemService
(Context.NOTIFICATION_SERVICE);
manager.notify(2, notification);
```

从上述内容来看，自定义通知的效果需要用到 RemoteViews，自定义通知的效果如图 5-1 所示。

图 5-1　自定义通知栏样式

RemoteViews 的使用也很简单，只要提供当前应用的包名和布局文件的资源 id 即可创建一个 RemoteViews 对象。如何更新 RemoteViews 呢？这一点和更新 View 有很大的不同，更新 RemoteViews 时，无法直接访问里面的 View，而必须通过 RemoteViews 所提供的一系列方法来更新 View。比如设置 TextView 的文本，要采用如下方式：remoteViews.setTextViewText(R.id.msg, "chapter_5")，其中 setTextViewText 的两个参数分别为 TextView 的 id 和要设置的文本。而设置 ImageView 的图片也不能直接访问 ImageView，必须通过如下方式：remoteViews.setImageViewResource(R.id.icon, R.drawable.icon1)，setImageViewResource 的两个参数分别为 ImageView 的 id 和要设置的图片资源的 id。如果要给一个控件加单击事件，则要使用 PendingIntent 并通过 setOnClickPendingIntent 方法来实现，比如 remoteViews.setOnClickPendingIntent(R.id.open_activity2, openActivity2Pending- Intent)这句代码会给 id 为 open_activity2 的 View 加上单击事件。关于 PendingIntent，它表示的是一种待定的 Intent，这个 Intent 中所包含的意图必须由用户来触发。为什么更新 RemoteViews 如此复杂呢？直观原因是因为 RemoteViews 并没有提供和 View 类似的 findViewById 这个方法，因此我们无法获取到 RemoteViews 中的子 View，当然实际原因绝非如此，具体会在 5.2 节中进行详细介绍。

5.1.2 RemoteViews 在桌面小部件上的应用

AppWidgetProvider 是 Android 中提供的用于实现桌面小部件的类，其本质是一个广播，即 BroadcastReceiver，图 5-2 所示的是它的类继承关系。所以，在实际的使用中，把 AppWidgetProvider 当成一个 BroadcastReceiver 就可以了，这样许多功能就很好理解了。

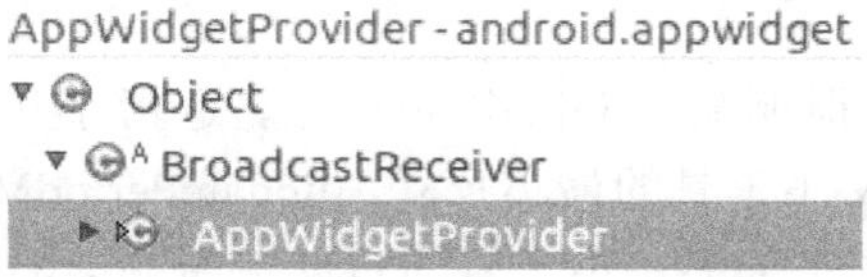

图 5-2　AppWidgetProvider 的类继承关系

为了更好地展示 RemoteViews 在桌面小部件上的应用，我们先简单介绍桌面小部件的开发步骤，分为如下几步。

1．定义小部件界面

在 res/layout/下新建一个 XML 文件，命名为 widget.xml，名称和内容可以自定义，看这个小部件要做成什么样子，内容如下所示。

```
<?xml version="1.0" encoding="utf-8"?>
<LinearLayout xmlns:android="http://schemas.android.com/apk/res/android"
    android:layout_width="match_parent"
    android:layout_height="match_parent"
    android:orientation="vertical" >

    <ImageView
```

```
            android:id="@+id/imageView1"
            android:layout_width="wrap_content"
            android:layout_height="wrap_content"
            android:src="@drawable/icon1" />

</LinearLayout>
```

2. 定义小部件配置信息

在 res/xml/ 下新建 appwidget_provider_info.xml，名称随意选择，添加如下内容：

```
<?xml version="1.0" encoding="utf-8"?>
<appwidget-provider xmlns:android="http://schemas.android.com/apk/res/android"
    android:initialLayout="@layout/widget"
    android:minHeight="84dp"
    android:minWidth="84dp"
    android:updatePeriodMillis="86400000" >

</appwidget-provider>
```

上面几个参数的意义很明确，initialLayout 就是指小工具所使用的初始化布局，minHeight 和 minWidth 定义小工具的最小尺寸，updatePeriodMillis 定义小工具的自动更新周期，毫秒为单位，每隔一个周期，小工具的自动更新就会触发。

3. 定义小部件的实现类

这个类需要继承 AppWidgetProvider，代码如下：

```
public class MyAppWidgetProvider extends AppWidgetProvider {
    public static final String TAG = "MyAppWidgetProvider";
    public static final String CLICK_ACTION = "com.ryg.chapter_5.action.CLICK";

    public MyAppWidgetProvider() {
        super();
    }

    @Override
    public void onReceive(final Context context, Intent intent) {
        super.onReceive(context, intent);
        Log.i(TAG, "onReceive : action = " + intent.getAction());
```

```java
// 这里判断是自己的action，做自己的事情，比如小部件被单击了要干什么，这里是做
一个动画效果
if (intent.getAction().equals(CLICK_ACTION)) {
    Toast.makeText(context, "clicked it", Toast.LENGTH_SHORT).show();

    new Thread(new Runnable() {
        @Override
        public void run() {
            Bitmap srcbBitmap = BitmapFactory.decodeResource(
                    context.getResources(), R.drawable.icon1);
            AppWidgetManager appWidgetManager = AppWidgetManager.
            getInstance(context);
            for (int i = 0; i < 37; i++) {
                float degree = (i * 10) % 360;
                RemoteViews remoteViews = new RemoteViews(context
                        .getPackageName(), R.layout.widget);
                remoteViews.setImageViewBitmap(R.id.imageView1,
                        rotateBitmap(context, srcbBitmap, degree));
                Intent intentClick = new Intent();
                intentClick.setAction(CLICK_ACTION);
                PendingIntent pendingIntent = PendingIntent
                        .getBroadcast(context, 0, intentClick, 0);
                remoteViews.setOnClickPendingIntent(R.id.imageView1,
                pendingIntent);
                appWidgetManager.updateAppWidget(new ComponentName(
                        context, MyAppWidgetProvider.class),
                        remoteViews);
                SystemClock.sleep(30);
            }

        }
    }).start();
}
}

/**
 * 每次桌面小部件更新时都调用一次该方法
 */
@Override
```

```java
public void onUpdate(Context context, AppWidgetManager appWidgetManager,
        int[] appWidgetIds) {
    super.onUpdate(context, appWidgetManager, appWidgetIds);
    Log.i(TAG, "onUpdate");

    final int counter = appWidgetIds.length;
    Log.i(TAG, "counter = " + counter);
    for (int i = 0; i < counter; i++) {
        int appWidgetId = appWidgetIds[i];
        onWidgetUpdate(context, appWidgetManager, appWidgetId);
    }

}

/**
 *桌面小部件更新
 *
 * @param context
 * @param appWidgeManger
 * @param appWidgetId
 */
private void onWidgetUpdate(Context context,
        AppWidgetManager appWidgeManger, int appWidgetId) {

    Log.i(TAG, "appWidgetId = " + appWidgetId);
    RemoteViews remoteViews = new RemoteViews(context.getPackageName(),
            R.layout.widget);

    // "桌面小部件"单击事件发送的Intent广播
    Intent intentClick = new Intent();
    intentClick.setAction(CLICK_ACTION);
    PendingIntent pendingIntent = PendingIntent.getBroadcast(context, 0,
            intentClick, 0);
    remoteViews.setOnClickPendingIntent(R.id.imageView1, pendingIntent);
    appWidgeManger.updateAppWidget(appWidgetId, remoteViews);
}

private Bitmap rotateBitmap(Context context, Bitmap srcbBitmap, float degree) {
    Matrix matrix = new Matrix();
```

```
        matrix.reset();
        matrix.setRotate(degree);
        Bitmap tmpBitmap = Bitmap.createBitmap(srcbBitmap, 0, 0,
                srcbBitmap.getWidth(), srcbBitmap.getHeight(), matrix, true);
        return tmpBitmap;
    }
}
```

上面的代码实现了一个简单的桌面小部件，在小部件上面显示一张图片，单击它后，这个图片就会旋转一周。当小部件被添加到桌面后，会通过 RemoteViews 来加载布局文件，而当小部件被单击后的旋转效果则是通过不断地更新 RemoteViews 来实现的，由此可见，桌面小部件不管是初始化界面还是后续的更新界面都必须使用 RemoteViews 来完成。

4．在 AndroidManifest.xml 中声明小部件

这是最后一步，因为桌面小部件本质上是一个广播组件，因此必须要注册，如下所示。

```xml
<receiver
    android:name=".MyAppWidgetProvider" >
    <meta-data
        android:name="android.appwidget.provider"
        android:resource="@xml/appwidget_provider_info" >
    </meta-data>

    <intent-filter>
        <action android:name="com.ryg.chapter_5.action.CLICK" />
        <action android:name="android.appwidget.action.APPWIDGET_UPDATE" />
    </intent-filter>
</receiver>
```

上面的代码中有两个 Action，其中第一个 Action 用于识别小部件的单击行为，而第二个 Action 则作为小部件的标识而必须存在，这是系统的规范，如果不加，那么这个 receiver 就不是一个桌面小部件并且也无法出现在手机的小部件列表里。

AppWidgetProvider 除了最常用的 onUpdate 方法，还有其他几个方法：onEnabled、onDisabled、onDeleted 以及 onReceive。这些方法会自动地被 onReceive 方法在合适的时间调用。确切来说，当广播到来以后，AppWidgetProvider 会自动根据广播的 Action 通过 onReceive 方法来自动分发广播，也就是调用上述几个方法。这几个方法的调用时机如下所示。

- **onEnable**：当该窗口小部件第一次添加到桌面时调用该方法，可添加多次但只在第一次调用。

- **onUpdate**：小部件被添加时或者每次小部件更新时都会调用一次该方法，小部件的更新时机由 updatePeriodMillis 来指定，每个周期小部件都会自动更新一次。

- **onDeleted**：每删除一次桌面小部件就调用一次。

- **onDisabled**：当最后一个该类型的桌面小部件被删除时调用该方法，注意是最后一个。

- **onReceive**：这是广播的内置方法，用于分发具体的事件给其他方法。

关于 AppWidgetProvider 的 onReceive 方法的具体分发过程，可以参看源码中的实现，如下所示。通过下面的代码可以看出，onReceive 中会根据不同的 Action 来分别调用 onEnable、onDisable 和 onUpdate 等方法。

```java
public void onReceive(Context context, Intent intent) {
    // Protect against rogue update broadcasts (not really a security issue,
    // just filter bad broacasts out so subclasses are less likely to crash).
    String action = intent.getAction();
    if (AppWidgetManager.ACTION_APPWIDGET_UPDATE.equals(action)) {
        Bundle extras = intent.getExtras();
        if (extras != null) {
            int[] appWidgetIds = extras.getIntArray(AppWidgetManager.
            EXTRA_APPWIDGET_IDS);
            if (appWidgetIds != null && appWidgetIds.length > 0) {
                this.onUpdate(context, AppWidgetManager.getInstance
                (context), appWidgetIds);
            }
        }
    } else if (AppWidgetManager.ACTION_APPWIDGET_DELETED.equals(action)) {
        Bundle extras = intent.getExtras();
        if (extras != null && extras.containsKey(AppWidgetManager.EXTRA_
        APPWIDGET_ID)) {
            final int appWidgetId = extras.getInt(AppWidgetManager.EXTRA_
            APPWIDGET_ID);
            this.onDeleted(context, new int[] { appWidgetId });
        }
```

```java
        } else if (AppWidgetManager.ACTION_APPWIDGET_OPTIONS_CHANGED.equals
(action)) {
            Bundle extras = intent.getExtras();
            if (extras != null && extras.containsKey(AppWidgetManager.EXTRA_
                APPWIDGET_ID)
                    && extras.containsKey(AppWidgetManager.EXTRA_APPWIDGET_
                    OPTIONS)) {
                int appWidgetId = extras.getInt(AppWidgetManager.EXTRA_
                APPWIDGET_ID);
                Bundle widgetExtras = extras.getBundle(AppWidgetManager.EXTRA_
                APPWIDGET_OPTIONS);
                this.onAppWidgetOptionsChanged(context, AppWidgetManager.
                getInstance(context),
                        appWidgetId, widgetExtras);
            }
        } else if (AppWidgetManager.ACTION_APPWIDGET_ENABLED.equals(action)) {
            this.onEnabled(context);
        } else if (AppWidgetManager.ACTION_APPWIDGET_DISABLED.equals(action)) {
            this.onDisabled(context);
        } else if (AppWidgetManager.ACTION_APPWIDGET_RESTORED.equals(action)) {
            Bundle extras = intent.getExtras();
            if (extras != null) {
                int[] oldIds = extras.getIntArray(AppWidgetManager.EXTRA_
                APPWIDGET_OLD_IDS);
                int[] newIds = extras.getIntArray(AppWidgetManager.EXTRA_
                APPWIDGET_IDS);
                if (oldIds != null && oldIds.length > 0) {
                    this.onRestored(context, oldIds, newIds);
                    this.onUpdate(context, AppWidgetManager.getInstance
                        (context), newIds);
                }
            }
        }
    }
}
```

上面描述了开发一个桌面小部件的典型过程，例子比较简单，实际开发中会稍微复杂一些，但是开发流程是一样的。可以发现，桌面小部件在界面上的操作都要通过 RemoteViews，不管是小部件的界面初始化还是界面更新都必须依赖它。

5.1.3 PendingIntent 概述

在 5.1.2 节中,我们多次提到 PendingIntent,那么 PendingIntent 到底是什么东西呢?它和 Intent 的区别是什么呢?在本节中将介绍 PendingIntent 的使用方法。

顾名思义,PendingIntent 表示一种处于 pending 状态的意图,而 pending 状态表示的是一种待定、等待、即将发生的意思,就是说接下来有一个 Intent(即意图)将在某个待定的时刻发生。可以看出 PendingIntent 和 Intent 的区别在于,PendingIntent 是在将来的某个不确定的时刻发生,而 Intent 是立刻发生。PendingIntent 典型的使用场景是给 RemoteViews 添加单击事件,因为 RemoteViews 运行在远程进程中,因此 RemoteViews 不同于普通的 View,所以无法直接向 View 那样通过 setOnClickListener 方法来设置单击事件。要想给 RemoteViews 设置单击事件,就必须使用 PendingIntent,PendingIntent 通过 send 和 cancel 方法来发送和取消特定的待定 Intent。

PendingIntent 支持三种待定意图:启动 Activity、启动 Service 和发送广播,对应着它的三个接口方法,如表 5-1 所示。

表 5-1 PendingIntent 的主要方法

static PendingIntent	getActivity(Context context, int requestCode, Intent intent, int flags) 获得一个 PendingIntent,该待定意图发生时,效果相当于 Context.startActivity(Intent)
static PendingIntent	getService(Context context, int requestCode, Intent intent, int flags) 获得一个 PendingIntent,该待定意图发生时,效果相当于 Context.startService(Intent)
static PendingIntent	getBroadcast(Context context, int requestCode, Intent intent, int flags) 获得一个 PendingIntent,该待定意图发生时,效果相当于 Context.sendBroadcast(Intent)

如表 5-1 所示,getActivity、getService 和 getBroadcast 这三个方法的参数意义都是相同的,第一个和第三个参数比较好理解,这里主要说下第二个参数 requestCode 和第四个参数 flags,其中 requestCode 表示 PendingIntent 发送方的请求码,多数情况下设为 0 即可,另外 requestCode 会影响到 flags 的效果。flags 常见的类型有:FLAG_ONE_SHOT、FLAG_NO_CREATE、FLAG_CANCEL_CURRENT 和 FLAG_UPDATE_CURRENT。在说明这四个标记位之前,必须要明白一个概念,那就是 PendingIntent 的匹配规则,即在什么情况下两个

PendingIntent 是相同的。

PendingIntent 的匹配规则为：如果两个 PendingIntent 它们内部的 Intent 相同并且 requestCode 也相同，那么这两个 PendingIntent 就是相同的。requestCode 相同比较好理解，那么什么情况下 Intent 相同呢？Intent 的匹配规则是：如果两个 Intent 的 ComponentName 和 intent-filter 都相同，那么这两个 Intent 就是相同的。需要注意的是 Extras 不参与 Intent 的匹配过程，只要 Intent 之间的 ComponentName 和 intent-filter 相同，即使它们的 Extras 不同，那么这两个 Intent 也是相同的。了解了 PendingIntent 的匹配规则后，就可以进一步理解 flags 参数的含义了，如下所示。

FLAG_ONE_SHOT

当前描述的 PendingIntent 只能被使用一次，然后它就会被自动 cancel，如果后续还有相同的 PendingIntent，那么它们的 send 方法就会调用失败。对于通知栏消息来说，如果采用此标记位，那么同类的通知只能使用一次，后续的通知单击后将无法打开。

FLAG_NO_CREATE

当前描述的 PendingIntent 不会主动创建，如果当前 PendingIntent 之前不存在，那么 getActivity、getService 和 getBroadcast 方法会直接返回 null，即获取 PendingIntent 失败。这个标记位很少见，它无法单独使用，因此在日常开发中它并没有太多的使用意义，这里就不再过多介绍了。

FLAG_CANCEL_CURRENT

当前描述的 PendingIntent 如果已经存在，那么它们都会被 cancel，然后系统会创建一个新的 PendingIntent。对于通知栏消息来说，那些被 cancel 的消息单击后将无法打开。

FLAG_UPDATE_CURRENT

当前描述的 PendingIntent 如果已经存在，那么它们都会被更新，即它们的 Intent 中的 Extras 会被替换成最新的。

从上面的分析来看还是不太好理解这四个标记位，下面结合通知栏消息再描述一遍。这里分两种情况，如下代码中：manager.notify(1, notification)，如果 notify 的第一个参数 id 是常量，那么多次调用 notify 只能弹出一个通知，后续的通知会把前面的通知完全替代掉，而如果每次 id 都不同，那么多次调用 notify 会弹出多个通知，下面一一说明。

如果 notify 方法的 id 是常量，那么不管 PendingIntent 是否匹配，后面的通知会直接替换前面的通知，这个很好理解。

如果 notify 方法的 id 每次都不同，那么当 PendingIntent 不匹配时，这里的匹配是指 PendingIntent 中的 Intent 相同并且 requestCode 相同，在这种情况下不管采用何种标记位，这些通知之间不会相互干扰。如果 PendingIntent 处于匹配状态时，这个时候要分情况讨论：如果采用了 FLAG_ONE_SHOT 标记位，那么后续通知中的 PendingIntent 会和第一条通知保持完全一致，包括其中的 Extras，单击任何一条通知后，剩下的通知均无法再打开，当所有的通知都被清除后，会再次重复这个过程；如果采用 FLAG_CANCEL_CURRENT 标记位，那么只有最新的通知可以打开，之前弹出的所有通知均无法打开；如果采用 FLAG_UPDATE_CURRENT 标记位，那么之前弹出的通知中的 PendingIntent 会被更新，最终它们和最新的一条通知保持完全一致，包括其中的 Extras，并且这些通知都是可以打开的。

5.2 RemoteViews 的内部机制

RemoteViews 的作用是在其他进程中显示并更新 View 界面，为了更好地理解它的内部机制，我们先来看一下它的主要功能。首先看一下它的构造方法，这里只介绍一个最常用的构造方法：public RemoteViews(String packageName, int layoutId)，它接受两个参数，第一个表示当前应用的包名，第二个参数表示待加载的布局文件，这个很好理解。RemoteViews 目前并不能支持所有的 View 类型，它所支持的所有类型如下：

Layout

FrameLayout、LinearLayout、RelativeLayout、GridLayout。

View

AnalogClock、Button、Chronometer、ImageButton、ImageView、ProgressBar、TextView、ViewFlipper、ListView、GridView、StackView、AdapterViewFlipper、ViewStub。

上面所描述的是 RemoteViews 所支持的所有的 View 类型，RemoteViews 不支持它们的子类以及其他 View 类型，也就是说 RemoteViews 中不能使用除了上述列表中以外的 View，也无法使用自定义 View。比如如果我们在通知栏的 RemoteViews 中使用系统的 EditText，那么通知栏消息将无法弹出并且会抛出如下异常：

```
E/StatusBar(765): couldn't inflate view for notification com.ryg.chapter_
5/0x2
E/StatusBar(765): android.view.InflateException: Binary XML file line #25:
Error inflating class android.widget.EditText
E/StatusBar(765): Caused by: android.view.InflateException: Binary XML file
line #25: Class not allowed to be inflated android.widget.EditText
E/StatusBar(765):     at android.view.LayoutInflater.failNotAllowed
(LayoutInflater.java:695)
E/StatusBar(765):     at android.view.LayoutInflater.createView
(LayoutInflater.java:628)
E/StatusBar(765):     ... 21 more
```

上面的异常信息很明确，android.widget.EditText 不允许在 RemoteViews 中使用。

RemoteViews 没有提供 findViewById 方法，因此无法直接访问里面的 View 元素，而必须通过 RemoteViews 所提供的一系列 set 方法来完成，当然这是因为 RemoteViews 在远程进程中显示，所以没办法直接 findViewById。表 5-2 列举了部分常用的 set 方法，更多的方法请查看相关资料。

表 5-2　RemoteViews 的部分 set 方法

方 法 名	作　　用
setTextViewText(int viewId, CharSequence text)	设置 TextView 的文本
setTextViewTextSize(int viewId, int units, float size)	设置 TextView 的字体大小
setTextColor(int viewId, int color)	设置 TextView 的字体颜色
setImageViewResource(int viewId, int srcId)	设置 ImageView 的图片资源
setImageViewBitmap(int viewId, Bitmap bitmap)	设置 ImageView 的图片
setInt(int viewId, String methodName, int value)	反射调用 View 对象的参数类型为 int 的方法
setLong(int viewId, String methodName, long value)	反射调用 View 对象的参数类型为 long 的方法
setBoolean(int viewId, String methodName, boolean value)	反射调用 View 对象的参数类型为 boolean 的方法
setOnClickPendingIntent(int viewId, PendingIntent pendingIntent)	为 View 添加单击事件，事件类型只能为 PendingIntent

从表 5-2 中可以看出，原本可以直接调用的 View 的方法，现在却必须要通过 RemoteViews 的一系列 set 方法才能完成，而且从方法的声明上来看，很像是通过反射来完成的，事实上大部分 set 方法的确是通过反射来完成的。

下面描述一下 RemoteViews 的内部机制，由于 RemoteViews 主要用于通知栏和桌面小部件之中，这里就通过它们来分析 RemoteViews 的工作过程。我们知道，通知栏和桌面小部件分别由 NotificationManager 和 AppWidgetManager 管理，而 NotificationManager 和 AppWidgetManager 通过 Binder 分别和 SystemServer 进程中的 NotificationManagerService 以及 AppWidgetService 进行通信。由此可见，通知栏和桌面小部件中的布局文件实际上是在 NotificationManagerService 以及 AppWidgetService 中被加载的，而它们运行在系统的 SystemServer 中，这就和我们的进程构成了跨进程通信的场景。

首先 RemoteViews 会通过 Binder 传递到 SystemServer 进程，这是因为 RemoteViews 实现了 Parcelable 接口，因此它可以跨进程传输，系统会根据 RemoteViews 中的包名等信息去得到该应用的资源。然后会通过 LayoutInflater 去加载 RemoteViews 中的布局文件。在 SystemServer 进程中加载后的布局文件是一个普通的 View，只不过相对于我们的进程它是一个 RemoteViews 而已。接着系统会对 View 执行一系列界面更新任务，这些任务就是之前我们通过 set 方法来提交的。set 方法对 View 所做的更新并不是立刻执行的，在 RemoteViews 内部会记录所有的更新操作，具体的执行时机要等到 RemoteViews 被加载以后才能执行，这样 RemoteViews 就可以在 SystemServer 进程中显示了，这就是我们所看到的通知栏消息或者桌面小部件。当需要更新 RemoteViews 时，我们需要调用一系列 set 方法并通过 NotificationManager 和 AppWidgetManager 来提交更新任务，具体的更新操作也是在 SystemServer 进程中完成的。

从理论上来说，系统完全可以通过 Binder 去支持所有的 View 和 View 操作，但是这样做的话代价太大，因为 View 的方法太多了，另外就是大量的 IPC 操作会影响效率。为了解决这个问题，系统并没有通过 Binder 去直接支持 View 的跨进程访问，而是提供了一个 Action 的概念，Action 代表一个 View 操作，Action 同样实现了 Parcelable 接口。系统首先将 View 操作封装到 Action 对象并将这些对象跨进程传输到远程进程，接着在远程进程中执行 Action 对象中的具体操作。在我们的应用中每调用一次 set 方法，RemoteViews 中就会添加一个对应的 Action 对象，当我们通过 NotificationManager 和 AppWidgetManager 来提交我们的更新时，这些 Action 对象就会传输到远程进程并在远程进程中依次执行，这个过程可以参看图 5-3。远程进程通过 RemoteViews 的 apply 方法来进行 View 的更新操作，RemoteViews 的 apply 方法内部则会去遍历所有的 Action 对象并调用它们的 apply 方法，具体的 View 更新操作是由 Action 对象的 apply 方法来完成的。上述做法的好处是显而易见的，首先不需要定义大量的 Binder 接口，其次通过在远程进程中批量执行 RemoteViews 的修改操作从而避免了大量的 IPC 操作，这就提高了程序的性能，由此可见，Android 系统在这方面的设计的确很精妙。

上面从理论上分析了 RemoteViews 的内部机制，接下来我们从源码的角度再来分析 RemoteViews 的工作流程。它的构造方法就不用多说了，这里我们首先看一下它提供的一系列 set 方法，比如 setTextViewText 方法，其源码如下所示。

```
public void setTextViewText(int viewId, CharSequence text) {
    setCharSequence(viewId, "setText", text);
}
```

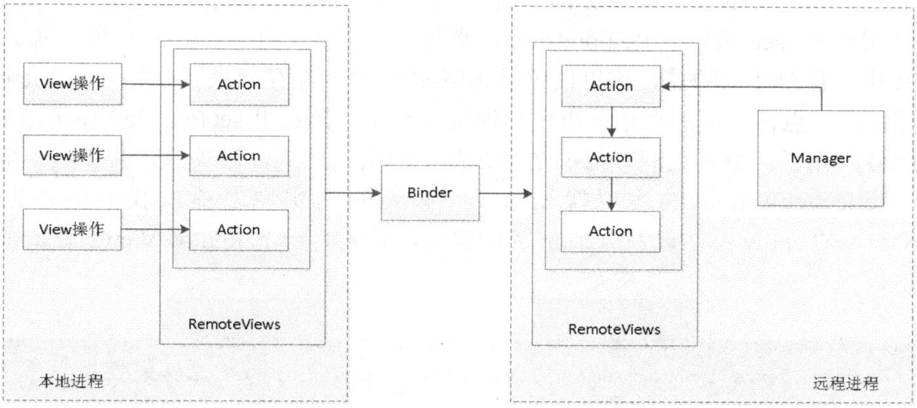

图 5-3　RemoteViews 的内部机制

在上面的代码中，viewId 是被操作的 View 的 id，"setText" 是方法名，text 是要给 TextView 设置的文本，这里可以联想一下 TextView 的 setText 方法，是不是很一致呢？接着再看 setCharSequence 的实现，如下所示。

```
public void setCharSequence(int viewId, String methodName, CharSequence 
value) {
    addAction(new ReflectionAction(viewId, methodName, ReflectionAction.
    CHAR_SEQUENCE, value));
}
```

从 setCharSequence 的实现可以看出，它的内部并没有对 View 进程直接的操作，而是添加了一个 ReflectionAction 对象，从名字来看，这应该是一个反射类型的动作。再看 addAction 的实现，如下所示。

```
private void addAction(Action a) {
    …
    if (mActions == null) {
        mActions = new ArrayList<Action>();
    }
    mActions.add(a);
```

233

```
    // update the memory usage stats
    a.updateMemoryUsageEstimate(mMemoryUsageCounter);
}
```

从上述代码可以知道，RemoteViews 内部有一个 mActions 成员，它是一个 ArrayList，外界每调用一次 set 方法，RemoteViews 就会为其创建一个 Action 对象并加入到这个 ArrayList 中。需要注意的是，这里仅仅是将 Action 对象保存起来了，并未对 View 进行实际的操作，这一点在上面的理论分析中已经提到过了。到这里 setTextViewText 这个方法的源码已经分析完了，但是我们好像还是什么都不知道的感觉，没关系，接着我们需要看一下这个 ReflectionAction 的实现就知道了。再看它的实现之前，我们需要先看一下 RemoteViews 的 apply 方法以及 Action 类的实现，首先看一下 RemoteViews 的 apply 方法，如下所示。

```
public View apply(Context context, ViewGroup parent,OnClickHandler handler) {
    RemoteViews rvToApply = getRemoteViewsToApply(context);

    View result;
    ...

    LayoutInflater inflater = (LayoutInflater)
            context.getSystemService(Context.LAYOUT_INFLATER_SERVICE);

    // Clone inflater so we load resources from correct context and
    // we don't add a filter to the static version returned by getSystem-
        Service.
    inflater = inflater.cloneInContext(inflationContext);
    inflater.setFilter(this);
    result = inflater.inflate(rvToApply.getLayoutId(), parent, false);

    rvToApply.performApply(result, parent, handler);

    return result;
}
```

从上面代码可以看出，首先会通过 LayoutInflater 去加载 RemoteViews 中的布局文件，RemoteViews 中的布局文件可以通过 getLayoutId 这个方法获得，加载完布局文件后会通过 performApply 去执行一些更新操作，代码如下所示。

```java
private void performApply(View v,ViewGroup parent,OnClickHandler handler) {
    if (mActions != null) {
        handler = handler == null ? DEFAULT_ON_CLICK_HANDLER : handler;
        final int count = mActions.size();
        for (int i = 0; i < count; i++) {
            Action a = mActions.get(i);
            a.apply(v, parent, handler);
        }
    }
}
```

performApply 的实现就比较好理解了,它的作用就是遍历 mActions 这个列表并执行每个 Action 对象的 apply 方法。还记得 mAction 吗？每一次的 set 操作都会对应着它里面的一个 Action 对象,因此我们可以断定,Action 对象的 apply 方法就是真正操作 View 的地方,实际上的确如此。

RemoteViews 在通知栏和桌面小部件中的工作过程和上面描述的过程是一致的,当我们调用 RemoteViews 的 set 方法时,并不会立刻更新它们的界面,而必须要通过 Notification-Manager 的 notify 方法以及 AppWidgetManager 的 updateAppWidget 才能更新它们的界面。实际上在 AppWidgetManager 的 updateAppWidget 的内部实现中,它们的确是通过 RemoteViews 的 apply 以及 reapply 方法来加载或者更新界面的,apply 和 reApply 的区别在于：apply 会加载布局并更新界面,而 reApply 则只会更新界面。通知栏和桌面小插件在初始化界面时会调用 apply 方法,而在后续的更新界面时则会调用 reapply 方法。这里先看一下 BaseStatusBar 的 updateNotificationViews 方法中,如下所示。

```java
private void updateNotificationViews(NotificationData.Entry entry,
        StatusBarNotification notification, boolean isHeadsUp) {
    final RemoteViews contentView = notification.getNotification().
    contentView;
    final RemoteViews bigContentView = isHeadsUp
            ? notification.getNotification().headsUpContentView
            : notification.getNotification().bigContentView;
    final Notification publicVersion = notification.getNotification().
    publicVersion;
    final RemoteViews publicContentView = publicVersion != null ? public-
    Version.contentView : null;

    // Reapply the RemoteViews
```

```
            contentView.reapply(mContext, entry.expanded, mOnClickHandler);
            ...
}
```

很显然，上述代码表示当通知栏界面需要更新时，它会通过 RemoteViews 的 reapply 方法来更新界面。

接着再看一下 AppWidgetHostView 的 updateAppWidget 方法，在它的内部有如下一段代码：

```
mRemoteContext = getRemoteContext();
int layoutId = remoteViews.getLayoutId();

// If our stale view has been prepared to match active, and the new
// layout matches, try recycling it
if (content == null && layoutId == mLayoutId) {
    try {
        remoteViews.reapply(mContext, mView, mOnClickHandler);
        content = mView;
        recycled = true;
        if (LOGD) Log.d(TAG, "was able to recycled existing layout");
    } catch (RuntimeException e) {
        exception = e;
    }
}

// Try normal RemoteView inflation
if (content == null) {
    try {
        content = remoteViews.apply(mContext, this, mOnClickHandler);
        if (LOGD) Log.d(TAG, "had to inflate new layout");
    } catch (RuntimeException e) {
        exception = e;
    }
}
```

从上述代码可以发现，桌面小部件在更新界面时也是通过 RemoteViews 的 reapply 方法来实现的。

了解了 apply 以及 reapply 的作用以后，我们再继续看一些 Action 的子类的具体实现，首先看一下 ReflectionAction 的具体实现，它的源码如下所示。

```
private final class ReflectionAction extends Action {
    ReflectionAction(int viewId,String methodName, int type, Object value) {
        this.viewId = viewId;
        this.methodName = methodName;
        this.type = type;
        this.value = value;
    }

    ...
    @Override
    public void apply(View root, ViewGroup rootParent, OnClickHandler
    handler) {
        final View view = root.findViewById(viewId);
        if (view == null) return;

        Class<?> param = getParameterType();
        if (param == null) {
            throw new ActionException("bad type: " + this.type);
        }

        try {
            getMethod(view, this.methodName, param).invoke(view, wrapArg
            (this.value));
        } catch (ActionException e) {
            throw e;
        } catch (Exception ex) {
            throw new ActionException(ex);
        }
    }
}
```

通过上述代码可以发现，ReflectionAction 表示的是一个反射动作，通过它对 View 的操作会以反射的方式来调用，其中 getMethod 就是根据方法名来得到反射所需的 Method 对象。使用 ReflectionAction 的 set 方法有：setTextViewText、setBoolean、setLong、setDouble 等。除了 ReflectionAction，还有其他 Action，比如 TextViewSizeAction、ViewPaddingAction、SetOnClickPendingIntent 等。这里再分析一下 TextViewSizeAction，它的实现如下所示。

```java
    private class TextViewSizeAction extends Action {
        public TextViewSizeAction(int viewId, int units, float size) {
            this.viewId = viewId;
            this.units = units;
            this.size = size;
        }
        ...

        @Override
        public void apply(View root, ViewGroup rootParent, OnClickHandler
            handler) {
            final TextView target = (TextView) root.findViewById(viewId);
            if (target == null) return;
            target.setTextSize(units, size);
        }

        public String getActionName() {
            return "TextViewSizeAction";
        }

        int units;
        float size;

        public final static int TAG = 13;
    }
```

TextViewSizeAction 的实现比较简单，它之所以不用反射来实现，是因为 setTextSize 这个方法有 2 个参数，因此无法复用 ReflectionAction，因为 ReflectionAction 的反射调用只有一个参数。其他 Action 这里就不一一进行分析了，读者可以查看 RemoteViews 的源代码。

关于单击事件，RemoteViews 中只支持发起 PendingIntent，不支持 onClickListener 那种模式。另外，我们需要注意 setOnClickPendingIntent、setPendingIntentTemplate 以及 setOnClickFillInIntent 它们之间的区别和联系。首先 setOnClickPendingIntent 用于给普通 View 设置单击事件，但是不能给集合（ListView 和 StackView）中的 View 设置单击事件，比如我们不能给 ListView 中的 item 通过 setOnClickPendingIntent 这种方式添加单击事件，因为开销比较大，所以系统禁止了这种方式；其次，如果要给 ListView 和 StackView 中的 item 添加单击事件，则必须将 setPendingIntentTemplate 和 setOnClickFillInIntent 组合使用才可以。

5.3 RemoteViews 的意义

在 5.2 节中我们分析了 RemoteViews 的内部机制，了解 RemoteViews 的内部机制可以让我们更加清楚通知栏和桌面小工具的底层实现原理，但是本章对 RemoteViews 的探索并没有停止，在本节中，我们将打造一个模拟的通知栏效果并实现跨进程的 UI 更新。

首先有 2 个 Activity 分别运行在不同的进程中，一个名字叫 A，另一个叫 B，其中 A 扮演着模拟通知栏的角色，而 B 则可以不停地发送通知栏消息，当然这是模拟的消息。为了模拟通知栏的效果，我们修改 A 的 process 属性使其运行在单独的进程中，这样 A 和 B 就构成了多进程通信的情形。我们在 B 中创建 RemoteViews 对象，然后通知 A 显示这个 RemoteViews 对象。如何通知 A 显示 B 中的 RemoteViews 呢？我们可以像系统一样采用 Binder 来实现，但是这里为了简单起见就采用了广播。B 每发送一次模拟通知，就会发送一个特定的广播，然后 A 接收到广播后就开始显示 B 中定义的 RemoteViews 对象，这个过程和系统的通知栏消息的显示过程几乎一致，或者说这里就是复制了通知栏的显示过程而已。

首先看 B 的实现，B 只要构造 RemoteViews 对象并将其传输给 A 即可，这一过程通知栏是采用 Binder 实现的，但是本例中采用广播来实现，RemoteViews 对象通过 Intent 传输到 A 中，代码如下所示。

```
RemoteViews remoteViews = new RemoteViews(getPackageName(), R.layout.
layout_simulated_notification);
remoteViews.setTextViewText(R.id.msg, "msg from process:" + Process.
myPid());
remoteViews.setImageViewResource(R.id.icon, R.drawable.icon1);
PendingIntent pendingIntent = PendingIntent.getActivity(this,
        0, new Intent(this, DemoActivity_1.class), PendingIntent.FLAG_
        UPDATE_CURRENT);
PendingIntent openActivity2PendingIntent = PendingIntent.getActivity(
        this, 0, new Intent(this, DemoActivity_2.class), PendingIntent.
        FLAG_UPDATE_CURRENT);
remoteViews.setOnClickPendingIntent(R.id.item_holder, pendingIntent);
remoteViews.setOnClickPendingIntent(R.id.open_activity2, openActivity2-
PendingIntent);
```

```
Intent intent = new Intent(MyConstants.REMOTE_ACTION);
intent.putExtra(MyConstants.EXTRA_REMOTE_VIEWS, remoteViews);
sendBroadcast(intent);
```

A 的代码也很简单，只需要接收 B 中的广播并显示 RemoteViews 即可，如下所示。

```
public class MainActivity extends Activity {
    private static final String TAG = "MainActivity";

    private LinearLayout mRemoteViewsContent;

    private BroadcastReceiver mRemoteViewsReceiver=new BroadcastReceiver() {
        @Override
        public void onReceive(Context context, Intent intent) {
            RemoteViews remoteViews = intent.getParcelableExtra(MyConstants.
            EXTRA_REMOTE_VIEWS);
            if (remoteViews != null) {
                updateUI(remoteViews);
            }
        }
    };

    @Override
    protected void onCreate(Bundle savedInstanceState) {
        super.onCreate(savedInstanceState);
        setContentView(R.layout.activity_main);
        initView();
    }

    private void initView() {
        mRemoteViewsContent = (LinearLayout) findViewById(R.id.remote_
        views_content);
        IntentFilter filter = new IntentFilter(MyConstants.REMOTE_ACTION);
        registerReceiver(mRemoteViewsReceiver, filter);
    }

    private void updateUI(RemoteViews remoteViews) {
        View view = remoteViews.apply(this, mRemoteViewsContent);
        mRemoteViewsContent.addView(view);
    }
```

```
    @Override
    protected void onDestroy() {
        unregisterReceiver(mRemoteViewsReceiver);
        super.onDestroy();
    }
}
```

上述代码很简单，除了注册和解除广播以外，最主要的逻辑其实就是 updateUI 方法。当 A 收到广播后，会从 Intent 中取出 RemoteViews 对象，然后通过它的 apply 方法加载布局文件并执行更新操作，最后将得到的 View 添加到 A 的布局中即可。可以发现，这个过程很简单，但是通知栏的底层就是这么实现的。

本节这个例子是可以在实际中使用的，比如现在有两个应用，一个应用需要能够更新另一个应用中的某个界面，这个时候我们当然可以选择 AIDL 去实现，但是如果对界面的更新比较频繁，这个时候就会有效率问题，同时 AIDL 接口就有可能会变得很复杂。这个时候如果采用 RemoteViews 来实现就没有这个问题了，当然 RemoteViews 也有缺点，那就是它仅支持一些常见的 View，对于自定义 View 它是不支持的。面对这种问题，到底是采用 AIDL 还是采用 RemoteViews，这个要看具体情况，如果界面中的 View 都是一些简单的且被 RemoteViews 支持的 View，那么可以考虑采用 RemoteViews，否则就不适合用 RemoteViews 了。

如果打算采用 RemoteViews 来实现两个应用之间的界面更新，那么这里还有一个问题，那就是布局文件的加载问题。在上面的代码中，我们直接通过 RemoteViews 的 apply 方法来加载并更新界面，如下所示。

```
View view = remoteViews.apply(this, mRemoteViewsContent);
mRemoteViewsContent.addView(view);
```

这种写法在同一个应用的多进程情形下是适用的，但是如果 A 和 B 属于不同应用，那么 B 中的布局文件的资源 id 传输到 A 以后很有可能是无效的，因为 A 中的这个布局文件的资源 id 不可能刚好和 B 中的资源 id 一样，面对这种情况，我们就要适当修改 RemoteViews 的显示过程的代码了。这里给出一种方法，既然资源 id 不相同，那我们就通过资源名称来加载布局文件。首先两个应用要提前约定好 RemoteViews 中的布局文件的资源名称，比如 "layout_simulated_notification"，然后在 A 中根据名称查找到对应的布局文件并加载，接着再调用 RemoteViews 的 reapply 方法即可将 B 中对 View 所做的一系列更新

操作全部作用到 A 中加载的 View 上面。关于 apply 和 reapply 方法的差别在前面已经提到过，这里就不多说了，这样整个跨应用更新界面的流程就走通了，具体效果如图 5-4 所示。可以发现 B 中的布局文件已经成功地在 A 中显示了出来。修改后的代码如下：

```
int         layoutId        =       getResources().getIdentifier("layout_simulated_
notification", "layout", getPackageName());
View  view = getLayoutInflater().inflate(layoutId, mRemoteViewsContent,
false);
remoteViews.reapply(this, view);
mRemoteViewsContent.addView(view);
```

图 5-4　模拟通知栏的效果

第 6 章 Android 的 Drawable

本章所讲述的话题是 Android 的 Drawable，Drawable 表示的是一种可以在 Canvas 上进行绘制的抽象的概念，它的种类有很多，最常见的颜色和图片都可以是一个 Drawable。在本章中，首先描述 Drawable 的层次关系，接着介绍 Drawable 的分类，最后介绍自定义 Drawable 相关的知识。本章的内容看起来稍微有点简单，但是由于 Drawable 的种类比较繁多，从而导致了开发者对不同 Drawable 的理解比较混乱。另外一点，熟练掌握各种类型的 Drawable 可以方便我们做出一些特殊的 UI 效果，这一点在 UI 相关的开发工作中尤其重要。Drawable 在开发中有着自己的优点：首先，它使用简单，比自定义 View 的成本要低；其次，非图片类型的 Drawable 占用空间较小，这对减小 apk 的大小也很有帮助。鉴于上述两点，全面理解 Drawable 的使用细节还是很有必要的，这也是本章的出发点。

6.1　Drawable 简介

Drawable 有很多种，它们都表示一种图像的概念，但是它们又不全是图片，通过颜色也可以构造出各式各样的图像的效果。在实际开发中，Drawable 常被用来作为 View 的背景使用。Drawable 一般都是通过 XML 来定义的，当然我们也可以通过代码来创建具体的 Drawable 对象，只是用代码创建会稍显复杂。在 Android 的设计中，Drawable 是一个抽象类，它是所有 Drawable 对象的基类，每个具体的 Drawable 都是它的子类，比如 ShapeDrawable、BitmapDrawable 等，Drawable 的层次关系如图 6-1 所示。

Drawable 的内部宽/高这个参数比较重要，通过 getIntrinsicWidth 和 getIntrinsicHeight 这两个方法可以获取到它们。但是并不是所有的 Drawable 都有内部宽/高，比如一张图片所形成的 Drawable，它的内部宽/高就是图片的宽/高，但是一个颜色所形成的 Drawable，它就没有内部宽/高的概念。另外需要注意的是，Drawable 的内部宽/高不等同于它的大小，一般来说，Drawable 是没有大小概念的，当用作 View 的背景时，Drawable 会被拉伸至 View 的同等大小。

图 6-1 Drawable 的层次关系

6.2 Drawable 的分类

Drawable 的种类繁多，常见的有 BitmapDrawable、ShapeDrawable、LayerDrawable 以及 StateListDrawable 等，这里就不一一列举了，下面会分别介绍它们的使用细节。

6.2.1 BitmapDrawable

这几乎是最简单的 Drawable 了，它表示的就是一张图片。在实际开发中，我们可以直接引用原始的图片即可，但是也可以通过 XML 的方式来描述它，通过 XML 来描述的 BitmapDrawable 可以设置更多的效果，如下所示：

```
<?xml version="1.0" encoding="utf-8"?>
<bitmap
    xmlns:android="http://schemas.android.com/apk/res/android"
    android:src="@[package:]drawable/drawable_resource"
    android:antialias=["true" | "false"]
    android:dither=["true" | "false"]
    android:filter=["true" | "false"]
    android:gravity=["top" | "bottom"|"left"|"right" | "center_vertical" |
                    "fill_vertical"|"center_horizontal" | "fill_horizontal" |
                    "center" | "fill" | "clip_vertical" | "clip_horizontal"]
    android:mipMap=["true" | "false"]
    android:tileMode=["disabled" | "clamp" | "repeat" | "mirror"] />
```

下面是它各个属性的含义。

android:src

这个很简单，就是图片的资源 id。

android:antialias

是否开启图片抗锯齿功能。开启后会让图片变得平滑，同时也会在一定程度上降低图片的清晰度，但是这个降低的幅度较低以至于可以忽略，因此抗锯齿选项应该开启。

android:dither

是否开启抖动效果。当图片的像素配置和手机屏幕的像素配置不一致时，开启这个选项可以让高质量的图片在低质量的屏幕上还能保持较好的显示效果，比如图片的色彩模式为 ARGB8888，但是设备屏幕所支持的色彩模式为 RGB555，这个时候开启抖动选项可以让图片显示不会过于失真。在 Android 中创建的 Bitmap 一般会选用 ARGB8888 这个模式，即 ARGB 四个通道各占 8 位，在这种色彩模式下，一个像素所占的大小为 4 个字节，一个像素的位数总和越高，图像也就越逼真。根据分析，抖动效果也应该开启。

android:filter

是否开启过滤效果。当图片尺寸被拉伸或者压缩时，开启过滤效果可以保持较好的显示效果，因此此选项也应该开启。

android:gravity

当图片小于容器的尺寸时，设置此选项可以对图片进行定位。这个属性的可选项比较多，不同的选项可以通过"|"来组合使用，如表 6-1 所示。

表 6-1　gravity 属性的可选项

可选项	含义
top	将图片放在容器的顶部，不改变图片的大小
bottom	将图片放在容器的底部，不改变图片的大小
left	将图片放在容器的左部，不改变图片的大小
right	将图片放在容器的右部，不改变图片的大小
center_vertical	使图片竖直居中，不改变图片的大小
fill_vertical	图片竖直方向填充容器
center_horizontal	使图片水平居中，不改变图片的大小
fill_horizontal	图片水平方向填充容器
center	使图片在水平和竖直方向同时居中，不改变图片的大小
fill	图片在水平和竖直方向均填充容器，这是默认值
clip_vertical	附加选项，表示竖直方向的裁剪，较少使用
clip_horizontal	附加选项，表示水平方向的裁剪，较少使用

android:mipMap

这是一种图像相关的处理技术，也叫纹理映射，比较抽象，这里也不对其深究了，默认值为 false，在日常开发中此选项不常用。

android:tileMode

平铺模式。这个选项有如下几个值：["disabled" | "clamp" | "repeat" | "mirror"]，其中 disable 表示关闭平铺模式，这也是默认值，当开启平铺模式后，gravity 属性会被忽略。这里主要说一下 repeat、mirror 和 clamp 的区别，这三者都表示平铺模式，但是它们的表现却有很大不同。repeat 表示的是简单的水平和竖直方向上的平铺效果；mirror 表示一种在水平和竖直方向上的镜面投影效果；而 clamp 表示的效果就更加奇特，图片四周的像素会扩展到周围区域。下面我们看一下这三者的实际效果，通过实际效果可以更好地理解不同的平铺模式的区别，如图 6-2 所示。

第 **6** 章　Android 的 Drawable

 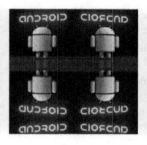 

　　（1）repeat　　　　　　　（2）mirror　　　　　　　（3）clamp

图 6-2　平铺模式下的图片显示效果

　　接下来介绍 NinePatchDrawable，它表示的是一张.9 格式的图片，.9 图片可以自动地根据所需的宽/高进行相应的缩放并保证不失真，之所以把它和 BitmapDrawable 放在一起介绍是因为它们都表示一张图片。和 BitmapDrawable 一样，在实际使用中直接引用图片即可，但是也可以通过 XML 来描述.9 图，如下所示。

```
<?xml version="1.0" encoding="utf-8"?>
<nine-patch
    xmlns:android="http://schemas.android.com/apk/res/android"
    android:src="@[package:]drawable/drawable_resource"
    android:dither=["true" | "false"] />
```

　　上述 XML 中的属性的含义和 BitmapDrawable 中的对应属性的含义是相同的，这里就不再描述了，另外，在实际使用中发现在 bitmap 标签中也可以使用.9 图，即 BitmapDrawable 也可以代表一个.9 格式的图片。

6.2.2　ShapeDrawable

　　ShapeDrawable 是一种很常见的 Drawable，可以理解为通过颜色来构造的图形，它既可以是纯色的图形，也可以是具有渐变效果的图形。ShapeDrawable 的语法稍显复杂，如下所示。

```
<?xml version="1.0" encoding="utf-8"?>
<shape
    xmlns:android="http://schemas.android.com/apk/res/android"
    android:shape=["rectangle" | "oval" | "line" | "ring"] >
    <corners
        android:radius="integer"
        android:topLeftRadius="integer"
```

```
        android:topRightRadius="integer"
        android:bottomLeftRadius="integer"
        android:bottomRightRadius="integer" />
    <gradient
        android:angle="integer"
        android:centerX="integer"
        android:centerY="integer"
        android:centerColor="integer"
        android:endColor="color"
        android:gradientRadius="integer"
        android:startColor="color"
        android:type=["linear" | "radial" | "sweep"]
        android:useLevel=["true" | "false"] />
    <padding
        android:left="integer"
        android:top="integer"
        android:right="integer"
        android:bottom="integer" />
    <size
        android:width="integer"
        android:height="integer" />
    <solid
        android:color="color" />
    <stroke
        android:width="integer"
        android:color="color"
        android:dashWidth="integer"
        android:dashGap="integer" />
</shape>
```

需要注意的是<shape>标签创建的 Drawable，其实体类实际上是 GradientDrawable，下面分别介绍各个属性的含义。

android:shape

表示图形的形状，有四个选项：rectangle（矩形）、oval（椭圆）、line（横线）和 ring（圆环）。它的默认值是矩形，另外 line 和 ring 这两个选项必须要通过<stroke>标签来指定线的宽度和颜色等信息，否则将无法达到预期的显示效果。

针对 ring 这个形状，有 5 个特殊的属性：android:innerRadius、android:thickness、android:innerRadiusRatio、android:thicknessRatio 和 android:useLevel，它们的含义如表 6-2 所示。

表 6-2　ring 的属性值

Value	Desciption
android:innerRadius	圆环的内半径，和 android:innerRadiusRatio 同时存在时，以 android:innerRadius 为准
android:thickness	圆环的厚度，即外半径减去内半径的大小，和 android:thicknessRatio 同时存在时，以 android:thickness 为准
android:innerRadiusRatio	内半径占整个 Drawable 宽度的比例，默认值为 9。如果为 n，那么内半径 = 宽度 / n
android:thicknessRatio	厚度占整个 Drawable 宽度的比例，默认值为 3。如果为 n，那么厚度 = 宽度 / n
android:useLevel	一般都应该使用 false，否则有可能无法到达预期的显示效果，除非它被当作 LevelListDrawable 来使用

<corners>

表示 shape 的四个角的角度。它只适用于矩形 shape，这里的角度是指圆角的程度，用 px 来表示，它有如下 5 个属性：

- android:radius——为四个角同时设定相同的角度，优先级较低，会被其他四个属性覆盖；

- android:topLeftRadius——设定最上角的角度；

- android:topRightRadius——设定右上角的角度；

- android:bottomLeftRadius——设定最下角的角度；

- android:bottomRightRadius——设定右下角的角度。

<gradient>

它与<solid>标签是互相排斥的，其中 solid 表示纯色填充，而 gradient 则表示渐变效果，gradient 有如下几个属性：

- android:angle——渐变的角度，默认为 0，其值必须为 45 的倍数，0 表示从左到右，90 表示从下到上，具体的效果需要自行体验，总之角度会影响渐变的方向；

- android:centerX——渐变的中心点的横坐标；

- android:centerY——渐变的中心点的纵坐标，渐变的中心点会影响渐变的具体效果；

- android:startColor——渐变的起始色；

- android:centerColor——渐变的中间色；

- android:endColor——渐变的结束色；

- android:gradientRadius——渐变半径，仅当 android:type= "radial"时有效；

- android:useLevel——一般为 false，当 Drawable 作为 StateListDrawable 使用时为 true；

- android:type——渐变的类别，有 linear（线性渐变）、radial（径向渐变）、sweep（扫描线渐变）三种，其中默认值为线性渐变，它们三者的区别如图 6-3 所示。

图 6-3　渐变的类别，从左到右依次为 linear、radial、sweep

<solid>

这个标签表示纯色填充，通过 android:color 即可指定 shape 中填充的颜色。

<stroke>

Shape 的描边，有如下几个属性：

- android:width——描边的宽度，越大则 shape 的边缘线就会看起来越粗；

- android:color——描边的颜色；

- android:dashWidth——组成虚线的线段的宽度；

- android:dashGap——组成虚线的线段之间的间隔，间隔越大则虚线看起来空隙就越大。

注意如果 android:dashWidth 和 android:dashGap 有任何一个为 0，那么虚线效果将不能生效。下面是一个具体的例子，效果图如图 6-4 所示。

```
<?xml version="1.0" encoding="utf-8"?>
<shape xmlns:android="http://schemas.android.com/apk/res/android"
    android:shape="rectangle" >

    <solid android:color="#ff0000" />

    <stroke
        android:dashGap="2dp"
        android:dashWidth="10dp"
        android:width="2dp"
        android:color="#00ff00" />
</shape>
```

\<padding\>

这个表示空白，但是它表示的不是 shape 的空白，而是包含它的 View 的空白，有四个属性：android:left、android:top、android:right 和 android:bottom。

\<size\>

shape 的大小，有两个属性：android:width 和 android:height，分别表示 shape 的宽/高。这个表示的是 shape 的固有大小，但是一般来说它并不是 shape 最终显示的大小，这个有点抽象，但是我们要明白，对于 shape 来说它并没有宽/高的概念，作为 View 的背景它会自适应 View 的宽/高。我们知道 Drawable 的两个方法 getIntrinsicWidth 和 getIntrinsicHeight 表示的是 Drawable 的固有宽/高，对于有些 Drawable 比如图片来说，它的固有宽/高就是图片的尺寸。而对于 shape 来说，默认情况下它是没有固有宽/高这个概念的，这个时候 getIntrinsicWidth 和 getIntrinsicHeight 会返回-1，但是如果通过\<size\>标签来指定宽/高信息，那么这个时候 shape 就有了所谓的固有宽/高。因此，总结来说，\<size\>标签设置的宽/高就是 ShapeDrawable 的固有宽/高，但是作为 View 的背景时，shape 还会被拉伸或者缩小为 View 的大小。

图 6-4　shape 的描边效果

6.2.3　LayerDrawable

LayerDrawable 对应的 XML 标签是\<layer-list\>，它表示一种层次化的 Drawable 集合，

通过将不同的 Drawable 放置在不同的层上面从而达到一种叠加后的效果。它的语法如下所示。

```xml
<?xml version="1.0" encoding="utf-8"?>
<layer-list
    xmlns:android="http://schemas.android.com/apk/res/android" >
    <item
        android:drawable="@[package:]drawable/drawable_resource"
        android:id="@[+][package:]id/resource_name"
        android:top="dimension"
        android:right="dimension"
        android:bottom="dimension"
        android:left="dimension" />
</layer-list>
```

一个 layer-list 中可以包含多个 item，每个 item 表示一个 Drawable。Item 的结构也比较简单，比较常用的属性有 android:top、android:bottom、android:left 和 android:right，它们分别表示 Drawable 相对于 View 的上下左右的偏移量，单位为像素。另外，我们可以通过 android:drawable 属性来直接引用一个已有的 Drawable 资源，也可以在 item 中自定义 Drawable。默认情况下，layer-list 中的所有的 Drawable 都会被缩放至 View 的大小，对于 bitmap 来说，需要使用 android:gravity 属性才能控制图片的显示效果。Layer-list 有层次的概念，下面的 item 会覆盖上面的 item，通过合理的分层，可以实现一些特殊的叠加效果。

下面是一个 layer-list 具体使用的例子，它实现了微信中的文本输入框的效果，如图 6-5 所示。当然它只适用于白色背景上的文本输入框，另外这种效果也可以采用 .9 图来实现。

```xml
<?xml version="1.0" encoding="utf-8"?>
<layer-list xmlns:android="http://schemas.android.com/apk/res/android" >

    <item>
        <shape android:shape="rectangle" >
            <solid android:color="#0ac39e" />
        </shape>
    </item>

    <item android:bottom="6dp">
        <shape android:shape="rectangle" >
            <solid android:color="#ffffff" />
        </shape>
```

```
        </item>

        <item
            android:bottom="1dp"
            android:left="1dp"
            android:right="1dp">
            <shape android:shape="rectangle" >
                <solid android:color="#ffffff" />
            </shape>
        </item>

</layer-list>
```

文本框

图 6-5 layer-list 的应用

6.2.4 StateListDrawable

StateListDrawable 对应于<selector>标签，它也是表示 Drawable 集合，每个 Drawable 都对应着 View 的一种状态，这样系统就会根据 View 的状态来选择合适的 Drawable。StateListDrawable 主要用于设置可单击的 View 的背景，最常见的是 Button，这个读者应该不陌生，它的语法如下所示。

```
<?xml version="1.0" encoding="utf-8"?>
<selector xmlns:android="http://schemas.android.com/apk/res/android"
    android:constantSize=["true" | "false"]
    android:dither=["true" | "false"]
    android:variablePadding=["true" | "false"] >
    <item
        android:drawable="@[package:]drawable/drawable_resource"
        android:state_pressed=["true" | "false"]
        android:state_focused=["true" | "false"]
        android:state_hovered=["true" | "false"]
        android:state_selected=["true" | "false"]
        android:state_checkable=["true" | "false"]
        android:state_checked=["true" | "false"]
        android:state_enabled=["true" | "false"]
        android:state_activated=["true" | "false"]
```

```
            android:state_window_focused=["true" | "false"] />
</selector>
```

针对上面的语法，下面做简单介绍。

android:constantSize

StateListDrawable 的固有大小是否不随着其状态的改变而改变的，因为状态的改变会导致 StateListDrawable 切换到具体的 Drawable，而不同的 Drawable 具有不同的固有大小。True 表示 StateListDrawable 的固有大小保持不变，这时它的固有大小是内部所有 Drawable 的固有大小的最大值，false 则会随着状态的改变而改变。此选项默认值为 false。

android:dither

是否开启抖动效果，这个在 BitmapDrawable 中也有提到，开启此选项可以让图片在低质量的屏幕上仍然获得较好的显示效果。此选项默认值为 true。

android:variablePadding

StateListDrawable 的 padding 表示是否随着其状态的改变而改变，true 表示会随着状态的改变而改变，false 表示 StateListDrawable 的 padding 是内部所有 Drawable 的 padding 的最大值。此选项默认值为 false，并且不建议开启此选项。

<item>标签表示一个具体的 Drawable，它的结构也比较简单，其中 android:drawable 是一个已有 Drawable 的资源 id，剩下的属性表示的是 View 的各种状态，每个 item 表示的都是一种状态下的 Drawable 信息。View 的常见状态如表 6-3 所示。

表 6-3 View 的常见状态

状　　态	含　　义
android:state_pressed	表示按下状态，比如 Button 被按下后仍没有松开时的状态
android:state_focused	表示 View 已经获取了焦点
android:state_selected	表示用户选择了 View
android:state_checked	表示用户选中了 View，一般适用于 CheckBox 这类在选中和非选中状态之间进行切换的 View
android:state_enabled	表示 View 当前处于可用状态

下面给出具体的例子，如下所示。

```
<?xml version="1.0" encoding="utf-8"?>
```

```xml
<selector xmlns:android="http://schemas.android.com/apk/res/android">
    <item android:state_pressed="true"
        android:drawable="@drawable/button_pressed" /> <!-- pressed -->
    <item android:state_focused="true"
        android:drawable="@drawable/button_focused" /> <!-- focused -->
    <item android:drawable="@drawable/button_normal" /> <!-- default -->
</selector>
```

系统会根据 View 当前的状态从 selector 中选择对应的 item，每个 item 对应着一个具体的 Drawable，系统按照从上到下的顺序查找，直至查找到第一条匹配的 item。一般来说，默认的 item 都应该放在 selector 的最后一条并且不附带任何的状态，这样当上面的 item 都无法匹配 View 的当前状态时，系统就会选择默认的 item，因为默认的 item 不附带状态，所以它可以匹配 View 的任何状态。

6.2.5 LevelListDrawable

LevelListDrawable 对应于<level-list>标签，它同样表示一个 Drawable 集合，集合中的每个 Drawable 都有一个等级（level）的概念。根据不同的等级，LevelListDrawable 会切换为对应的 Drawable，它的语法如下所示。

```xml
<?xml version="1.0" encoding="utf-8"?>
<level-list
    xmlns:android="http://schemas.android.com/apk/res/android" >
    <item
        android:drawable="@drawable/drawable_resource"
        android:maxLevel="integer"
        android:minLevel="integer" />
</level-list>
```

上面的语法中，每个 item 表示一个 Drawable，并且有对应的等级范围，由 android:minLevel 和 android:maxLevel 来指定，在最小值和最大值之间的等级会对应此 item 中的 Drawable。下面是一个实际的例子，当它作为 View 的背景时，可以通过 Drawable 的 setLevel 方法来设置不同的等级从而切换具体的 Drawable。如果它被用来作为 ImageView 的前景 Drawable，那么还可以通过 ImageView 的 setImageLevel 方法来切换 Drawable。最后，Drawable 的等级是有范围的，即 0~10000，最小等级是 0，这也是默认值，最大等级是 10000。

```xml
<?xml version="1.0" encoding="utf-8"?>
<level-list xmlns:android="http://schemas.android.com/apk/res/android" >
```

```
    <item
        android:drawable="@drawable/status_off"
        android:maxLevel="0" />
    <item
        android:drawable="@drawable/status_on"
        android:maxLevel="1" />
</level-list>
```

6.2.6 TransitionDrawable

TransitionDrawable 对应于<transition>标签，它用于实现两个 Drawable 之间的淡入淡出效果，它的语法如下所示。

```
<?xml version="1.0" encoding="utf-8"?>
<transition
xmlns:android="http://schemas.android.com/apk/res/android" >
    <item
        android:drawable="@[package:]drawable/drawable_resource"
        android:id="@[+][package:]id/resource_name"
        android:top="dimension"
        android:right="dimension"
        android:bottom="dimension"
        android:left="dimension" />
</transition>
```

上面语法中的属性前面已经都介绍过了，其中 android:top、android:bottom、android:left 和 android:right 仍然表示的是 Drawable 四周的偏移量，这里就不多介绍了。下面给出一个实际的例子。

首先定义 TransitionDrawable，如下所示。

```
// res/drawable/transition_drawable.xml
<?xml version="1.0" encoding="utf-8"?>
<transition xmlns:android="http://schemas.android.com/apk/res/android">
    <item android:drawable="@drawable/drawable1" />
    <item android:drawable="@drawable/drawable2" />
</transition>
```

接着将上面的 TransitionDrawable 设置为 View 的背景，如下所示。当然也可以在 ImageView 中直接作为 Drawable 来使用。

```xml
<TextView
    android:id="@+id/button"
    android:layout_height="wrap_content"
    android:layout_width="wrap_content"
android:background="@drawable/transition_drawable" />
```

最后，通过它的 startTransition 和 reverseTransition 方法来实现淡入淡出的效果以及它的逆过程，如下所示。

```java
TextView textView = (TextView) findViewById(R.id.test_transition);
TransitionDrawable drawable = (TransitionDrawable) textView.getBackground();
drawable.startTransition(1000);
```

6.2.7　InsetDrawable

InsetDrawable 对应于<inset>标签，它可以将其他 Drawable 内嵌到自己当中，并可以在四周留出一定的间距。当一个 View 希望自己的背景比自己的实际区域小的时候，可以采用 InsetDrawable 来实现，同时我们知道，通过 LayerDrawable 也可以实现这种效果。InsetDrawable 的语法如下所示。

```xml
<?xml version="1.0" encoding="utf-8"?>
<inset
    xmlns:android="http://schemas.android.com/apk/res/android"
    android:drawable="@drawable/drawable_resource"
    android:insetTop="dimension"
    android:insetRight="dimension"
    android:insetBottom="dimension"
    android:insetLeft="dimension" />
```

上面的属性都比较好理解，其中 android:insetTop、android:insetBottom、android:insetLeft 和 android:insetRight 分别表示顶部、底部、左边和右边内凹的大小。在下面的例子中，inset 中的 shape 距离 View 的边界为 15dp。

```xml
<?xml version="1.0" encoding="utf-8"?>
<inset xmlns:android="http://schemas.android.com/apk/res/android"
    android:insetBottom="15dp"
    android:insetLeft="15dp"
    android:insetRight="15dp"
    android:insetTop="15dp" >
```

```
    <shape android:shape="rectangle" >
        <solid android:color="#ff0000" />
    </shape>
</inset>
```

6.2.8 ScaleDrawable

ScaleDrawable 对应于<scale>标签，它可以根据自己的等级（level）将指定的 Drawable 缩放到一定比例，它的语法如下所示。

```
<?xml version="1.0" encoding="utf-8"?>
<scale
    xmlns:android="http://schemas.android.com/apk/res/android"
    android:drawable="@drawable/drawable_resource"
    android:scaleGravity=["top" | "bottom" | "left" | "right" | "center_vertical" |"fill_vertical" | "center_horizontal" | "fill_horizontal" | "center" | "fill" | "clip_vertical" | "clip_horizontal"]
    android:scaleHeight="percentage"
    android:scaleWidth="percentage" />
```

在上面的属性中，android:scaleGravity 的含义等同于 shape 中的 android:gravity，而 android:scaleWidth 和 android:scaleHeight 分别表示对指定 Drawable 宽和高的缩放比例，以百分比的形式表示，比如 25%。

ScaleDrawable 有点费解，要理解它，我们首先要明白等级对 ScaleDrawable 的影响。等级 0 表示 ScaleDrawable 不可见，这是默认值，要想 ScaleDrawable 可见，需要等级不能为 0，这一点从源码中可以得出。来看一下 ScaleDrawable 的 draw 方法，如下所示。

```
public void draw(Canvas canvas) {
    if (mScaleState.mDrawable.getLevel() != 0)
        mScaleState.mDrawable.draw(canvas);
}
```

很显然，由于 ScaleDrawable 的等级和 mDrawable 的等级是保持一致的，所以如果 ScaleDrawable 的等级为 0，那么它内部的 mDrawable 的等级也必然为 0，这时 mDrawable 就无法绘制出来，也就是 ScaleDrawable 不可见。下面再看一下 ScaleDrawable 的 onBoundsChange 方法，如下所示。

```
protected void onBoundsChange(Rect bounds) {
    final Rect r = mTmpRect;
    final boolean min = mScaleState.mUseIntrinsicSizeAsMin;
    int level = getLevel();
    int w = bounds.width();
    if (mScaleState.mScaleWidth > 0) {
        final int iw = min ? mScaleState.mDrawable.getIntrinsicWidth() : 0;
        w -= (int) ((w - iw) * (10000 - level) * mScaleState.mScaleWidth /
            10000);
    }
    int h = bounds.height();
    if (mScaleState.mScaleHeight > 0) {
        final int ih = min ? mScaleState.mDrawable.getIntrinsicHeight() : 0;
        h -= (int) ((h - ih) * (10000 - level) * mScaleState.mScaleHeight
            / 10000);
    }
    final int layoutDirection = getLayoutDirection();
    Gravity.apply(mScaleState.mGravity, w, h, bounds, r, layoutDirection);

    if (w > 0 && h > 0) {
        mScaleState.mDrawable.setBounds(r.left, r.top, r.right, r.bottom);
    }
}
```

在 ScaleDrawable 的 onBoundsChange 方法中，我们可以看出 mDrawable 的大小和等级以及缩放比例的关系，这里拿宽度来说，如下所示。

```
final int iw = min ? mScaleState.mDrawable.getIntrinsicWidth() : 0;
w -= (int) ((w - iw) * (10000 - level) * mScaleState.mScaleWidth / 10000);
```

由于 iw 一般都为 0，所以上面的代码可以简化为：w -= (int) (w * (10000 - level) * mScaleState.mScaleWidth / 10000)。由此可见，如果 ScaleDrawable 的级别为最大值 10000，那么就没有缩放的效果；如果 ScaleDrawable 的级别（level）越大，那么内部的 Drawable 看起来就越大；如果 ScaleDrawable 的 XML 中所定义的缩放比例越大，那么内部的 Drawable 看起来就越小。另外，从 ScaleDrawable 的内部实现来看，ScaleDrawable 的作用更偏向于缩小一个特定的 Drawable。在下面的例子中，可以近似地将一张图片缩小为原大小的 30%，代码如下所示。

```
// res/drawable/scale_drawable.xml
```

```xml
<?xml version="1.0" encoding="utf-8"?>
<scale xmlns:android="http://schemas.android.com/apk/res/android"
    android:drawable="@drawable/image1"
    android:scaleHeight="70%"
    android:scaleWidth="70%"
    android:scaleGravity="center" />
```

直接使用上面的 drawable 资源是不行的，还必须设置 ScaleDrawable 的等级为大于 0 且小于等于 10000 的值，如下所示。

```
View testScale = findViewById(R.id.test_scale);
ScaleDrawable testScaleDrawable = (ScaleDrawable) testScale.getBackground();
testScaleDrawable.setLevel(1);
```

经过上面的两步可以正确地缩放一个 Drawable，如果少了设置等级这一步，由于 Drawable 的默认等级为 0，那么 ScaleDrawable 将无法显示出来。我们可以武断地将 Drawable 的等级设置为大于 10000 的值，比如 20000，虽然也能正常工作，但是不推荐这么做，这是因为系统内部约定 Drawable 等级的范围为 0 到 10000。

6.2.9 ClipDrawable

ClipDrawable 对应于<clip>标签，它可以根据自己当前的等级（level）来裁剪另一个 Drawable，裁剪方向可以通过 android:clipOrientation 和 android:gravity 这两个属性来共同控制，它的语法如下所示。

```xml
<?xml version="1.0" encoding="utf-8"?>
<clip
    xmlns:android="http://schemas.android.com/apk/res/android"
    android:drawable="@drawable/drawable_resource"
    android:clipOrientation=["horizontal" | "vertical"]
    android:gravity=["top" | "bottom"|"left" | "right"|"center_vertical" |
               "fill_vertical"|"center_horizontal"|"fill_horizontal" |
               "center" | "fill"|"clip_vertical"|"clip_horizontal"] />
```

其中 clipOrientation 表示裁剪方向，有水平和竖直两个方向，gravity 比较复杂，需要和 clipOrientation 一起才能发挥作用，如表 6-4 所示。另外 gravity 的各种选项是可以通过 "|" 来组合使用的。

表 6-4 ClipDrawable 的 gravity 属性

选项	含义
top	将内部的 Drawable 放在容器的顶部,不改变它的大小。如果为竖直裁剪,那么从底部开始裁剪
bottom	将内部的 Drawable 放在容器的底部,不改变它的大小。如果为竖直裁剪,那么从顶部开始裁剪
left	将内部的 Drawable 放在容器的左边,不改变它的大小。如果为水平裁剪,那么从右边开始裁剪,这是默认值
right	将内部的 Drawable 放在容器的右边,不改变它的大小。如果为水平裁剪,那么从左边开始裁剪
center_vertical	使内部的 Drawable 在容器中竖直居中,不改变它的大小。如果为竖直裁剪,那么从上下同时开始裁剪
fill_vertical	使内部的 Drawable 在竖直方向上填充容器。如果为竖直裁剪,那么仅当 ClipDrawable 的等级为 0(0 表示 ClipDrawable 被完全裁剪,即不可见)时,才能有裁剪行为
center_horizontal	使内部的 Drawable 在容器中水平居中,不改变它的大小。如果为水平裁剪,那么从左右两边同时开始裁剪
fill_horizontal	使内部的 Drawable 在水平方向上填充容器。如果为水平裁剪,那么仅当 ClipDrawable 的等级为 0 时,才能有裁剪行为
center	使内部的 Drawable 在容器中水平和竖直方向都居中,不改变它的大小。如果为竖直裁剪,那么从上下同时开始裁剪;如果为水平裁剪,那么从左右同时开始裁剪
fill	使内部的 Drawable 在水平和竖直方向上同时填充容器。仅当 ClipDrawable 的等级为 0 时,才能有裁剪行为
clip_vertical	附加选项,表示竖直方向的裁剪,较少使用
clip_horizontal	附加选项,表示水平方向的裁剪,较少使用

下面举个例子,我们实现将一张图片从上往下进行裁剪的效果,首先定义 ClipDrawable,xml 如下:

```xml
<?xml version="1.0" encoding="utf-8"?>
<clip xmlns:android="http://schemas.android.com/apk/res/android"
    android:clipOrientation="vertical"
    android:drawable="@drawable/image1"
    android:gravity="bottom" />
```

在上面的 XML 中,因为我们要实现顶部的裁剪效果,所以裁剪方向应该为竖直方向,同时从表 6-4 可以知道,gravity 属性应该选择 bottom。有了 ClipDrawable 如何使用呢?也是很简单的,首先将它设置给 ImageView,当然也可以作为普通 View 的背景,如下所示。

```xml
<ImageView
    android:id="@+id/test_clip"
    android:layout_width="100dp"
    android:layout_height="100dp"
```

```
            android:src="@drawable/clip_drawable"
            android:gravity="center" />
```

接着在代码中设置 ClipDrawable 的等级,如下所示。

```
ImageView testClip = (ImageView) findViewById(R.id.test_clip);
ClipDrawable testClipDrawable = (ClipDrawable) testClip.getDrawable();
testClipDrawable.setLevel(5000);
```

在 6.2.5 节中已经提到,Drawable 的等级(level)是有范围的,即 0~10000,最小等级是 0,最大等级是 10000,对于 ClipDrawable 来说,等级 0 表示完全裁剪,即整个 Drawable 都不可见了,而等级 10000 表示不裁剪。在上面的代码中将等级设置为 8000 表示裁剪了 2000,即在顶部裁剪掉 20%的区域,被裁剪的区域就相当于不存在了,具体效果如图 6-6 所示。

对于本例来说,等级越大,表示裁剪的区域越小,因此等级 10000 表示不裁剪,这个时候整个图片都可以完全显示出来;而等级 0 则表示裁剪全部区域,这个时候整个图片将不可见。另外裁剪效果还受裁剪方向和 gravity 属性的影响,表 6-4 中的选项读者可以自行尝试一下,这样就能比较好地理解不同属性对裁剪效果的影响了。

图 6-6 ClipDrawable 的裁剪效果

6.3 自定义 Drawable

Drawable 的使用范围很单一,一个是作为 ImageView 中的图像来显示,另外一个就是作为 View 的背景,大多数情况下 Drawable 都是以 View 的背景这种形式出现的。Drawable 的工作原理很简单,其核心就是 draw 方法。在第 5 章中,我们分析了 View 的工作原理,我们知道系统会调用 Drawable 的 draw 方法来绘制 View 的背景,从这一点我们明白,可以通过重写 Drawable 的 draw 方法来自定义 Drawable。

通常我们没有必要去自定义 Drawable,这是因为自定义的 Drawable 无法在 XML 中使用,这就降低了自定义 Drawable 的使用范围。某些特殊情况下我们的确想自定义 Drawable,这也是可以的。下面演示一个自定义 Drawable 的实现过程,我们通过自定义 Drawable 来绘制一个圆形的 Drawable,并且它的半径会随着 View 的变化而变化,这种 Drawable 可以作为 View 的通用背景,代码如下所示。

```java
public class CustomDrawable extends Drawable {
    private Paint mPaint;

    public CustomDrawable(int color) {
        mPaint = new Paint(Paint.ANTI_ALIAS_FLAG);
        mPaint.setColor(color);
    }

    @Override
    public void draw(Canvas canvas) {
        final Rect r = getBounds();
        float cx = r.exactCenterX();
        float cy = r.exactCenterY();
        canvas.drawCircle(cx, cy, Math.min(cx, cy), mPaint);
    }

    @Override
    public void setAlpha(int alpha) {
        mPaint.setAlpha(alpha);
        invalidateSelf();
    }

    @Override
    public void setColorFilter(ColorFilter cf) {
        mPaint.setColorFilter(cf);
        invalidateSelf();
    }

    @Override
    public int getOpacity() {
        // not sure, so be safe
        return PixelFormat.TRANSLUCENT;
    }
}
```

在上面的代码中，draw、setAlpha、setColorFilter 和 getOpacity 这几个方法都是必须要实现的，其中 draw 是最主要的方法，这个方法就和 View 的 draw 方法类似，而 setAlpha、setColorFilter 和 getOpacity 这三个方法的实现都比较简单，这里不再多说了。在上面的例子中，参考了 ShapeDrawable 和 BitmapDrawable 的源码，所以说，源码是一个很好的学习资料，有些技术细节我们不清楚，就可以查看源码中类似功能的实现以及相应的文档，这样就可以更有针对性地解决一些问题。

上面的例子比较简单，但是流程是完整的，读者可以根据自己的需要实现更复杂的自定义 Drawable。另外 getIntrinsicWidth 和 getIntrinsicHeight 这两个方法需要注意一下，当自定义的 Drawable 有固有大小时最好重写这两个方法，因为它会影响到 View 的 wrap_content 布局，比如自定义 Drawable 是绘制一张图片，那么这个 Drawable 的内部大小就可以选用图片的大小。在上面的例子中，自定义的 Drawable 是由颜色填充的圆形并且没有固定的大小，因此没有重写这两个方法，这个时候它的内部大小为-1，即内部宽度和内部高度都为-1。需要注意的是，内部大小不等于 Drawable 的实际区域大小，Drawable 的实际区域大小可以通过它的 getBounds 方法来得到，一般来说它和 View 的尺寸相同。

第 7 章 Android 动画深入分析

Android 的动画可以分为三种：View 动画、帧动画和属性动画，其实帧动画也属于 View 动画的一种，只不过它和平移、旋转等常见的 View 动画在表现形式上略有不同而已。View 动画通过对场景里的对象不断做图像变换（平移、缩放、旋转、透明度）从而产生动画效果，它是一种渐近式动画，并且 View 动画支持自定义。帧动画通过顺序播放一系列图像从而产生动画效果，可以简单理解为图片切换动画，很显然，如果图片过多过大就会导致 OOM。属性动画通过动态地改变对象的属性从而达到动画效果，属性动画为 API 11 的新特性，在低版本无法直接使用属性动画，但是我们仍然可以通过兼容库来使用它。在本章中，首先简单介绍 View 动画以及自定义 View 动画的方式，接着介绍 View 动画的一些特殊的使用场景，最后对属性动画做一个全面性的介绍，另外还介绍使用动画的一些注意事项。

7.1 View 动画

View 动画的作用对象是 View，它支持 4 种动画效果，分别是平移动画、缩放动画、旋转动画和透明度动画。除了这四种典型的变换效果外，帧动画也属于 View 动画，但是帧动画的表现形式和上面的四种变换效果不太一样。为了更好地区分这四种变换和帧动画，在本章中如果没有特殊说明，那么所提到的 View 动画均指这四种变换，帧动画会单独介绍。本节将介绍 View 动画的四种效果以及帧动画，同时还会介绍自定义 View 动画的方法。

7.1.1 View 动画的种类

View 动画的四种变换效果对应着 Animation 的四个子类：TranslateAnimation、

ScaleAnimation、RotateAnimation 和 AlphaAnimation，如表 7-1 所示。这四种动画既可以通过 XML 来定义，也可以通过代码来动态创建，对于 View 动画来说，建议采用 XML 来定义动画，这是因为 XML 格式的动画可读性更好。

表 7-1 View 动画的四种变换

名称	标签	子类	效果
平移动画	\<translate\>	TranslateAnimation	移动 View
缩放动画	\<scale\>	ScaleAnimation	放大或缩小 View
旋转动画	\<rotate\>	RotateAnimation	旋转 View
透明度动画	\<alpha\>	AlphaAnimation	改变 View 的透明度

要使用 View 动画，首先要创建动画的 XML 文件，这个文件的路径为：res/anim/filename.xml。View 动画的描述文件是有固定的语法的，如下所示。

```xml
<?xml version="1.0" encoding="utf-8"?>
<set xmlns:android="http://schemas.android.com/apk/res/android"
    android:interpolator="@[package:]anim/interpolator_resource"
    android:shareInterpolator=["true" | "false"] >
    <alpha
        android:fromAlpha="float"
        android:toAlpha="float" />
    <scale
        android:fromXScale="float"
        android:toXScale="float"
        android:fromYScale="float"
        android:toYScale="float"
        android:pivotX="float"
        android:pivotY="float" />
    <translate
        android:fromXDelta="float"
        android:toXDelta="float"
        android:fromYDelta="float"
        android:toYDelta="float" />
    <rotate
        android:fromDegrees="float"
        android:toDegrees="float"
        android:pivotX="float"
        android:pivotY="float" />
    <set>
```

```
    ...
    </set>
</set>
```

从上面的语法可以看出，View 动画既可以是单个动画，也可以由一系列动画组成。

<set>标签表示动画集合，对应 AnimationSet 类，它可以包含若干个动画，并且它的内部也是可以嵌套其他动画集合的，它的两个属性的含义如下：

android:interpolator

表示动画集合所采用的插值器，插值器影响动画的速度，比如非匀速动画就需要通过插值器来控制动画的播放过程。这个属性可以不指定，默认为@android:anim/accelerate_decelerate_interpolator，即加速减速插值器，关于插值器的概念会在 7.3.2 节中进行具体介绍。

android:shareInterpolator

表示集合中的动画是否和集合共享同一个插值器。如果集合不指定插值器，那么子动画就需要单独指定所需的插值器或者使用默认值。

<translate>标签标示平移动画，对应 TranslateAnimation 类，它可以使一个 View 在水平和竖直方向完成平移的动画效果，它的一系列属性的含义如下：

- android:fromXDelta——表示 x 的起始值，比如 0；
- android:toXDelta——表示 x 的结束值，比如 100；
- android:fromYDelta——表示 y 的起始值；
- android:toYDelta——表示 y 的结束值。

<scale>标签表示缩放动画，对应 ScaleAnimation，它可以使 View 具有放大或者缩小的动画效果，它的一系列属性的含义如下：

- android:fromXScale——水平方向缩放的起始值，比如 0.5；
- android:toXScale——水平方向缩放的结束值，比如 1.2；

- android:fromYScale——竖直方向缩放的起始值；
- android:toYScale——竖直方向缩放的起始值；
- android:pivotX——缩放的轴点的 x 坐标，它会影响缩放的效果；
- android:pivotY——缩放的轴点的 y 坐标，它会影响缩放的效果。

在<scale>标签中提到了轴点的概念，这里举个例子，默认情况下轴点是 View 的中心点，这个时候在水平方向进行缩放的话会导致 View 向左右两个方向同时进行缩放，但是如果把轴点设为 View 的右边界，那么 View 就只会向左边进行缩放，反之则向右边进行缩放，具体效果读者可以自己测试一下。

<rotate>标签表示旋转动画，对于 RotateAnimation，它可以使 View 具有旋转的动画效果，它的属性的含义如下：

- android:fromDegrees——旋转开始的角度，比如 0；
- android:toDegrees——旋转结束的角度，比如 180；
- android:pivotX——旋转的轴点的 x 坐标；
- android:pivotY——旋转的轴点的 y 坐标。

在旋转动画中也有轴点的概念，它也会影响到旋转的具体效果。在旋转动画中，轴点扮演着旋转轴的角色，即 View 是围绕着轴点进行旋转的，默认情况下轴点为 View 的中心点。考虑一种情况，View 围绕着自己的中心点和围绕着自己的左上角旋转 90 度显然是不同的旋转轨迹，不同轴点对旋转效果的影响读者可以自己测试一下。

<alpha>标签表示透明度动画，对应 AlphaAnimation，它可以改变 View 的透明度，它的属性的含义如下：

- android:fromAlpha——表示透明度的起始值，比如 0.1；
- android:toAlpha——表示透明度的结束值，比如 1。

上面简单介绍了 View 动画的 XML 格式，具体的使用方法查看相关文档。除了上面介绍的属性以外，View 动画还有一些常用的属性，如下所示。

- android:duration——动画的持续时间；

- android:fillAfter——动画结束以后 View 是否停留在结束位置，true 表示 View 停留在结束位置，false 则不停留。

下面是一个实际的例子：

```xml
// res/anim/animation_test.xml
<?xml version="1.0" encoding="utf-8"?>
<set xmlns:android="http://schemas.android.com/apk/res/android"
    android:fillAfter="true"
    android:zAdjustment="normal" >

    <translate
        android:duration="100"
        android:fromXDelta="0"
        android:fromYDelta="0"
        android:interpolator="@android:anim/linear_interpolator"
        android:toXDelta="100"
        android:toYDelta="100" />

    <rotate
        android:duration="400"
        android:fromDegrees="0"
        android:toDegrees="90" />

</set>
```

如何应用上面的动画呢？也很简单，如下所示。

```java
Button mButton = (Button) findViewById(R.id.button1);
Animation animation = AnimationUtils.loadAnimation(this, R.anim.animation_test);
mButton.startAnimation(animation);
```

除了在 XML 中定义动画外，还可以通过代码来应用动画，这里举个例子，如下所示。

```java
AlphaAnimation alphaAnimation = new AlphaAnimation(0, 1);
alphaAnimation.setDuration(300);
mButton.startAnimation(alphaAnimation);
```

在上面的代码中，创建了一个透明度动画，将一个 Button 的透明度在 300ms 内由 0 变为 1，其他类型的 View 动画也可以通过代码来创建，这里就不做介绍了。另外，通过 Animation 的 setAnimationListener 方法可以给 View 动画添加过程监听，接口如下所示。从接口的定义可以很清楚地看出每个方法的含义。

```
public static interface AnimationListener {
    void onAnimationStart(Animation animation);
    void onAnimationEnd(Animation animation);
    void onAnimationRepeat(Animation animation);
}
```

7.1.2 自定义 View 动画

除了系统提供的四种 View 动画外，我们还可以自定义 View 动画。自定义动画是一件既简单又复杂的事情，说它简单，是因为派生一种新动画只需要继承 Animation 这个抽象类，然后重写它的 initialize 和 applyTransformation 方法，在 initialize 方法中做一些初始化工作，在 applyTransformation 中进行相应的矩阵变换即可，很多时候需要采用 Camera 来简化矩阵变换的过程。说它复杂，是因为自定义 View 动画的过程主要是矩阵变换的过程，而矩阵变换是数学上的概念，如果对这方面的知识不熟悉的话，就会觉得这个过程比较复杂了。

本节不打算详细地讲解自定义 View 动画的细节，因为这都是数学中的矩阵变换的细节，读者只需要知道自定义 View 的方法并在需要的时候参考矩阵变换的细节即可写出特定的自定义 View 动画。一般来说，在实际开发中很少用到自定义 View 动画。这里提供一个自定义 View 动画的例子，这个例子来自于 Android 的 ApiDemos 中的一个自定义 View 动画 Rotate3dAnimation。Rotate3dAnimation 可以围绕 y 轴旋转并且同时沿着 z 轴平移从而实现一种类似于 3D 的效果，它的代码如下：

```
public class Rotate3dAnimation extends Animation {
    private final float mFromDegrees;
    private final float mToDegrees;
    private final float mCenterX;
    private final float mCenterY;
    private final float mDepthZ;
    private final boolean mReverse;
    private Camera mCamera;

    /**
     * Creates a new 3D rotation on the Y axis. The rotation is defined by its
```

```
 * start angle and its end angle. Both angles are in degrees. The rotation
 * is performed around a center point on the 2D space, definied by a pair
 * of X and Y coordinates, called centerX and centerY. When the animation
 * starts, a translation on the Z axis (depth) is performed. The length
 * of the translation can be specified, as well as whether the translation
 * should be reversed in time.
 *
 * @param fromDegrees the start angle of the 3D rotation
 * @param toDegrees the end angle of the 3D rotation
 * @param centerX the X center of the 3D rotation
 * @param centerY the Y center of the 3D rotation
 * @param reverse true if the translation should be reversed, false
   otherwise
 */
public Rotate3dAnimation(float fromDegrees, float toDegrees,
        float centerX, float centerY, float depthZ, boolean reverse) {
    mFromDegrees = fromDegrees;
    mToDegrees = toDegrees;
    mCenterX = centerX;
    mCenterY = centerY;
    mDepthZ = depthZ;
    mReverse = reverse;
}


@Override
public void initialize(int width, int height, int parentWidth, int
parentHeight) {
    super.initialize(width, height, parentWidth, parentHeight);
    mCamera = new Camera();
}


@Override
protected void applyTransformation(float interpolatedTime, Transfor-
mation t) {
    final float fromDegrees = mFromDegrees;
    float degrees = fromDegrees + ((mToDegrees - fromDegrees) * interpo-
    latedTime);

    final float centerX = mCenterX;
    final float centerY = mCenterY;
```

```java
        final Camera camera = mCamera;

        final Matrix matrix = t.getMatrix();

        camera.save();
        if (mReverse) {
            camera.translate(0.0f, 0.0f, mDepthZ * interpolatedTime);
        } else {
            camera.translate(0.0f, 0.0f, mDepthZ * (1.0f - interpolated-
            Time));
        }
        camera.rotateY(degrees);
        camera.getMatrix(matrix);
        camera.restore();

        matrix.preTranslate(-centerX, -centerY);
        matrix.postTranslate(centerX, centerY);
    }
}
```

7.1.3 帧动画

帧动画是顺序播放一组预先定义好的图片，类似于电影播放。不同于 View 动画，系统提供了另外一个类 AnimationDrawable 来使用帧动画。帧动画的使用比较简单，首先需要通过 XML 来定义一个 AnimationDrawable，如下所示。

```xml
// res/drawable/frame_animation.xml
<?xml version="1.0" encoding="utf-8"?>
<animation-list xmlns:android="http://schemas.android.com/apk/res/android"
    android:oneshot="false">
    <item android:drawable="@drawable/image1" android:duration="500" />
    <item android:drawable="@drawable/image2" android:duration="500" />
    <item android:drawable="@drawable/image3" android:duration="500" />
</animation-list>
```

然后将上述的 Drawable 作为 View 的背景并通过 Drawable 来播放动画即可：

```java
Button mButton = (Button)findViewById(R.id.button1);
mButton.setBackgroundResource(R.drawable.frame_animation);
AnimationDrawable drawable = (AnimationDrawable) mButton.getBackground();
```

```
drawable.start();
```

帧动画的使用比较简单，但是比较容易引起 OOM，所以在使用帧动画时应尽量避免使用过多尺寸较大的图片。

7.2 View 动画的特殊使用场景

在 7.1 节中我们介绍了 View 动画的四种形式，除了这四种形式外，View 动画还可以在一些特殊的场景下使用，比如在 ViewGroup 中可以控制子元素的出场效果，在 Activity 中可以实现不同 Activity 之间的切换效果。

7.2.1 LayoutAnimation

LayoutAnimation 作用于 ViewGroup，为 ViewGroup 指定一个动画，这样当它的子元素出场时都会具有这种动画效果。这种效果常常被用在 ListView 上，我们时常会看到一种特殊的 ListView，它的每个 item 都以一定的动画的形式出现，其实这并非什么高深的技术，它使用的就是 LayoutAnimation。LayoutAnimation 也是一个 View 动画，为了给 ViewGroup 的子元素加上出场效果，遵循如下几个步骤。

（1）定义 LayoutAnimation，如下所示。

```
// res/anim/anim_layout.xml
<layoutAnimation
    xmlns:android="http://schemas.android.com/apk/res/android"
    android:delay="0.5"
    android:animationOrder="normal"
android:animation="@anim/anim_item"/>
```

它的属性的含义如下所示。

android:delay

表示子元素开始动画的时间延迟，比如子元素入场动画的时间周期为 300ms，那么 0.5 表示每个子元素都需要延迟 150ms 才能播放入场动画。总体来说，第一个子元素延迟 150ms 开始播放入场动画，第 2 个子元素延迟 300ms 开始播放入场动画，依次类推。

android:animationOrder

表示子元素动画的顺序,有三种选项:normal、reverse 和 random,其中 normal 表示顺序显示,即排在前面的子元素先开始播放入场动画;reverse 表示逆向显示,即排在后面的子元素先开始播放入场动画;random 则是随机播放入场动画。

android:animation

为子元素指定具体的入场动画。

(2)为子元素指定具体的入场动画,如下所示。

```
// res/anim/anim_item.xml
<?xml version="1.0" encoding="utf-8"?>
<set xmlns:android="http://schemas.android.com/apk/res/android"
    android:duration="300"
    android:interpolator="@android:anim/accelerate_interpolator"
    android:shareInterpolator="true" >

    <alpha
        android:fromAlpha="0.0"
        android:toAlpha="1.0" />

    <translate
        android:fromXDelta="500"
        android:toXDelta="0" />

</set>
```

(3)为 ViewGroup 指定 android:layoutAnimation 属性:android:layoutAnimation= "@anim/anim_layout"。对于 ListView 来说,这样 ListView 的 item 就具有出场动画了,这种方式适用于所有的 ViewGroup,如下所示。

```
<ListView
    android:id="@+id/list"
    android:layout_width="match_parent"
    android:layout_height="match_parent"
    android:layoutAnimation="@anim/anim_layout"
    android:background="#fff4f7f9"
```

```
android:cacheColorHint="#00000000"
android:divider="#dddbdb"
android:dividerHeight="1.0px"
android:listSelector="@android:color/transparent" />
```

除了在 XML 中指定 LayoutAnimation 外，还可以通过 LayoutAnimationController 来实现，具体代码如下所示。

```
ListView listView = (ListView) layout.findViewById(R.id.list);
Animation animation = AnimationUtils.loadAnimation(this, R.anim.anim_item);
LayoutAnimationController controller = new LayoutAnimationController(animation);
controller.setDelay(0.5f);
controller.setOrder(LayoutAnimationController.ORDER_NORMAL);
listView.setLayoutAnimation(controller);
```

7.2.2 Activity 的切换效果

Activity 有默认的切换效果，但是这个效果我们是可以自定义的，主要用到 overridePendingTransition(int enterAnim, int exitAnim) 这个方法，这个方法必须在 startActivity(Intent)或者 finish()之后被调用才能生效，它的参数含义如下：

- enterAnim——Activity 被打开时，所需的动画资源 id；
- exitAnim——Activity 被暂停时，所需的动画资源 id。

当启动一个 Activity 时，可以按照如下方式为其添加自定义的切换效果：

```
Intent intent = new Intent(this, TestActivity.class);
startActivity(intent);
overridePendingTransition(R.anim.enter_anim, R.anim.exit_anim);
```

当 Activity 退出时，也可以为其指定自己的切换效果，如下所示。

```
@Override
public void finish() {
    super.finish();
    overridePendingTransition(R.anim.enter_anim, R.anim.exit_anim);
}
```

需要注意的是，overridePendingTransition 这个方法必须位于 startActivity 或者 finish 的后面，否则动画效果将不起作用。

Fragment 也可以添加切换动画，由于 Fragment 是在 API 11 中新引入的类，因此为了兼容性我们需要使用 support-v4 这个兼容包，在这种情况下我们可以通过 FragmentTransaction 中的 setCustomAnimations() 方法来添加切换动画。这个切换动画需要是 View 动画，之所以不能采用属性动画是因为属性动画也是 API 11 新引入的。还有其他方式可以给 Activity 和 Fragment 添加切换动画，但是它们大多都有兼容性问题，在低版本上无法使用，因此不具有很高的使用价值，这里就不再一一介绍了。

7.3 属性动画

属性动画是 API 11 新加入的特性，和 View 动画不同，它对作用对象进行了扩展，属性动画可以对任何对象做动画，甚至还可以没有对象。除了作用对象进行了扩展以外，属性动画的效果也得到了加强，不再像 View 动画那样只能支持四种简单的变换。属性动画中有 ValueAnimator、ObjectAnimator 和 AnimatorSet 等概念，通过它们可以实现绚丽的动画。

7.3.1 使用属性动画

属性动画可以对任意对象的属性进行动画而不仅仅是 View，动画默认时间间隔 300ms，默认帧率 10ms/帧。其可以达到的效果是：在一个时间间隔内完成对象从一个属性值到另一个属性值的改变。因此，属性动画几乎是无所不能的，只要对象有这个属性，它都能实现动画效果。但是属性动画从 API 11 才有，这就严重制约了属性动画的使用。可以采用开源动画库 nineoldandroids 来兼容以前的版本，采用 nineoldandroids，可以在 API 11 以前的系统上使用属性动画，nineoldandroids 的网址为：http://nineoldandroids.com。

Nineoldandroids 对属性动画做了兼容，在 API 11 以前的版本其内部是通过代理 View 动画来实现的，因此在 Android 低版本上，它的本质还是 View 动画，尽管使用方法看起来是属性动画。Nineoldandroids 的功能和系统原生对的 android.animation.* 中类的功能完全一致，使用方法也完全一样，只要我们用 nineoldandroids 来编写动画，就可以在所有的 Android 系统上运行。比较常用的几个动画类是：ValueAnimator、ObjectAnimator 和 AnimatorSet，其中 ObjectAnimator 继承自 ValueAnimator，AnimatorSet 是动画集合，可以定义一组动画，它们使用起来也是极其简单的。如何使用属性动画呢？下面简单举几个小例子，读者一看就明白了。

（1）改变一个对象（myObject）的 translationY 属性，让其沿着 Y 轴向上平移一段距离：它的高度，该动画在默认时间内完成，动画的完成时间是可以定义的。想要更灵活的效果我们还可以定义插值器和估值算法，但是一般来说我们不需要自定义，系统已经预置了一些，能够满足常用的动画。

```
ObjectAnimator.ofFloat(myObject, "translationY", -myObject.getHeight()).start();
```

（2）改变一个对象的背景色属性，典型的情形是改变 View 的背景色，下面的动画可以让背景色在 3 秒内实现从 0xFFFF8080 到 0xFF8080FF 的渐变，动画会无限循环而且会有反转的效果。

```
ValueAnimator colorAnim = ObjectAnimator.ofInt(this, "backgroundColor",
    /*Red*/0xFFFF8080, /*Blue*/0xFF8080FF);
colorAnim.setDuration(3000);
colorAnim.setEvaluator(new ArgbEvaluator());
colorAnim.setRepeatCount(ValueAnimator.INFINITE);
colorAnim.setRepeatMode(ValueAnimator.REVERSE);
colorAnim.start();
```

（3）动画集合，5 秒内对 View 的旋转、平移、缩放和透明度都进行了改变。

```
AnimatorSet set = new AnimatorSet();
set.playTogether(
    ObjectAnimator.ofFloat(myView, "rotationX", 0, 360),
    ObjectAnimator.ofFloat(myView, "rotationY", 0, 180),
    ObjectAnimator.ofFloat(myView, "rotation", 0, -90),
    ObjectAnimator.ofFloat(myView, "translationX", 0, 90),
    ObjectAnimator.ofFloat(myView, "translationY", 0, 90),
    ObjectAnimator.ofFloat(myView, "scaleX", 1, 1.5f),
    ObjectAnimator.ofFloat(myView, "scaleY", 1, 0.5f),
    ObjectAnimator.ofFloat(myView, "alpha", 1, 0.25f, 1)
);
set.setDuration(5 * 1000).start();
```

属性动画除了通过代码实现以外，还可以通过 XML 来定义。属性动画需要定义在 res/animator/ 目录下，它的语法如下所示。

```xml
<set
  android:ordering=["together" | "sequentially"]>

  <objectAnimator
      android:propertyName="string"
      android:duration="int"
      android:valueFrom="float | int | color"
      android:valueTo="float | int | color"
      android:startOffset="int"
      android:repeatCount="int"
      android:repeatMode=["restart" | "reverse"]
      android:valueType=["intType" | "floatType"]/>

  <animator
      android:duration="int"
      android:valueFrom="float | int | color"
      android:valueTo="float | int | color"
      android:startOffset="int"
      android:repeatCount="int"
      android:repeatMode=["restart" | "reverse"]
      android:valueType=["intType" | "floatType"]/>

  <set>
     ...
  </set>
</set>
```

属性动画的各种参数都比较好理解，在 XML 中可以定义 ValueAnimator、ObjectAnimator 以及 AnimatorSet，其中<set>标签对应 AnimatorSet，<animator>标签对应 ValueAnimator，而<objectAnimator>则对应 ObjectAnimator。<set>标签的 android:ordering 属性有两个可选值："together" 和 "sequentially"，其中 "together" 表示动画集合中的子动画同时播放，"sequentially" 则表示动画集合中的子动画按照前后顺序依次播放，android:ordering 属性的默认值是 "together"。

对于<objectAnimator>标签的各个属性的含义，下面简单说明一下，对于<animator>标签这里就不再介绍了，因为它只是比<objectAnimator>少了一个 android:propertyName 属性而已，其他都是一样的。

- android:propertyName——表示属性动画的作用对象的属性的名称；

- android:duration——表示动画的时长；

- android:valueFrom——表示属性的起始值；

- android:valueTo——表示属性的结束值；

- android:startOffset——表示动画的延迟时间，当动画开始后，需要延迟多少毫秒才会真正播放此动画；

- android:repeatCount——表示动画的重复次数；

- android:repeatMode——表示动画的重复模式；

- android:valueType——表示 android:propertyName 所指定的属性的类型，有"intType"和"floatType"两个可选项，分别表示属性的类型为整型和浮点型。另外，如果 android:propertyName 所指定的属性表示的是颜色，那么不需要指定 android:valueType，系统会自动对颜色类型的属性做处理。

对于一个动画来说，有两个属性这里要特殊说明一下，一个是 android:repeatCount，它表示动画循环的次数，默认值为 0，其中-1 表示无限循环；另一个是 android:repeatMode，它表示动画循环的模式，有两个选项："restart"和"reverse"，分别表示连续重复和逆向重复。连续重复比较好理解，就是动画每次都重新开始播放，而逆向重复是指第一次播放完以后，第二次会倒着播放动画，第三次再重头开始播放动画，第四次再倒着播放动画，如此反复。

下面是一个具体的例子，我们通过 XML 定义一个属性动画并将其作用在 View 上，如下所示。

```
// res/animator/property_animator.xml
<set android:ordering="together">
    <objectAnimator
        android:propertyName="x"
        android:duration="300"
        android:valueTo="200"
        android:valueType="intType"/>
    <objectAnimator
        android:propertyName="y"
        android:duration="300"
```

```
            android:valueTo="300"
            android:valueType="intType"/>
</set>
```

如何使用上面的属性动画呢？也很简单，如下所示。

```
AnimatorSet set = (AnimatorSet) AnimatorInflater.loadAnimator(myContext,
R.anim.property_animator);
set.setTarget(mButton);
set.start();
```

在实际开发中建议采用代码来实现属性动画，这是因为通过代码来实现比较简单。更重要的是，很多时候一个属性的起始值是无法提前确定的，比如让一个 Button 从屏幕左边移动到屏幕的右边，由于我们无法提前知道屏幕的宽度，因此无法将属性动画定义在 XML 中，在这种情况下就必须通过代码来动态地创建属性动画。

7.3.2 理解插值器和估值器

TimeInterpolator 中文翻译为时间插值器，它的作用是根据时间流逝的百分比来计算出当前属性值改变的百分比，系统预置的有 LinearInterpolator（线性插值器：匀速动画）、AccelerateDecelerateInterpolator（加速减速插值器：动画两头慢中间快）和 DecelerateInterpolator（减速插值器：动画越来越慢）等。TypeEvaluator 的中文翻译为类型估值算法，也叫估值器，它的作用是根据当前属性改变的百分比来计算改变后的属性值，系统预置的有 IntEvaluator（针对整型属性）、FloatEvaluator（针对浮点型属性）和 ArgbEvaluator（针对 Color 属性）。属性动画中的插值器（Interpolator）和估值器（TypeEvaluator）很重要，它们是实现非匀速动画的重要手段。可能这么说还有点晦涩，没关系，下面给出一个实例就很好理解了。

如图 7-1 所示，它表示一个匀速动画，采用了线性插值器和整型估值算法，在 40ms 内，View 的 x 属性实现从 0 到 40 的变换。

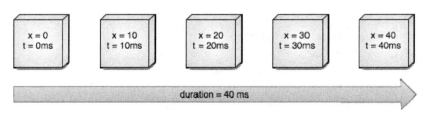

图 7-1　插值器的工作原理（注：此图来自 Android 官方文档）

由于动画的默认刷新率为 10ms/帧,所以该动画将分 5 帧进行,我们来考虑第三帧($x=20$, $t=20ms$),当时间 $t=20ms$ 的时候,时间流逝的百分比是 0.5(20/40=0.5),意味着现在时间过了一半,那 x 应该改变多少呢?这个就由插值器和估值算法来确定。拿线性插值器来说,当时间流逝一半的时候,x 的变换也应该是一半,即 x 的改变是 0.5,为什么呢?因为它是线性插值器,是实现匀速动画的,下面看它的源码:

```
public class LinearInterpolator implements Interpolator {

    public LinearInterpolator() {
    }

    public LinearInterpolator(Context context, AttributeSet attrs) {
    }

    public float getInterpolation(float input) {
        return input;
    }
}
```

很显然,线性插值器的返回值和输入值一样,因此插值器返回的值是 0.5,这意味着 x 的改变是 0.5,这个时候插值器的工作就完成了。具体 x 变成了什么值,这个需要估值算法来确定,我们来看看整型估值算法的源码:

```
public class IntEvaluator implements TypeEvaluator<Integer> {

    public Integer evaluate(float fraction, Integer startValue, Integer endValue) {
        int startInt = startValue;
        return (int)(startInt + fraction * (endValue - startInt));
    }
}
```

上述算法很简单,evaluate 的三个参数分别表示估值小数、开始值和结束值,对应于我们的例子就分别是 0.5、0、40。根据上述算法,整型估值返回给我们的结果是 20,这就是($x=20$, $t=20ms$)的由来。

属性动画要求对象的该属性有 set 方法和 get 方法(可选)。插值器和估值算法除了系统提供的外,我们还可以自定义。实现方式也很简单,因为插值器和估值算法都是一个接口,且内部都只有一个方法,我们只要派生一个类实现接口就可以了,然后就可以做出千奇百怪的动画效果了。具体一点就是：自定义插值器需要实现 Interpolator 或者 TimeInterpolator，自定义估值算法需要实现 TypeEvaluator。另外就是如果要对其他类型（非 int、float、Color）做动画，那么必须要自定义类型估值算法。

7.3.3 属性动画的监听器

属性动画提供了监听器用于监听动画的播放过程，主要有如下两个接口：AnimatorUpdateListener 和 AnimatorListener。

AnimatorListener 的定义如下：

```
public static interface AnimatorListener {
    void onAnimationStart(Animator animation);
    void onAnimationEnd(Animator animation);
    void onAnimationCancel(Animator animation);
    void onAnimationRepeat(Animator animation);
}
```

从 AnimatorListener 的定义可以看出，它可以监听动画的开始、结束、取消以及重复播放。同时为了方便开发，系统还提供了 AnimatorListenerAdapter 这个类，它是 AnimatorListener 的适配器类，这样我们就可以有选择地实现上面的 4 个方法了，毕竟不是所有方法都是我们感兴趣的。

下面再看一下 AnimatorUpdateListener 的定义，如下所示。

```
public static interface AnimatorUpdateListener {
    void onAnimationUpdate(ValueAnimator animation);
}
```

AnimatorUpdateListener 比较特殊，它会监听整个动画过程，动画是由许多帧组成的，每播放一帧，onAnimationUpdate 就会被调用一次，利用这个特性，我们可以做一些特殊的事情。

7.3.4 对任意属性做动画

这里先提出一个问题：给 Button 加一个动画，让这个 Button 的宽度从当前宽度增加到 500px。也许你会说，这很简单，用 View 动画就可以搞定，我们可以来试试，你能写出来吗？很快你就会恍然大悟，原来 View 动画根本不支持对宽度进行动画。没错，View 动画只支持四种类型：平移（Translate）、旋转（Rotate）、缩放（Scale）、不透明度（Alpha）。
当然用 x 方向缩放（scaleX）可以让 Button 在 x 方向放大，看起来好像是宽度增加了，实际上不是，只是 Button 被放大了而已，而且由于只 x 方向被放大，这个时候 Button 的背景以及上面的文本都被拉伸了，甚至有可能 Button 会超出屏幕，如图 7-2 所示。

图 7-2　属性动画的缩放效果

图 7-2 中的效果显然是很差的，而且也不是真正地对宽度做动画，不过，所幸我们还有属性动画，我们用属性动画试试，如下所示。

```
private void performAnimate() {
    ObjectAnimator.ofInt(mButton, "width", 500).setDuration(5000).start();
}

@Override
public void onClick(View v) {
    if (v == mButton) {
        performAnimate();
    }
}
```

上述代码运行后发现没效果，其实没效果是对的，如果随便传递一个属性过去，轻则没动画效果，重则程序直接 Crash。

下面分析属性动画的原理：属性动画要求动画作用的对象提供该属性的 get 和 set 方法，属性动画根据外界传递的该属性的初始值和最终值，以动画的效果多次去调用 set 方法，每次传递给 set 方法的值都不一样，确切来说是随着时间的推移，所传递的值越来越接近最终值。总结一下，我们对 object 的属性 abc 做动画，如果想让动画生效，要同时满足两个条件：

（1）object 必须要提供 setAbc 方法，如果动画的时候没有传递初始值，那么还要提供 getAbc 方法，因为系统要去取 abc 属性的初始值（如果这条不满足，程序直接 Crash）。

（2）object 的 setAbc 对属性 abc 所做的改变必须能够通过某种方法反映出来，比如会带来 UI 的改变之类的（如果这条不满足，动画无效果但不会 Crash）。

以上条件缺一不可。那么为什么我们对 Button 的 width 属性做动画会没有效果？这是因为 Button 内部虽然提供了 getWidth 和 setWidth 方法，但是这个 setWidth 方法并不是改变视图的大小，它是 TextView 新添加的方法，View 是没有这个 setWidth 方法的，由于 Button

继承了 TextView，所以 Button 也就有了 setWidth 方法。下面看一下这个 getWidth 和 setWidth 方法的源码：

```
/**
 * Makes the TextView exactly this many pixels wide.
 * You could do the same thing by specifying this number in the
 * LayoutParams.
 *
 * @see #setMaxWidth(int)
 * @see #setMinWidth(int)
 * @see #getMinWidth()
 * @see #getMaxWidth()
 *
 * @attr ref android.R.styleable#TextView_width
 */
@android.view.RemotableViewMethod
public void setWidth(int pixels) {
    mMaxWidth = mMinWidth = pixels;
    mMaxWidthMode = mMinWidthMode = PIXELS;

    requestLayout();
    invalidate();
}

/**
 * Return the width of the your view.
 *
 * @return The width of your view, in pixels.
 */
@ViewDebug.ExportedProperty(category = "layout")
public final int getWidth() {
    return mRight - mLeft;
}
```

从上述源码可以看出，getWidth 的确是获取 View 的宽度的，而 setWidth 是 TextView 和其子类的专属方法，它的作用不是设置 View 的宽度，而是设置 TextView 的最大宽度和最小宽度的，这个和 TextView 的宽度不是一个东西。具体来说，TextView 的宽度对应 XML 中的 android:layout_width 属性，而 TextView 还有一个属性 android:width，这个 android:width 属性就对应了 TextView 的 setWidth 方法。总之，TextView 和 Button 的 setWidth、getWidth

干的不是同一件事情,通过 setWidth 无法改变控件的宽度,所以对 width 做属性动画没有效果。对应于属性动画的两个条件来说,本例中动画不生效的原因是只满足了条件 1 而未满足条件 2。

针对上述问题,官方文档上告诉我们有 3 种解决方法:

- 给你的对象加上 get 和 set 方法,如果你有权限的话;
- 用一个类来包装原始对象,间接为其提供 get 和 set 方法;
- 采用 ValueAnimator,监听动画过程,自己实现属性的改变。

针对上面提出的三种解决方法,下面给出具体的介绍。

1. 给你的对象加上 get 和 set 方法,如果你有权限的话

这个的意思很好理解,如果你有权限的话,加上 get 和 set 就搞定了。但是很多时候我们没权限去这么做。比如本文开头所提到的问题,你无法给 Button 加上一个合乎要求的 setWidth 方法,因为这是 Android SDK 内部实现的。这个方法最简单,但是往往是不可行的,这里就不对其进行更多的分析了。

2. 用一个类来包装原始对象,间接为其提供 get 和 set 方法

这是一个很有用的解决方法,是笔者最喜欢用的,因为用起来很方便,也很好理解,下面将通过一个具体的例子来介绍它。

```
private void performAnimate() {
    ViewWrapper wrapper = new ViewWrapper(mButton);
    ObjectAnimator.ofInt(wrapper, "width", 500).setDuration(5000).start();
}

@Override
public void onClick(View v) {
    if (v == mButton) {
        performAnimate();
    }
}

private static class ViewWrapper {
    private View mTarget;
```

```
public ViewWrapper(View target) {
    mTarget = target;
}

public int getWidth() {
    return mTarget.getLayoutParams().width;
}

public void setWidth(int width) {
    mTarget.getLayoutParams().width = width;
    mTarget.requestLayout();
}
}
```

上述代码在 5s 内让 Button 的宽度增加到了 500px，为了达到这个效果，我们提供了 ViewWrapper 类专门用于包装 View，具体到本例是包装 Button。然后我们对 ViewWrapper 的 width 属性做动画，并且在 setWidth 方法中修改其内部的 target 的宽度，而 target 实际上就是我们包装的 Button。这样一个间接属性动画就搞定了，上述代码同样适用于一个对象的其他属性。如图 7-3 所示，很显然效果达到了，真正实现了对宽度做动画。

图 7-3 属性动画对宽度做动画的效果

3. 采用 ValueAnimator，监听动画过程，自己实现属性的改变

首先说说什么是 ValueAnimator，ValueAnimator 本身不作用于任何对象，也就是说直接使用它没有任何动画效果。它可以对一个值做动画，然后我们可以监听其动画过程，在动画过程中修改我们的对象的属性值，这样也就相当于我们的对象做了动画。下面用例子来说明：

```
private void performAnimate(final View target, final int start, final int end) {
    ValueAnimator valueAnimator = ValueAnimator.ofInt(1, 100);
    valueAnimator.addUpdateListener(new AnimatorUpdateListener() {
```

```java
        // 持有一个 IntEvaluator 对象，方便下面估值的时候使用
        private IntEvaluator mEvaluator = new IntEvaluator();

        @Override
        public void onAnimationUpdate(ValueAnimator animator) {
            // 获得当前动画的进度值，整型，1~100 之间
            int currentValue = (Integer) animator.getAnimatedValue();
            Log.d(TAG, "current value: " + currentValue);

            // 获得当前进度占整个动画过程的比例，浮点型，0~1 之间
            float fraction = animator.getAnimatedFraction();
            // 直接调用整型估值器，通过比例计算出宽度，然后再设给 Button
            target.getLayoutParams().width = mEvaluator.evaluate(fraction,
            start, end);
            target.requestLayout();
        }
    });

    valueAnimator.setDuration(5000).start();
}

@Override
public void onClick(View v) {
    if (v == mButton) {
        performAnimate(mButton, mButton.getWidth(), 500);
    }
}
```

上述代码的效果图和采用 ViewWrapper 是一样的，请参看图 7-3。关于这个 ValueAnimator 要再说一下，拿上面的例子来说，它会在 5000ms 内将一个数从 1 变到 100，然后动画的每一帧会回调 onAnimationUpdate 方法。在这个方法里，我们可以获取当前的值（1~100）和当前值所占的比例，我们可以计算出 Button 现在的宽度应该是多少。比如时间过了一半，当前值是 50，比例为 0.5，假设 Button 的起始宽度是 100px，最终宽度是 500px，那么 Button 增加的宽度也应该占总增加宽度的一半，总增加宽度是 500－100=400，所以这个时候 Button 应该增加的宽度是 400×0.5=200，那么当前 Button 的宽度应该为初始宽度 + 增加宽度（100+200=300）。上述计算过程很简单，其实它就是整型估值器 IntEvaluator 的内部实现，所以我们不用自己写了，直接用吧。

7.3.5 属性动画的工作原理

属性动画要求动画作用的对象提供该属性的 set 方法，属性动画根据你传递的该属性的初始值和最终值，以动画的效果多次去调用 set 方法。每次传递给 set 方法的值都不一样，确切来说是随着时间的推移，所传递的值越来越接近最终值。如果动画的时候没有传递初始值，那么还要提供 get 方法，因为系统要去获取属性的初始值。对于属性动画来说，其动画过程中所做的就是这么多，下面看源码分析。

首先我们要找一个入口，就从 ObjectAnimator.ofInt(mButton, "width", 500).setDuration(5000).start() 开始，其他动画都是类似的。先看 ObjectAnimator 的 start 方法：

```
public void start() {
    // See if any of the current active/pending animators need to be canceled
    AnimationHandler handler = sAnimationHandler.get();
    if (handler != null) {
        int numAnims = handler.mAnimations.size();
        for (int i = numAnims - 1; i >= 0; i--) {
            if (handler.mAnimations.get(i) instanceof ObjectAnimator) {
                ObjectAnimator anim = (ObjectAnimator) handler.mAnimations.get(i);
                if (anim.mAutoCancel && hasSameTargetAndProperties(anim)) {
                    anim.cancel();
                }
            }
        }
        numAnims = handler.mPendingAnimations.size();
        for (int i = numAnims - 1; i >= 0; i--) {
            if (handler.mPendingAnimations.get(i) instanceof ObjectAnimator) {
                ObjectAnimator anim = (ObjectAnimator) handler.mPending-
                    Animations.get(i);
                if (anim.mAutoCancel && hasSameTargetAndProperties(anim)) {
                    anim.cancel();
                }
            }
        }
        numAnims = handler.mDelayedAnims.size();
        for (int i = numAnims - 1; i >= 0; i--) {
            if (handler.mDelayedAnims.get(i) instanceof ObjectAnimator) {
                ObjectAnimator anim = (ObjectAnimator) handler.mDelayed-
```

```
                    Anims.get(i);
                if (anim.mAutoCancel && hasSameTargetAndProperties(anim)) {
                    anim.cancel();
                }
            }
        }
    }
    if (DBG) {
        Log.d(LOG_TAG, "Anim target, duration: " + getTarget() + ", " +
        getDuration());
        for (int i = 0; i < mValues.length; ++i) {
            PropertyValuesHolder pvh = mValues[i];
            Log.d(LOG_TAG, "   Values[" + i + "]: " +
                pvh.getPropertyName() + ", " + pvh.mKeyframes.getValue(0)
                + ", " +
                pvh.mKeyframes.getValue(1));
        }
    }
    super.start();
}
```

上面的代码别看那么长，其实做的事情很简单，首先会判断如果当前动画、等待的动画（Pending）和延迟的动画（Delay）中有和当前动画相同的动画，那么就把相同的动画给取消掉，接下来那一段是 log，再接着就调用了父类的 super.start()方法。因为 ObjectAnimator 继承了 ValueAnimator，所以接下来我们看一下 ValueAnimator 的 Start 方法：

```
private void start(boolean playBackwards) {
    if (Looper.myLooper() == null) {
        throw new AndroidRuntimeException("Animators may only be run on
        Looper threads");
    }
    mPlayingBackwards = playBackwards;
    mCurrentIteration = 0;
    mPlayingState = STOPPED;
    mStarted = true;
    mStartedDelay = false;
    mPaused = false;
    updateScaledDuration(); // in case the scale factor has changed since
    creation time
    AnimationHandler animationHandler = getOrCreateAnimationHandler();
```

```
        animationHandler.mPendingAnimations.add(this);
        if (mStartDelay == 0) {
            // This sets the initial value of the animation, prior to actually
            starting it running
            setCurrentPlayTime(0);
            mPlayingState = STOPPED;
            mRunning = true;
            notifyStartListeners();
        }
        animationHandler.start();
}
```

可以看出属性动画需要运行在有 Looper 的线程中。上述代码最终会调用 Animation-Handler 的 start 方法，这个 AnimationHandler 并不是 Handler，它是一个 Runnable。看一下它的代码，通过代码我们发现，很快就调到了 JNI 层，不过 JNI 层最终还是要调回来的。它的 run 方法会被调用，这个 Runnable 涉及和底层的交互，我们就忽略这部分，直接看重点：ValueAnimator 中的 doAnimationFrame 方法，如下所示。

```
final boolean doAnimationFrame(long frameTime) {
    if (mPlayingState == STOPPED) {
        mPlayingState = RUNNING;
        if (mSeekTime < 0) {
            mStartTime = frameTime;
        } else {
            mStartTime = frameTime - mSeekTime;
            // Now that we're playing, reset the seek time
            mSeekTime = -1;
        }
    }
    if (mPaused) {
        if (mPauseTime < 0) {
            mPauseTime = frameTime;
        }
        return false;
    } else if (mResumed) {
        mResumed = false;
        if (mPauseTime > 0) {
            // Offset by the duration that the animation was paused
            mStartTime += (frameTime - mPauseTime);
```

```
        }
    }
    // The frame time might be before the start time during the first frame of
    // an animation. The "current time" must always be on or after the start
    // time to avoid animating frames at negative time intervals.  In practice,this
    // is very rare and only happens when seeking backwards.
    final long currentTime = Math.max(frameTime, mStartTime);
    return animationFrame(currentTime);
}
```

注意到上述代码末尾调用了 animationFrame 方法，而 animationFrame 内部调用了 animateValue，下面看 animateValue 的代码：

```
void animateValue(float fraction) {
    fraction = mInterpolator.getInterpolation(fraction);
    mCurrentFraction = fraction;
    int numValues = mValues.length;
    for (int i = 0; i < numValues; ++i) {
        mValues[i].calculateValue(fraction);
    }
    if (mUpdateListeners != null) {
        int numListeners = mUpdateListeners.size();
        for (int i = 0; i < numListeners; ++i) {
            mUpdateListeners.get(i).onAnimationUpdate(this);
        }
    }
}
```

上述代码中的 calculateValue 方法就是计算每帧动画所对应的属性的值，下面着重看一下到底是在哪里调用属性的 get 和 set 方法的，毕竟这个才是我们最关心的。

在初始化的时候，如果属性的初始值没有提供，则 get 方法将会被调用，请看 PropertyValuesHolder 的 setupValue 方法，可以发现 get 方法是通过反射来调用的，如下所示。

```
private void setupValue(Object target, Keyframe kf) {
    if (mProperty != null) {
        Object value = convertBack(mProperty.get(target));
        kf.setValue(value);
    }
    try {
        if (mGetter == null) {
```

```
            Class targetClass = target.getClass();
            setupGetter(targetClass);
            if (mGetter == null) {
                // Already logged the error - just return to avoid NPE
                return;
            }
        }
        Object value = convertBack(mGetter.invoke(target));
        kf.setValue(value);
    } catch (InvocationTargetException e) {
        Log.e("PropertyValuesHolder", e.toString());
    } catch (IllegalAccessException e) {
        Log.e("PropertyValuesHolder", e.toString());
    }
}
```

当动画的下一帧到来的时候，PropertyValuesHolder 中的 setAnimatedValue 方法会将新的属性值设置给对象，调用其 set 方法。从下面的源码可以看出，set 方法也是通过反射来调用的：

```
void setAnimatedValue(Object target) {
    if (mProperty != null) {
        mProperty.set(target, getAnimatedValue());
    }
    if (mSetter != null) {
        try {
            mTmpValueArray[0] = getAnimatedValue();
            mSetter.invoke(target, mTmpValueArray);
        } catch (InvocationTargetException e) {
            Log.e("PropertyValuesHolder", e.toString());
        } catch (IllegalAccessException e) {
            Log.e("PropertyValuesHolder", e.toString());
        }
    }
}
```

7.4 使用动画的注意事项

通过动画可以实现一些比较绚丽的效果，但是在使用过程中，也需要注意一些事情，主要分为下面几类。

1．OOM 问题

这个问题主要出现在帧动画中，当图片数量较多且图片较大时就极易出现 OOM，这个在实际的开发中要尤其注意，尽量避免使用帧动画。

2．内存泄露

在属性动画中有一类无限循环的动画，这类动画需要在 Activity 退出时及时停止，否则将导致 Activity 无法释放从而造成内存泄露,通过验证后发现 View 动画并不存在此问题。

3．兼容性问题

动画在 3.0 以下的系统上有兼容性问题，在某些特殊场景可能无法正常工作，因此要做好适配工作。

4．View 动画的问题

View 动画是对 View 的影像做动画，并不是真正地改变 View 的状态，因此有时候会出现动画完成后 View 无法隐藏的现象，即 setVisibility(View.GONE)失效了，这个时候只要调用 view.clearAnimation()清除 View 动画即可解决此问题。

5．不要使用 px

在进行动画的过程中，要尽量使用 dp，使用 px 会导致在不同的设备上有不同的效果。

6．动画元素的交互

将 view 移动（平移）后，在 Android 3.0 以前的系统上，不管是 View 动画还是属性动画，新位置均无法触发单击事件，同时，老位置仍然可以触发单击事件。尽管 View 已经在视觉上不存在了，将 View 移回原位置以后，原位置的单击事件继续生效。从 3.0 开始，属性动画的单击事件触发位置为移动后的位置，但是 View 动画仍然在原位置。

7．硬件加速

使用动画的过程中，建议开启硬件加速，这样会提高动画的流畅性。

第 8 章　理解 Window 和 WindowManager

　　Window 表示一个窗口的概念,在日常开发中直接接触 Window 的机会并不多,但是在某些特殊时候我们需要在桌面上显示一个类似悬浮窗的东西,那么这种效果就需要用到 Window 来实现。Window 是一个抽象类,它的具体实现是 PhoneWindow。创建一个 Window 是很简单的事,只需要通过 WindowManager 即可完成。WindowManager 是外界访问 Window 的入口,Window 的具体实现位于 WindowManagerService 中,WindowManager 和 WindowManagerService 的交互是一个 IPC 过程。Android 中所有的视图都是通过 Window 来呈现的,不管是 Activity、Dialog 还是 Toast,它们的视图实际上都是附加在 Window 上的,因此 Window 实际是 View 的直接管理者。从第 4 章中所讲述的 View 的事件分发机制也可以知道,单击事件由 Window 传递给 DecorView,然后再由 DecorView 传递给我们的 View,就连 Activity 的设置视图的方法 setContentView 在底层也是通过 Window 来完成的。

8.1　Window 和 WindowManager

　　为了分析 Window 的工作机制,我们需要先了解如何使用 WindowManager 添加一个 Window。下面的代码演示了通过 WindowManager 添加 Window 的过程,是不是很简单呢?

```
mFloatingButton = new Button(this);
mFloatingButton.setText("button");
mLayoutParams = new WindowManager.LayoutParams(
```

```
            LayoutParams.WRAP_CONTENT, LayoutParams.WRAP_CONTENT, 0, 0,
            PixelFormat.TRANSPARENT);
mLayoutParams.flags = LayoutParams.FLAG_NOT_TOUCH_MODAL
        | LayoutParams.FLAG_NOT_FOCUSABLE
        | LayoutParams. FLAG_SHOW_WHEN_LOCKED
mLayoutParams.gravity = Gravity.LEFT | Gravity.TOP;
mLayoutParams.x = 100;
mLayoutParams.y = 300;
mWindowManager.addView(mFloatingButton, mLayoutParams);
```

上述代码可以将一个 Button 添加到屏幕坐标为 (100，300) 的位置上。WindowManager. LayoutParams 中的 flags 和 type 这两个参数比较重要，下面对其进行说明。

Flags 参数表示 Window 的属性，它有很多选项，通过这些选项可以控制 Window 的显示特性，这里主要介绍几个比较常用的选项，剩下的请查看官方文档。

FLAG_NOT_FOCUSABLE

表示 Window 不需要获取焦点，也不需要接收各种输入事件，此标记会同时启用 FLAG_NOT_TOUCH_MODAL，最终事件会直接传递给下层的具有焦点的 Window。

FLAG_NOT_TOUCH_MODAL

在此模式下，系统会将当前 Window 区域以外的单击事件传递给底层的 Window，当前 Window 区域以内的单击事件则自己处理。这个标记很重要，一般来说都需要开启此标记，否则其他 Window 将无法收到单击事件。

FLAG_SHOW_WHEN_LOCKED

开启此模式可以让 Window 显示在锁屏的界面上。

Type 参数表示 Window 的类型，Window 有三种类型，分别是应用 Window、子 Window 和系统 Window。应用类 Window 对应着一个 Activity。子 Window 不能单独存在，它需要附属在特定的父 Window 之中，比如常见的一些 Dialog 就是一个子 Window。系统 Window 是需要声明权限才能创建的 Window，比如 Toast 和系统状态栏这些都是系统 Window。

Window 是分层的，每个 Window 都有对应的 z-ordered，层级大的会覆盖在层级小的 Window 的上面，这和 HTML 中的 z-index 的概念是完全一致的。在三类 Window 中，应用

Window 的层级范围是 1~99，子 Window 的层级范围是 1000~1999，系统 Window 的层级范围是 2000~2999，这些层级范围对应着 WindowManager.LayoutParams 的 type 参数。如果想要 Window 位于所有 Window 的最顶层，那么采用较大的层级即可。很显然系统 Window 的层级是最大的，而且系统层级有很多值，一般我们可以选用 TYPE_SYSTEM_OVERLAY 或者 TYPE_SYSTEM_ERROR，如果采用 TYPE_SYSTEM_ERROR，只需要为 type 参数指定这个层级即可：mLayoutParams.type = LayoutParams.TYPE_SYSTEM_ERROR；同时声明权限：<uses-permission android:name="android.permission.SYSTEM_ALERT_WINDOW" />。因为系统类型的 Window 是需要检查权限的，如果不在 AndroidManifest 中使用相应的权限，那么创建 Window 的时候就会报错，错误如下所示。

```
E/AndroidRuntime(8071): Caused by: android.view.WindowManager$BadToken-
Exception: Unable to add window android.view.ViewRootImpl$W@42882fe8 --
permission denied for this window type
E/AndroidRuntime(8071): at android.view.ViewRootImpl.setView(ViewRootImpl.
java:677)
E/AndroidRuntime(8071): at android.view.WindowManagerImpl.addView(Window-
ManagerImpl.java:326)
E/AndroidRuntime(8071): at android.view.WindowManagerImpl.addView(Window-
ManagerImpl.java:224)
E/AndroidRuntime(8071):   at   android.view.WindowManagerImpl$CompatMode-
Wrapper.addView(WindowManagerImpl.java:149)
E/AndroidRuntime(8071): at android.view.Window$LocalWindowManager.addView
(Window.java:558)
E/AndroidRuntime(8071):  at  com.ryg.chapter_8.TestActivity.onButtonClick
(TestActivity.java:60)
E/AndroidRuntime(8071): ... 14 more
W/ActivityManager( 514):   Force finishing activity com.ryg.chapter_8/.
TestActivity
```

WindowManager 所提供的功能很简单，常用的只有三个方法，即添加 View、更新 View 和删除 View，这三个方法定义在 ViewManager 中，而 WindowManager 继承了 ViewManager。

```
public interface ViewManager
{
    public void addView(View view, ViewGroup.LayoutParams params);
    public void updateViewLayout(View view, ViewGroup.LayoutParams params);
    public void removeView(View view);
}
```

对开发者来说，WindowManager 常用的就只有这三个功能而已，但是这三个功能已经足够我们使用了。它可以创建一个 Window 并向其添加 View，还可以更新 Window 中的 View，另外如果想要删除一个 Window，那么只需要删除它里面的 View 即可。由此来看，WindowManager 操作 Window 的过程更像是在操作 Window 中的 View。我们时常见到那种可以拖动的 Window 效果，其实是很好实现的，只需要根据手指的位置来设定 LayoutParams 中的 x 和 y 的值即可改变 Window 的位置。首先给 View 设置 onTouchListener：mFloatingButton.setOnTouchListener(this)。然后在 onTouch 方法中不断更新 View 的位置即可：

```
public boolean onTouch(View v, MotionEvent event) {
    int rawX = (int) event.getRawX();
    int rawY = (int) event.getRawY();
    switch (event.getAction()) {
    case MotionEvent.ACTION_MOVE: {
        mLayoutParams.x = rawX;
        mLayoutParams.y = rawY;
        mWindowManager.updateViewLayout(mFloatingButton, mLayoutParams);
        break;
    }
    default:
        break;
    }

    return false;
}
```

8.2 Window 的内部机制

Window 是一个抽象的概念，每一个 Window 都对应着一个 View 和一个 ViewRootImpl，Window 和 View 通过 ViewRootImpl 来建立联系，因此 Window 并不是实际存在的，它是以 View 的形式存在。这点从 WindowManager 的定义也可以看出，它提供的三个接口方法 addView、updateViewLayout 以及 removeView 都是针对 View 的，这说明 View 才是 Window 存在的实体。在实际使用中无法直接访问 Window，对 Window 的访问必须通过 WindowManager。为了分析 Window 的内部机制，这里从 Window 的添加、删除以及更新说起。

8.2.1 Window 的添加过程

Window 的添加过程需要通过 WindowManager 的 addView 来实现，WindowManager 是一个接口，它的真正实现是 WindowManagerImpl 类。在 WindowManagerImpl 中 Window 的三大操作的实现如下：

```
@Override
public void addView(View view, ViewGroup.LayoutParams params) {
    mGlobal.addView(view, params, mDisplay, mParentWindow);
}

@Override
public void updateViewLayout(View view, ViewGroup.LayoutParams params) {
    mGlobal.updateViewLayout(view, params);
}

@Override
public void removeView(View view) {
    mGlobal.removeView(view, false);
}
```

可以发现，WindowManagerImpl 并没有直接实现 Window 的三大操作，而是全部交给了 WindowManagerGlobal 来处理，WindowManagerGlobal 以工厂的形式向外提供自己的实例，在 WindowManagerGlobal 中有如下一段代码：private final WindowManagerGlobal mGlobal = WindowManagerGlobal.getInstance()。WindowManagerImpl 这种工作模式是典型的桥接模式，将所有的操作全部委托给 WindowManagerGlobal 来实现。WindowManager-Global 的 addView 方法主要分为如下几步。

1. 检查参数是否合法，如果是子 Window 那么还需要调整一些布局参数

```
if (view == null) {
    throw new IllegalArgumentException("view must not be null");
}
if (display == null) {
    throw new IllegalArgumentException("display must not be null");
}
if (!(params instanceof WindowManager.LayoutParams)) {
```

```
    throw new IllegalArgumentException("Params must be WindowManager.
    LayoutParams");
}

final WindowManager.LayoutParams wparams = (WindowManager.LayoutParams)
params;
if (parentWindow != null) {
    parentWindow.adjustLayoutParamsForSubWindow(wparams);
}
```

2. 创建 ViewRootImpl 并将 View 添加到列表中

在 WindowManagerGlobal 内部有如下几个列表比较重要：

```
    private final ArrayList<View> mViews = new ArrayList<View>();
    private final ArrayList<ViewRootImpl> mRoots = new ArrayList
    <ViewRootImpl>();
    private final ArrayList<WindowManager.LayoutParams> mParams =
        new ArrayList<WindowManager.LayoutParams>();
private final ArraySet<View> mDyingViews = new ArraySet<View>();
```

在上面的声明中，mViews 存储的是所有 Window 所对应的 View，mRoots 存储的是所有 Window 所对应的 ViewRootImpl，mParams 存储的是所有 Window 所对应的布局参数，而 mDyingViews 则存储了那些正在被删除的 View 对象，或者说是那些已经调用 removeView 方法但是删除操作还未完成的 Window 对象。在 addView 中通过如下方式将 Window 的一系列对象添加到列表中：

```
root = new ViewRootImpl(view.getContext(), display);
view.setLayoutParams(wparams);

mViews.add(view);
mRoots.add(root);
mParams.add(wparams);
```

3. 通过 ViewRootImpl 来更新界面并完成 Window 的添加过程

这个步骤由 ViewRootImpl 的 setView 方法来完成，从第 4 章可以知道，View 的绘制过程是由 ViewRootImpl 来完成的，这里当然也不例外，在 setView 内部会通过 requestLayout 来完成异步刷新请求。在下面的代码中，scheduleTraversals 实际是 View 绘制的入口：

```java
public void requestLayout() {
    if (!mHandlingLayoutInLayoutRequest) {
        checkThread();
        mLayoutRequested = true;
        scheduleTraversals();
    }
}
```

接着会通过 WindowSession 最终来完成 Window 的添加过程。在下面的代码中，mWindowSession 的类型是 IWindowSession，它是一个 Binder 对象，真正的实现类是 Session，也就是 Window 的添加过程是一次 IPC 调用。

```java
try {
    mOrigWindowType = mWindowAttributes.type;
    mAttachInfo.mRecomputeGlobalAttributes = true;
    collectViewAttributes();
    res = mWindowSession.addToDisplay(mWindow, mSeq, mWindowAttributes,
            getHostVisibility(), mDisplay.getDisplayId(),
            mAttachInfo.mContentInsets, mInputChannel);
} catch (RemoteException e) {
    mAdded = false;
    mView = null;
    mAttachInfo.mRootView = null;
    mInputChannel = null;
    mFallbackEventHandler.setView(null);
    unscheduleTraversals();
    setAccessibilityFocus(null, null);
    throw new RuntimeException("Adding window failed", e);
}
```

在 Session 内部会通过 WindowManagerService 来实现 Window 的添加，代码如下所示。

```java
public int addToDisplay(IWindow window, int seq, WindowManager.LayoutParams attrs,
        int viewVisibility, int displayId, Rect outContentInsets,
        InputChannel outInputChannel) {
    return mService.addWindow(this, window, seq, attrs, viewVisibility,
        displayId,
            outContentInsets, outInputChannel);
}
```

如此一来，Window 的添加请求就交给 WindowManagerService 去处理了，在 Window-ManagerService 内部会为每一个应用保留一个单独的 Session。具体 Window 在 Window-ManagerService 内部是怎么添加的，本章不对其进行进一步的分析，这是因为到此为止我们对 Window 的添加这一流程已经清楚了，而在 WindowManagerService 内部主要是代码细节，深入进去没有太大的意义，读者可以自行阅读源码或者参考相关的源码分析书籍，本书对源码的分析侧重的是整体流程，会尽量避免出现深入代码逻辑无法自拔的情形。

8.2.2　Window 的删除过程

Window 的删除过程和添加过程一样，都是先通过 WindowManagerImpl 后，再进一步通过 WindowManagerGlobal 来实现的。下面是 WindowManagerGlobal 的 removeView 的实现：

```
public void removeView(View view, boolean immediate) {
    if (view == null) {
        throw new IllegalArgumentException("view must not be null");
    }

    synchronized (mLock) {
        int index = findViewLocked(view, true);
        View curView = mRoots.get(index).getView();
        removeViewLocked(index, immediate);
        if (curView == view) {
            return;
        }

        throw new IllegalStateException("Calling with view " + view
                + " but the ViewAncestor is attached to " + curView);
    }
}
```

removeView 的逻辑很清晰，首先通过 findViewLocked 来查找待删除的 View 的索引，这个查找过程就是建立的数组遍历，然后再调用 removeViewLocked 来做进一步的删除，如下所示。

```
private void removeViewLocked(int index, boolean immediate) {
    ViewRootImpl root = mRoots.get(index);
    View view = root.getView();
```

```
        if (view != null) {
            InputMethodManager imm = InputMethodManager.getInstance();
            if (imm != null) {
                imm.windowDismissed(mViews.get(index).getWindowToken());
            }
        }
        boolean deferred = root.die(immediate);
        if (view != null) {
            view.assignParent(null);
            if (deferred) {
                mDyingViews.add(view);
            }
        }
    }
```

removeViewLocked 是通过 ViewRootImpl 来完成删除操作的。在 WindowManager 中提供了两种删除接口 removeView 和 removeViewImmediate，它们分别表示异步删除和同步删除，其中 removeViewImmediate 使用起来需要特别注意，一般来说不需要使用此方法来删除 Window 以免发生意外的错误。这里主要说异步删除的情况，具体的删除操作由 ViewRootImpl 的 die 方法来完成。在异步删除的情况下，die 方法只是发送了一个请求删除的消息后就立刻返回了，这个时候 View 并没有完成删除操作，所以最后会将其添加到 mDyingViews 中，mDyingViews 表示待删除的 View 列表。ViewRootImpl 的 die 方法如下所示。

```
boolean die(boolean immediate) {
    // Make sure we do execute immediately if we are in the middle of a traversal
    //   or the damage
    // done by dispatchDetachedFromWindow will cause havoc on return.
    if (immediate && !mIsInTraversal) {
        doDie();
        return false;
    }

    if (!mIsDrawing) {
        destroyHardwareRenderer();
    } else {
        Log.e(TAG, "Attempting to destroy the window while drawing!\n" +
                "  window=" + this + ", title=" + mWindowAttributes.
            getTitle());
```

```
    }
    mHandler.sendEmptyMessage(MSG_DIE);
    return true;
}
```

在 die 方法内部只是做了简单的判断，如果是异步删除，那么就发送一个 MSG_DIE 的消息，ViewRootImpl 中的 Handler 会处理此消息并调用 doDie 方法，如果是同步删除（立即删除），那么就不发消息直接调用 doDie 方法，这就是这两种删除方式的区别。在 doDie 内部会调用 dispatchDetachedFromWindow 方法，真正删除 View 的逻辑在 dispatchDetachedFromWindow 方法的内部实现。dispatchDetachedFromWindow 方法主要做四件事：

（1）垃圾回收相关的工作，比如清除数据和消息、移除回调。

（2）通过 Session 的 remove 方法删除 Window：mWindowSession.remove(mWindow)，这同样是一个 IPC 过程，最终会调用 WindowManagerService 的 removeWindow 方法。

（3）调用 View 的 dispatchDetachedFromWindow 方法，在内部会调用 View 的 onDetachedFromWindow()以及 onDetachedFromWindowInternal()。对于 onDetachedFromWindow()大家一定不陌生，当 View 从 Window 中移除时，这个方法就会被调用，可以在这个方法内部做一些资源回收的工作，比如终止动画、停止线程等。

（4）调用 WindowManagerGlobal 的 doRemoveView 方法刷新数据，包括 mRoots、mParams 以及 mDyingViews，需要将当前 Window 所关联的这三类对象从列表中删除。

8.2.3　Window 的更新过程

到这里，Window 的删除过程已经分析完毕了，下面分析 Window 的更新过程，还是要看 WindowManagerGlobal 的 updateViewLayout 方法，如下所示。

```
public void updateViewLayout(View view, ViewGroup.LayoutParams params) {
    if (view == null) {
        throw new IllegalArgumentException("view must not be null");
    }
    if (!(params instanceof WindowManager.LayoutParams)) {
        throw new IllegalArgumentException("Params must be WindowManager.
            LayoutParams");
    }
```

```
    final WindowManager.LayoutParams wparams = (WindowManager.Layout-
Params)params;

view.setLayoutParams(wparams);

synchronized (mLock) {
    int index = findViewLocked(view, true);
    ViewRootImpl root = mRoots.get(index);
    mParams.remove(index);
    mParams.add(index, wparams);
    root.setLayoutParams(wparams, false);
}
}
```

updateViewLayout 方法做的事情就比较简单了，首先它需要更新 View 的 LayoutParams 并替换掉老的 LayoutParams，接着再更新 ViewRootImpl 中的 LayoutParams，这一步是通过 ViewRootImpl 的 setLayoutParams 方法来实现的。在 ViewRootImpl 中会通过 scheduleTraversals 方法来对 View 重新布局，包括测量、布局、重绘这三个过程。除了 View 本身的重绘以外，ViewRootImpl 还会通过 WindowSession 来更新 Window 的视图，这个过程最终是由 WindowManagerService 的 relayoutWindow() 来具体实现的，它同样是一个 IPC 过程。

8.3 Window 的创建过程

通过上面的分析可以看出，View 是 Android 中的视图的呈现方式，但是 View 不能单独存在，它必须附着在 Window 这个抽象的概念上面，因此有视图的地方就有 Window。哪些地方有视图呢？这个读者都比较清楚，Android 中可以提供视图的地方有 Activity、Dialog、Toast，除此之外，还有一些依托 Window 而实现的视图，比如 PopUpWindow、菜单，它们也是视图，有视图的地方就有 Window，因此 Activity、Dialog、Toast 等视图都对应着一个 Window。本节将分析这些视图元素中的 Window 的创建过程，通过本节可以使读者进一步加深对 Window 的理解。

8.3.1 Activity 的 Window 创建过程

要分析 Activity 中的 Window 的创建过程就必须了解 Activity 的启动过程，详细的过程

第 8 章　理解 Window 和 WindowManager

会在第 9 章进行介绍，这里先大概了解即可。Activity 的启动过程很复杂，最终会由 ActivityThread 中的 performLaunchActivity()来完成整个启动过程，在这个方法内部会通过类加载器创建 Activity 的实例对象，并调用其 attach 方法为其关联运行过程中所依赖的一系列上下文环境变量，代码如下所示。

```
java.lang.ClassLoader cl = r.packageInfo.getClassLoader();
activity = mInstrumentation.newActivity(
        cl, component.getClassName(), r.intent);
...
if (activity != null) {
    Context appContext = createBaseContextForActivity(r, activity);
    CharSequence title = r.activityInfo.loadLabel(appContext.getPackage-
    Manager());
    Configuration config = new Configuration(mCompatConfiguration);
    if (DEBUG_CONFIGURATION) Slog.v(TAG, "Launching activity "
            + r.activityInfo.name + " with config " + config);
    activity.attach(appContext, this, getInstrumentation(), r.token,
            r.ident, app, r.intent, r.activityInfo, title, r.parent,
            r.embeddedID, r.lastNonConfigurationInstances, config,
            r.voiceInteractor);
    ...
}
```

在 Activity 的 attach 方法里，系统会创建 Activity 所属的 Window 对象并为其设置回调接口，Window 对象的创建是通过 PolicyManager 的 makeNewWindow 方法实现的。由于 Activity 实现了 Window 的 Callback 接口，因此当 Window 接收到外界的状态改变时就会回调 Activity 的方法。Callback 接口中的方法很多，但是有几个却是我们都非常熟悉的，比如 onAttachedToWindow、onDetachedFromWindow、dispatchTouchEvent，等等，代码如下所示。

```
mWindow = PolicyManager.makeNewWindow(this);
mWindow.setCallback(this);
mWindow.setOnWindowDismissedCallback(this);
mWindow.getLayoutInflater().setPrivateFactory(this);
if (info.softInputMode != WindowManager.LayoutParams.SOFT_INPUT_STATE_
UNSPECIFIED) {
    mWindow.setSoftInputMode(info.softInputMode);
}
if (info.uiOptions != 0) {
    mWindow.setUiOptions(info.uiOptions);
```

}

从上面的分析可以看出，Activity 的 Window 是通过 PolicyManager 的一个工厂方法来创建的，但是从 PolicyManager 的类名可以看出，它不是一个普通的类，它是一个策略类。PolicyManager 中实现的几个工厂方法全部在策略接口 IPolicy 中声明了，IPolicy 的定义如下：

```
public interface IPolicy {
    public Window makeNewWindow(Context context);
    public LayoutInflater makeNewLayoutInflater(Context context);
    public WindowManagerPolicy makeNewWindowManager();
    public FallbackEventHandler makeNewFallbackEventHandler(Context context);
}
```

在实际的调用中，PolicyManager 的真正实现是 Policy 类，Policy 类中的 makeNewWindow 方法的实现如下，由此可以发现，Window 的具体实现的确是 PhoneWindow。

```
public Window makeNewWindow(Context context) {
    return new PhoneWindow(context);
}
```

关于策略类 PolicyManager 是如何关联到 Policy 上面的，这个无法从源码中的调用关系来得出，这里猜测可能是由编译环节动态控制的。到这里 Window 已经创建完成了，下面分析 Activity 的视图是怎么附属在 Window 上的。由于 Activity 的视图由 setContentView 方法提供，我们只需要看 setContentView 方法的实现即可。

```
public void setContentView(int layoutResID) {
    getWindow().setContentView(layoutResID);
    initWindowDecorActionBar();
}
```

从 Activity 的 setContentView 的实现可以看出，Activity 将具体实现交给了 Window 处理，而 Window 的具体实现是 PhoneWindow，所以只需要看 PhoneWindow 的相关逻辑即可。PhoneWindow 的 setContentView 方法大致遵循如下几个步骤。

1. 如果没有 DecorView，那么就创建它

DecorView 是一个 FrameLayout，在第 4 章已经做了初步的介绍，这里再简单说一下。DecorView 是 Activity 中的顶级 View，一般来说它的内部包含标题栏和内部栏，但是这个

会随着主题的变换而发生改变。不管怎么样，内容栏是一定要存在的，并且内容栏具体固定的 id，那就是 "content"，它的完整 id 是 android.R.id.content。DecorView 的创建过程由 installDecor 方法来完成，在方法内部会通过 generateDecor 方法来直接创建 DecorView，这个时候 DecorView 还只是一个空白的 FrameLayout：

```
protected DecorView generateDecor() {
    return new DecorView(getContext(), -1);
}
```

为了初始化 DecorView 的结构，PhoneWindow 还需要通过 generateLayout 方法来加载具体的布局文件到 DecorView 中，具体的布局文件和系统版本以及主题有关，这个过程如下所示。

```
View in = mLayoutInflater.inflate(layoutResource, null);
decor.addView(in, new ViewGroup.LayoutParams(MATCH_PARENT, MATCH_PARENT));
mContentRoot = (ViewGroup) in;
ViewGroup contentParent = (ViewGroup)findViewById(ID_ANDROID_CONTENT);
```

其中 ID_ANDROID_CONTENT 的定义如下，这个 id 所对应的 ViewGroup 就是 mContentParent：

```
public static final int ID_ANDROID_CONTENT = com.android.internal.R.id.content
```

2. 将 View 添加到 DecorView 的 mContentParent 中

这个过程就比较简单了，由于在步骤 1 中已经创建并初始化了 DecorView，因此这一步直接将 Activity 的视图添加到 DecorView 的 mContentParent 中即可：mLayoutInflater.inflate(layoutResID, mContentParent)。到此为止，Activity 的布局文件已经添加到 DecorView 里面了，由此可以理解 Activity 的 setContentView 这个方法的来历了。不知道读者是否曾经怀疑过：为什么不叫 setView 呢？它明明是给 Activity 设置视图的啊！从这里来看，它的确不适合叫 setView，因为 Activity 的布局文件只是被添加到 DecorView 的 mContentParent 中，因此叫 setContentView 更加准确。

3. 回调 Activity 的 onContentChanged 方法通知 Activity 视图已经发生改变

这个过程就更简单了，由于 Activity 实现了 Window 的 Callback 接口，这里表示 Activity 的布局文件已经被添加到 DecorView 的 mContentParent 中了，于是需要通知 Activity，使其可以做相应的处理。Activity 的 onContentChanged 方法是个空实现，我们可以在子 Activity

中处理这个回调。这个过程的代码如下所示。

```
final Callback cb = getCallback();
if (cb != null && !isDestroyed()) {
    cb.onContentChanged();
}
```

经过了上面的三个步骤，到这里为止 DecorView 已经被创建并初始化完毕，Activity 的布局文件也已经成功添加到了 DecorView 的 mContentParent 中，但是这个时候 DecorView 还没有被 WindowManager 正式添加到 Window 中。这里需要正确理解 Window 的概念，Window 更多表示的是一种抽象的功能集合，虽然说早在 Activity 的 attach 方法中 Window 就已经被创建了，但是这个时候由于 DecorView 并没有被 WindowManager 识别，所以这个时候的 Window 无法提供具体功能，因为它还无法接收外界的输入信息。在 ActivityThread 的 handleResumeActivity 方法中，首先会调用 Activity 的 onResume 方法，接着会调用 Activity 的 makeVisible()，正是在 makeVisible 方法中，DecorView 真正地完成了添加和显示这两个过程，到这里 Activity 的视图才能被用户看到，如下所示。

```
void makeVisible() {
    if (!mWindowAdded) {
        ViewManager wm = getWindowManager();
        wm.addView(mDecor, getWindow().getAttributes());
        mWindowAdded = true;
    }
    mDecor.setVisibility(View.VISIBLE);
}
```

到这里，Activity 中的 Window 的创建过程已经分析完了，读者对整个过程是不是有了更进一步的理解了呢？

8.3.2 Dialog 的 Window 创建过程

Dialog 的 Window 的创建过程和 Activity 类似，有如下几个步骤。

1. 创建 Window

Dialog 中 Window 的创建同样是通过 PolicyManager 的 makeNewWindow 方法来完成的，从 8.3.1 节中可以知道，创建后的对象实际上就是 PhoneWindow，这个过程和 Activity 的 Window 的创建过程是一致的，这里就不再详细说明了。

```
Dialog(Context context, int theme, boolean createContextThemeWrapper) {
    ...
    mWindowManager = (WindowManager)context.getSystemService(Context.
    WINDOW_SERVICE);
    Window w = PolicyManager.makeNewWindow(mContext);
    mWindow = w;
    w.setCallback(this);
    w.setOnWindowDismissedCallback(this);
    w.setWindowManager(mWindowManager, null, null);
    w.setGravity(Gravity.CENTER);
    mListenersHandler = new ListenersHandler(this);
}
```

2. 初始化 DecorView 并将 Dialog 的视图添加到 DecorView 中

这个过程也和 Activity 的类似，都是通过 Window 去添加指定的布局文件。

```
public void setContentView(int layoutResID) {
    mWindow.setContentView(layoutResID);
}
```

3. 将 DecorView 添加到 Window 中并显示

在 Dialog 的 show 方法中，会通过 WindowManager 将 DecorView 添加到 Window 中，如下所示。

```
mWindowManager.addView(mDecor, l);
mShowing = true;
```

从上面三个步骤可以发现，Dialog 的 Window 创建和 Activity 的 Window 创建过程很类似，二者几乎没有什么区别。当 Dialog 被关闭时，它会通过 WindowManager 来移除 DecorView：mWindowManager.removeViewImmediate(mDecor)。

普通的 Dialog 有一个特殊之处，那就是必须采用 Activity 的 Context，如果采用 Application 的 Context，那么就会报错。

```
Dialog dialog = new Dialog(this.getApplicationContext());
TextView textView = new TextView(this);
textView.setText("this is toast!");
dialog.setContentView(textView);
dialog.show();
```

上述代码运行时会报错，错误信息如下所示。

```
E/AndroidRuntime(1185): Caused by: android.view.WindowManager$BadToken-
Exception: Unable to add window -- token null is not for an application
E/AndroidRuntime(1185): at android.view.ViewRootImpl.setView(ViewRoot-
Impl.java:657)
E/AndroidRuntime(1185): at android.view.WindowManagerImpl.addView(Window-
ManagerImpl.java:326)
E/AndroidRuntime(1185): at android.view.WindowManagerImpl.addView(Window-
ManagerImpl.java:224)
E/AndroidRuntime(1185): at android.view.WindowManagerImpl$CompatMode-
Wrapper.addView(WindowManagerImpl.java:149)
E/AndroidRuntime(1185): at android.app.Dialog.show(Dialog.java:316)
E/AndroidRuntime(1185): at com.ryg.chapter_8.DemoActivity_1.initView
(DemoActivity_1.java:26)
E/AndroidRuntime(1185): at com.ryg.chapter_8.DemoActivity_1.onCreate
(DemoActivity_1.java:18)
E/AndroidRuntime(1185): at android.app.Activity.performCreate(Activity.
java:5086)
E/AndroidRuntime(1185): at android.app.Instrumentation.callActivityOn-
Create(Instrumentation.java:1079)
E/AndroidRuntime(1185): at android.app.ActivityThread.performLaunch-
Activity(ActivityThread.java:2056)
```

上面的错误信息很明确，是没有应用 token 所导致的，而应用 token 一般只有 Activity 拥有，所以这里只需要用 Activity 作为 Context 来显示对话框即可。另外，系统 Window 比较特殊，它可以不需要 token，因此在上面的例子中，只需要指定对话框的 Window 为系统类型就可以正常弹出对话框。在本章一开始讲到，WindowManager.LayoutParams 中的 type 表示 Window 的类型，而系统 Window 的层级范围是 2000～2999，这些层级范围就对应着 type 参数。系统 Window 的层级有很多值，对于本例来说，可以选用 TYPE_SYSTEM_OVERLAY 来指定对话框的 Window 类型为系统 Window，如下所示。

```
dialog.getWindow().setType(LayoutParams.TYPE_SYSTEM_ERROR)
```

然后别忘了在 AndroidManifest 文件中声明权限从而可以使用系统 Window，如下所示。

```
<uses-permission android:name="android.permission.SYSTEM_ALERT_WINDOW" />
```

8.3.3 Toast 的 Window 创建过程

Toast 和 Dialog 不同，它的工作过程就稍显复杂。首先 Toast 也是基于 Window 来实现的，但是由于 Toast 具有定时取消这一功能，所以系统采用了 Handler。在 Toast 的内部有两类 IPC 过程，第一类是 Toast 访问 NotificationManagerService，第二类是 NotificationManagerService 回调 Toast 里的 TN 接口。关于 IPC 的一些知识，请读者参考第 2 章的相关内容。为了便于描述，下面将 NotificationManagerService 简称为 NMS。

Toast 属于系统 Window，它内部的视图由两种方式指定，一种是系统默认的样式，另一种是通过 setView 方法来指定一个自定义 View，不管如何，它们都对应 Toast 的一个 View 类型的内部成员 mNextView。Toast 提供了 show 和 cancel 分别用于显示和隐藏 Toast，它们的内部是一个 IPC 过程，show 方法和 cancel 方法的实现如下：

```java
public void show() {
    if (mNextView == null) {
        throw new RuntimeException("setView must have been called");
    }

    INotificationManager service = getService();
    String pkg = mContext.getOpPackageName();
    TN tn = mTN;
    tn.mNextView = mNextView;

    try {
        service.enqueueToast(pkg, tn, mDuration);
    } catch (RemoteException e) {
        // Empty
    }
}

public void cancel() {
    mTN.hide();

    try {
        getService().cancelToast(mContext.getPackageName(), mTN);
    } catch (RemoteException e) {
        // Empty
    }
}
```

从上面的代码可以看到，显示和隐藏 Toast 都需要通过 NMS 来实现，由于 NMS 运行在系统的进程中，所以只能通过远程调用的方式来显示和隐藏 Toast。需要注意的是 TN 这个类，它是一个 Binder 类，在 Toast 和 NMS 进行 IPC 的过程中，当 NMS 处理 Toast 的显示或隐藏请求时会跨进程回调 TN 中的方法，这个时候由于 TN 运行在 Binder 线程池中，所以需要通过 Handler 将其切换到当前线程中。这里的当前线程是指发送 Toast 请求所在的线程。注意，由于这里使用了 Handler，所以这意味着 Toast 无法在没有 Looper 的线程中弹出，这是因为 Handler 需要使用 Looper 才能完成切换线程的功能，关于 Handler 和 Looper 的具体介绍请参看第 10 章。

首先看 Toast 的显示过程，它调用了 NMS 中的 enqueueToast 方法，如下所示。

```
INotificationManager service = getService();
String pkg = mContext.getOpPackageName();
TN tn = mTN;
tn.mNextView = mNextView;
try {
    service.enqueueToast(pkg, tn, mDuration);
} catch (RemoteException e) {
    // Empty
}
```

NMS 的 enqueueToast 方法的第一个参数表示当前应用的包名，第二个参数 tn 表示远程回调，第三个参数表示 Toast 的时长。enqueueToast 首先将 Toast 请求封装为 ToastRecord 对象并将其添加到一个名为 mToastQueue 的队列中。mToastQueue 其实是一个 ArrayList。对于非系统应用来说，mToastQueue 中最多能同时存在 50 个 ToastRecord，这样做是为了防止 DOS（Denial of Service）。如果不这么做，试想一下，如果我们通过大量的循环去连续弹出 Toast，这将会导致其他应用没有机会弹出 Toast，那么对于其他应用的 Toast 请求，系统的行为就是拒绝服务，这就是拒绝服务攻击的含义，这种手段常用于网络攻击中。

```
// Limit the number of toasts that any given package except the android
// package can enqueue.  Prevents DOS attacks and deals with leaks.
if (!isSystemToast) {
    int count = 0;
    final int N = mToastQueue.size();
    for (int i=0; i<N; i++) {
        final ToastRecord r = mToastQueue.get(i);
        if (r.pkg.equals(pkg)) {
```

```
            count++;
            if (count >= MAX_PACKAGE_NOTIFICATIONS) {
                Slog.e(TAG, "Package has already posted " + count
                        + " toasts. Not showing more. Package=" + pkg);
                return;
            }
        }
    }
}
```

正常情况下，一个应用不可能达到上限，当 ToastRecord 被添加到 mToastQueue 中后，NMS 就会通过 showNextToastLocked 方法来显示当前的 Toast。下面的代码很好理解，需要注意的是，Toast 的显示是由 ToastRecord 的 callback 来完成的，这个 callback 实际上就是 Toast 中的 TN 对象的远程 Binder，通过 callback 来访问 TN 中的方法是需要跨进程来完成的，最终被调用的 TN 中的方法会运行在发起 Toast 请求的应用的 Binder 线程池中。

```
void showNextToastLocked() {
    ToastRecord record = mToastQueue.get(0);
    while (record != null) {
        if (DBG) Slog.d(TAG, "Show pkg=" + record.pkg + " callback=" + record.callback);
        try {
            record.callback.show();
            scheduleTimeoutLocked(record);
            return;
        } catch (RemoteException e) {
            Slog.w(TAG, "Object died trying to show notification " + record.callback
                    + " in package " + record.pkg);
            // remove it from the list and let the process die
            int index = mToastQueue.indexOf(record);
            if (index >= 0) {
                mToastQueue.remove(index);
            }
            keepProcessAliveLocked(record.pid);
            if (mToastQueue.size() > 0) {
                record = mToastQueue.get(0);
            } else {
                record = null;
```

            }
        }
    }
}
```

Toast 显示以后，NMS 还会通过 scheduleTimeoutLocked 方法来发送一个延时消息，具体的延时取决于 Toast 的时长，如下所示。

```
private void scheduleTimeoutLocked(ToastRecord r)
{
    mHandler.removeCallbacksAndMessages(r);
    Message m = Message.obtain(mHandler, MESSAGE_TIMEOUT, r);
    long delay = r.duration == Toast.LENGTH_LONG ? LONG_DELAY : SHORT_DELAY;
    mHandler.sendMessageDelayed(m, delay);
}
```

在上面的代码中，LONG_DELAY 是 3.5s，而 SHORT_DELAY 是 2s。延迟相应的时间后，NMS 会通过 cancelToastLocked 方法来隐藏 Toast 并将其从 mToastQueue 中移除，这个时候如果 mToastQueue 中还有其他 Toast，那么 NMS 就继续显示其他 Toast。

Toast 的隐藏也是通过 ToastRecord 的 callback 来完成的，这同样也是一次 IPC 过程，它的工作方式和 Toast 的显示过程是类似的，如下所示。

```
try {
    record.callback.hide();
} catch (RemoteException e) {
    Slog.w(TAG, "Object died trying to hide notification " + record.callback
            + " in package " + record.pkg);
    // don't worry about this, we're about to remove it from
    // the list anyway
}
```

通过上面的分析，大家知道 Toast 的显示和影响过程实际上是通过 Toast 中的 TN 这个类来实现的，它有两个方法 show 和 hide，分别对应 Toast 的显示和隐藏。由于这两个方法是被 NMS 以跨进程的方式调用的，因此它们运行在 Binder 线程池中。为了将执行环境切换到 Toast 请求所在的线程，在它们的内部使用了 Handler，如下所示。

```
/**
 * schedule handleShow into the right thread
 */
```
```

第 8 章 理解 Window 和 WindowManager

```
@Override
public void show() {
    if (localLOGV) Log.v(TAG, "SHOW: " + this);
    mHandler.post(mShow);
}

/**
 * schedule handleHide into the right thread
 */
@Override
public void hide() {
    if (localLOGV) Log.v(TAG, "HIDE: " + this);
    mHandler.post(mHide);
}
```

上述代码中，mShow 和 mHide 是两个 Runnable，它们内部分别调用了 handleShow 和 handleHide 方法。由此可见，handleShow 和 handleHide 才是真正完成显示和隐藏 Toast 的地方。TN 的 handleShow 中会将 Toast 的视图添加到 Window 中，如下所示。

```
mWM = (WindowManager)context.getSystemService(Context.WINDOW_SERVICE);
mWM.addView(mView, mParams)
```

而 NT 的 handleHide 中会将 Toast 的视图从 Window 中移除，如下所示。

```
if (mView.getParent() != null) {
    if (localLOGV) Log.v(TAG, "REMOVE! " + mView + " in " + this);
    mWM.removeView(mView);
}
```

到这里 Toast 的 Window 的创建过程已经分析完了，相信读者对 Toast 的工作过程有了一个更加全面的理解了。除了上面已经提到的 Activity、Dialog 和 Toast 以外，PopupWindow、菜单以及状态栏等都是通过 Window 来实现的，这里就不一一介绍了，读者可以找自己感兴趣的内容来分析。

本章的意义在于让读者对 Window 有一个更加清晰的认识，同时能够深刻理解 Window 和 View 的依赖关系，这有助于理解其他更深层次的概念，比如 SurfaceFlinger。通过本章读者应该知道，任何 View 都是附属在一个 Window 上面的，那么这里问一个问题：一个应用中到底有多少个 Window 呢？相信读者都已经清楚了。

第 9 章 四大组件的工作过程

本章讲述 Android 中的四大组件的工作过程。说到四大组件，开发者都再熟悉不过了，它们是 Activity、Service、BroadcastReceiver 和 ContentProvider。如何使用四大组件，这不是本章关心的，毕竟这是开发者都熟悉的内容，本章按照如下的逻辑来分析 Android 的四大组件：首先会对四大组件的运行状态和工作方式做一个概括化的描述，接着对四大组件的工作过程进行分析，通过本章的分析读者可以对四大组件有一个更深刻的认识。

本章主要侧重于四大组件工作过程的分析，通过分析它们的工作过程我们可以更好地理解系统内部的运行机制。本章的意义在于加深读者对四大组件的工作方式的认识，由于四大组件的特殊性，我们有必要对它们的工作过程有一定的了解，这也有助于加深对 Android 整体的体系结构的认识。很多情况下，只有对 Android 体系结构有一定认识，在实际的开发中才能写出优秀的代码。

9.1 四大组件的运行状态

Android 的四大组件中除了 BroadcastReceiver 以外，其他三种组件都必须在 AndroidManifest 中注册，对于 BroadcastReceiver 来说，它既可以在 AndroidManifest 中注册也可以通过代码来注册。在调用方式上，Activity、Service 和 BroadcastReceiver 需要借助 Intent，而 ContentProvider 则无须借助 Intent。

Activity 是一种展示型组件，用于向用户直接地展示一个界面，并且可以接收用户的输入信息从而进行交互。Activity 是最重要的一种组件，对用户来说，Activity 就是一个 Android

应用的全部，这是因为其他三大组件对用户来说都是不可感知的。Activity 的启动由 Intent 触发，其中 Intent 可以分为显式 Intent 和隐式 Intent，显式 Intent 可以明确地指向一个 Activity 组件，隐式 Intent 则指向一个或多个目标 Activity 组件，当然也可能没有任何一个 Activity 组件可以处理这个隐式 Intent。一个 Activity 组件可以具有特定的启动模式。关于 Activity 的启动模式在第 1 章中已经做了介绍，同一个 Activity 组件在不同的启动模式下会有不同的效果。Activity 组件是可以停止的，在实际开发中可以通过 Activity 的 finish 方法来结束一个 Activity 组件的运行。由此来看，Activity 组件的主要作用是展示一个界面并和用户交互，它扮演的是一种前台界面的角色。

Service 是一种计算型组件，用于在后台执行一系列计算任务。由于 Service 组件工作在后台，因此用户无法直接感知到它的存在。Service 组件和 Activity 组件略有不同，Activity 组件只有一种运行模式，即 Activity 处于启动状态，但是 Service 组件却有两种状态：启动状态和绑定状态。当 Service 组件处于启动状态时，这个时候 Service 内部可以做一些后台计算，并且不需要和外界有直接的交互。尽管 Service 组件是用于执行后台计算的，但是它本身是运行在主线程中的，因此耗时的后台计算仍然需要在单独的线程中去完成。当 Service 组件处于绑定状态时，这个时候 Service 内部同样可以进行后台计算，但是处于这种状态时外界可以很方便地和 Service 组件进行通信。Service 组件也是可以停止的，停止一个 Service 组件稍显复杂，需要灵活采用 stopService 和 unBindService 这两个方法才能完全停止一个 Service 组件。

BroadcastReceiver 是一种消息型组件，用于在不同的组件乃至不同的应用之间传递消息。BroadcastReceiver 同样无法被用户直接感知，因为它工作在系统内部。BroadcastReceiver 也叫广播，广播的注册有两种方式：静态注册和动态注册。静态注册是指在 AndroidManifest 中注册广播，这种广播在应用安装时会被系统解析，此种形式的广播不需要应用启动就可以收到相应的广播。动态注册广播需要通过 Context.registerReceiver() 来实现，并且在不需要的时候要通过 Context.unRegisterReceiver() 来解除广播，此种形态的广播必须要应用启动才能注册并接收广播，因为应用不启动就无法注册广播，无法注册广播就无法收到相应的广播。在实际开发中通过 Context 的一系列 send 方法来发送广播，被发送的广播会被系统发送给感兴趣的广播接收者，发送和接收过程的匹配是通过广播接收者的<intent-filter>来描述的。可以发现，BroadcastReceiver 组件可以用来实现低耦合的观察者模式，观察者和被观察者之间可以没有任何耦合。由于 BroadcastReceiver 的特性，它不适合用来执行耗时操作。BroadcastReceiver 组件一般来说不需要停止，它也没有停止的概念。

ContentProvider 是一种数据共享型组件，用于向其他组件乃至其他应用共享数据。和 BroadcastReceiver 一样，ContentProvider 同样无法被用户直接感知。对于一个 ContentProvider

组件来说，它的内部需要实现增删改查这四种操作，在它的内部维持着一份数据集合，这个数据集合既可以通过数据库来实现，也可以采用其他任何类型来实现，比如 List 和 Map，ContentProvider 对数据集合的具体实现并没有任何要求。需要注意的是，ContentProvider 内部的 insert、delete、update 和 query 方法需要处理好线程同步，因为这几个方法是在 Binder 线程池中被调用的，另外 ContentProvider 组件也不需要手动停止。

9.2 Activity 的工作过程

本节讲述的内容是 Activity 的工作过程。为了方便日常的开发工作，系统对四大组件的工作过程进行了很大程度的封装，这使得开发者无须关注实现细节即可快速地使用四大组件。Activity 作为很重要的一个组件，其内部工作过程系统当然也是做了很多的封装，这种封装使得启动一个 Activity 变得异常简单。在显式调用的情形下，只需要通过如下代码即可完成：

```
Intent intent = new Intent(this, TestActivity.class);
startActivity(intent);
```

通过上面的代码即可启动一个具体的 Activity，然后新 Activity 就会被系统启动并展示在用户的眼前。这个过程对于 Android 开发者来说最普通不过了，这也是很理所应当的事，但是有没有想过系统内部到底是如何启动一个 Activity 的呢？比如新 Activity 的对象是在何时创建的？Activity 的 onCreate 方法又是在何时被系统回调的呢？读者可能会有疑问：在日常开发中并不需要了解 Activity 底层到底是怎么工作的，那么了解它们又有什么意义呢？没错，在日常开发中是不需要了解系统的底层工作原理，但是如果想要在技术上有进一步的提高，那么就必须了解一些系统的工作原理，这是一个开发人员日后成长为高级工程师乃至架构师所必须具备的技术能力。从另外一个角度来说，Android 作为一个优秀的基于 Linux 的移动操作系统，其内部一定有很多值得我们学习和借鉴的地方，因此了解系统的工作过程就是学习 Android 操作系统。通过对 Android 操作系统的学习可以提高我们对操作系统在技术实现上的理解，这对于加强开发人员的内功是很有帮助的。但是有一点，由于 Android 的内部实现多数都比较复杂，在研究内部实现上应该更加侧重于对整体流程的把握，而不能深入代码细节不能自拔，太深入代码细节往往会导致"只见树木不见森林"的状态。处于这种状态下，无法对整体流程建立足够的认识，取而代之的是烦琐的代码细节，但是代码细节本身并不具有太多的指导意义，因此这种学习状态是要极力避免的。鉴于这一点，本章对 Activity 以及其他三个组件的工作过程的分析将会侧重于整体流程的讲解，目的是为了让读者对四大组件的工作过程有一个感性的认识并能够给予上层开发一些指导意义。但凡事不是绝对的，如果开发者从事的工作是 Android Rom 开发，那底层代码细节还是要有所涉猎的。

本节主要分析 Activity 的启动过程，通过本节读者可以对 Activity 的启动过程有一个感性的认识，至于启动模式以及任务栈等概念本节中并未涉及，读者感兴趣的话可以查看相应的代码细节即可。

我们从 Activity 的 startActivity 方法开始分析，startActivity 方法有好几种重载方式，但它们最终都会调用 startActivityForResult 方法，它的实现如下所示。

```java
public void startActivityForResult(Intent intent, int requestCode,
@Nullable Bundle options) {
    if (mParent == null) {
        Instrumentation.ActivityResult ar =
            mInstrumentation.execStartActivity(
                this, mMainThread.getApplicationThread(), mToken, this,
                intent, requestCode, options);
        if (ar != null) {
            mMainThread.sendActivityResult(
                mToken, mEmbeddedID, requestCode, ar.getResultCode(),
                ar.getResultData());
        }
        if (requestCode >= 0) {
            // If this start is requesting a result, we can avoid making
            // the activity visible until the result is received.  Setting
            // this code during onCreate(Bundle savedInstanceState) or
            //    onResume() will keep the
            // activity hidden during this time, to avoid flickering.
            // This can only be done when a result is requested because
            // that guarantees we will get information back when the
            // activity is finished, no matter what happens to it.
            mStartedActivity = true;
        }

        final View decor = mWindow != null ? mWindow.peekDecorView() : null;
        if (decor != null) {
            decor.cancelPendingInputEvents();
        }
```

```
            // TODO Consider clearing/flushing other event sources and events
               for child windows.
        } else {
            if (options != null) {
                mParent.startActivityFromChild(this,intent,requestCode,options);
            } else {
                // Note we want to go through this method for compatibility with
                // existing applications that may have overridden it.
                mParent.startActivityFromChild(this, intent, requestCode);
            }
        }
        if (options != null && !isTopOfTask()) {
            mActivityTransitionState.startExitOutTransition(this, options);
        }
    }
}
```

在上面的代码中，我们只需要关注 mParent == null 这部分逻辑即可。mParent 代表的是 ActivityGroup，ActivityGroup 最开始被用来在一个界面中嵌入多个子 Activity，但是其在 API 13 中已经被废弃了，系统推荐采用 Fragment 来代替 ActivityGroup，Fragment 的好处就不用多说了。在上面的代码中需要注意 mMainThread.getApplicationThread()这个参数，它的类型是 ApplicationThread，ApplicationThread 是 ActivityThread 的一个内部类，通过后面的分析可以发现，ApplicationThread 和 ActivityThread 在 Activity 的启动过程中发挥着很重要的作用。接着看一下 Instrumentation 的 execStartActivity 方法，如下所示。

```
    public ActivityResult execStartActivity(
            Context who, IBinder contextThread, IBinder token, Activity target,
            Intent intent, int requestCode, Bundle options) {
        IApplicationThread whoThread = (IApplicationThread) contextThread;
        if (mActivityMonitors != null) {
            synchronized (mSync) {
                final int N = mActivityMonitors.size();
                for (int i=0; i<N; i++) {
                    final ActivityMonitor am = mActivityMonitors.get(i);
                    if (am.match(who, null, intent)) {
                        am.mHits++;
                        if (am.isBlocking()) {
                            return requestCode >= 0 ? am.getResult() : null;
                        }
                        break;
```

```
                }
            }
        }
    }
    try {
        intent.migrateExtraStreamToClipData();
        intent.prepareToLeaveProcess();
        int result = ActivityManagerNative.getDefault()
            .startActivity(whoThread, who.getBasePackageName(), intent,
                    intent.resolveTypeIfNeeded(who.getContentResolver()),
                    token, target != null ? target.mEmbeddedID : null,
                    requestCode, 0, null, options);
        checkStartActivityResult(result, intent);
    } catch (RemoteException e) {
    }
    return null;
}
```

从上面的代码可以看出，启动 Activity 真正的实现由 ActivityManagerNative.getDefault() 的 startActivity 方法来完成。ActivityManagerService（下面简称为 AMS）继承自 ActivityManagerNative，而 ActivityManagerNative 继承自 Binder 并实现了 IActivityManager 这个 Binder 接口，因此 AMS 也是一个 Binder，它是 IActivityManager 的具体实现。由于 ActivityManagerNative.getDefault()其实是一个 IActivityManager 类型的 Binder 对象，因此它的具体实现是 AMS。可以发现，在 ActivityManagerNative 中，AMS 这个 Binder 对象采用单例模式对外提供，Singleton 是一个单例的封装类，第一次调用它的 get 方法时它会通过 create 方法来初始化 AMS 这个 Binder 对象，在后续的调用中则直接返回之前创建的对象，这个过程的源码如下所示。

```
static public IActivityManager getDefault() {
    return gDefault.get();
}

private static final Singleton<IActivityManager> gDefault = new Singleton
<IActivityManager>() {
    protected IActivityManager create() {
        IBinder b = ServiceManager.getService("activity");
        if (false) {
            Log.v("ActivityManager", "default service binder = " + b);
        }
```

```
            IActivityManager am = asInterface(b);
            if (false) {
                Log.v("ActivityManager", "default service = " + am);
            }
            return am;
        }
    };
```

从上面的分析可以知道，Activity 由 ActivityManagerNative.getDefault()来启动，而 ActivityManagerNative.getDefault()实际上是 AMS，因此 Activity 的启动过程又转移到了 AMS 中，为了继续分析这个过程，只需要查看 AMS 的 startActivity 方法即可。在分析 AMS 的的 startActivity 方法之前，我们先回过头来看一下 Instrumentation 的 execStartActivity 方法，其中有一行代码：checkStartActivityResult(result, intent)，直观上看起来这个方法的作用像是在检查启动 Activity 的结果，它的具体实现如下所示。

```
/** @hide */
public static void checkStartActivityResult(int res, Object intent) {
    if (res >= ActivityManager.START_SUCCESS) {
        return;
    }

    switch (res) {
        case ActivityManager.START_INTENT_NOT_RESOLVED:
        case ActivityManager.START_CLASS_NOT_FOUND:
            if (intent instanceof Intent && ((Intent)intent).getComponent() !=
            null)
                throw new ActivityNotFoundException(
                        "Unable to find explicit activity class "
                        + ((Intent)intent).getComponent().toShortString()
                        + "; have you declared this activity in your
                        AndroidManifest.xml?");
            throw new ActivityNotFoundException(
                    "No Activity found to handle " + intent);
        case ActivityManager.START_PERMISSION_DENIED:
            throw new SecurityException("Not allowed to start activity "
                    + intent);
        case ActivityManager.START_FORWARD_AND_REQUEST_CONFLICT:
            throw new AndroidRuntimeException(
                    "FORWARD_RESULT_FLAG used while also requesting a
```

```
            result");
        case ActivityManager.START_NOT_ACTIVITY:
            throw new IllegalArgumentException(
                    "PendingIntent is not an activity");
        case ActivityManager.START_NOT_VOICE_COMPATIBLE:
            throw new SecurityException(
                    "Starting under voice control not allowed for: " +
                    intent);
        default:
            throw new AndroidRuntimeException("Unknown error code "
                + res + " when starting " + intent);
    }
}
```

从上面的代码可以看出，checkStartActivityResult 的作用很明显，就是检查启动 Activity 的结果。当无法正确地启动一个 Activity 时，这个方法会抛出异常信息，其中最熟悉不过的就是"Unable to find explicit activity class; have you declared this activity in your AndroidManifest.xml?"这个异常了，当待启动的 Activity 没有在 AndroidManifest 中注册时，就会抛出这个异常。

接着我们继续分析 AMS 的 startActivity 方法，如下所示。

```
public final int startActivity(IApplicationThread caller, String callingPackage,
        Intent intent, String resolvedType, IBinder resultTo, String
        resultWho, int requestCode,
        int startFlags, ProfilerInfo profilerInfo, Bundle options) {
    return startActivityAsUser(caller, callingPackage, intent, resolved-
    Type, resultTo,
        resultWho, requestCode, startFlags, profilerInfo, options,
        UserHandle.getCallingUserId());
}

public final int startActivityAsUser(IApplicationThread caller, String
callingPackage,
        Intent intent, String resolvedType, IBinder resultTo, String
        resultWho, int requestCode,
        int startFlags, ProfilerInfo profilerInfo,Bundle options,int userId) {
    enforceNotIsolatedCaller("startActivity");
    userId = handleIncomingUser(Binder.getCallingPid(), Binder.getCalling-
    Uid(), userId,
```

```
            false, ALLOW_FULL_ONLY, "startActivity", null);
    // TODO: Switch to user app stacks here.
    return mStackSupervisor.startActivityMayWait(caller, -1, calling
Package, intent,
            resolvedType, null, null, resultTo, resultWho, requestCode,
            startFlags,
            profilerInfo, null, null, options, userId, null, null);
}
```

可以看出，Activity 的启动过程又转移到了 ActivityStackSupervisor 的 startActivity-MayWait 方法中了，在 startActivityMayWait 中又调用了 startActivityLocked 方法，然后 startActivityLocked 方法又调用了 startActivityUncheckedLocked 方法，接着 startActivityUncheckedLocked 又调用了 ActivityStack 的 resumeTopActivitiesLocked 方法，这个时候启动过程已经从 ActivityStackSupervisor 转移到了 ActivityStack。

ActivityStack 的 resumeTopActivitiesLocked 方法的实现如下所示。

```
final boolean resumeTopActivityLocked(ActivityRecord prev, Bundle options) {
    if (inResumeTopActivity) {
        // Don't even start recursing.
        return false;
    }

    boolean result = false;
    try {
        // Protect against recursion.
        inResumeTopActivity = true;
        result = resumeTopActivityInnerLocked(prev, options);
    } finally {
        inResumeTopActivity = false;
    }
    return result;
}
```

从上面的代码可以看出，resumeTopActivityLocked 调用了 resumeTopActivityInnerLocked 方法，resumeTopActivityInnerLocked 方法又调用了 ActivityStackSupervisor 的 startSpecificActivityLocked 方法，startSpecificActivityLocked 的源码如下所示。

```
void startSpecificActivityLocked(ActivityRecord r,
        boolean andResume, boolean checkConfig) {
```

```
        // Is this activity's application already running?
        ProcessRecord app = mService.getProcessRecordLocked(r.processName,
                r.info.applicationInfo.uid, true);

        r.task.stack.setLaunchTime(r);

        if (app != null && app.thread != null) {
            try {
                if ((r.info.flags&ActivityInfo.FLAG_MULTIPROCESS) == 0
                        || !"android".equals(r.info.packageName)) {
                    // Don't add this if it is a platform component that is marked
                    // to run in multiple processes, because this is actually
                    // part of the framework so doesn't make sense to track as a
                    // separate apk in the process.
                    app.addPackage(r.info.packageName, r.info.applicationInfo.
                    versionCode,
                            mService.mProcessStats);
                }
                realStartActivityLocked(r, app, andResume, checkConfig);
                return;
            } catch (RemoteException e) {
                Slog.w(TAG, "Exception when starting activity "
                        + r.intent.getComponent().flattenToShortString(), e);
            }

            // If a dead object exception was thrown -- fall through to
            // restart the application.
        }

        mService.startProcessLocked(r.processName, r.info.applicationInfo,true,0,
                "activity", r.intent.getComponent(), false, false, true);
    }
```

从上面代码可以看出，startSpecificActivityLocked 方法调用了 realStartActivityLocked 方法。为了更清晰地说明 Activity 的启动过程在 ActivityStackSupervisor 和 ActivityStack 之间的传递顺序，这里给出了一张流程图，如图9-1所示。

在 ActivityStackSupervisor 的 realStartActivityLocked 方法中有如下一段代码：

```
app.thread.scheduleLaunchActivity(new Intent(r.intent), r.appToken,
        System.identityHashCode(r), r.info, new Configuration(mService.
```

```
                    mConfiguration),
            r.compat, r.task.voiceInteractor, app.repProcState, r.icicle,
            r.persistentState,
            results, newIntents, !andResume, mService.isNextTransitionForward(),
            profilerInfo);
```

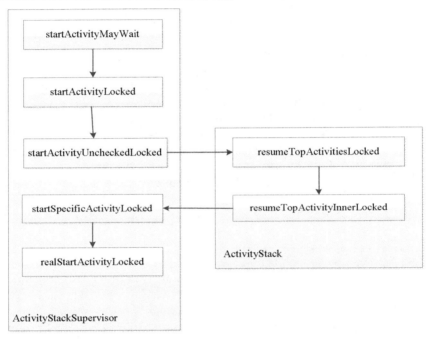

图 9-1　Activity 的启动过程在 ActivityStackSupervisor 和 ActivityStack 之间的传递顺序

上面的这段代码很重要，其中 app.thread 的类型为 IApplicationThread，IApplicationThread 的声明如下：

```
public interface IApplicationThread extends IInterface {
void schedulePauseActivity(IBinder token, boolean finished, boolean userLeaving,
        int configChanges, boolean dontReport) throws RemoteException;
void scheduleStopActivity(IBinder token, boolean showWindow,
        int configChanges) throws RemoteException;
void scheduleWindowVisibility(IBinder token, boolean showWindow) throws
RemoteException;
void scheduleSleeping(IBinder token, boolean sleeping) throws RemoteException;
    void scheduleResumeActivity(IBinder token, int procState, boolean
```

```
        isForward, Bundle resumeArgs)
        throws RemoteException;
void scheduleSendResult(IBinder token, List<ResultInfo> results) throws
RemoteException;
void scheduleLaunchActivity(Intent intent, IBinder token, int ident,
        ActivityInfo info, Configuration curConfig, CompatibilityInfo
        compatInfo,
        IVoiceInteractor voiceInteractor, int procState, Bundle state,
        PersistableBundle persistentState, List<ResultInfo> pendingResults,
        List<Intent> pendingNewIntents, boolean notResumed, boolean
        isForward,
        ProfilerInfo profilerInfo) throws RemoteException;
void scheduleRelaunchActivity(IBinder token, List<ResultInfo> pending-
Results,
        List<Intent> pendingNewIntents, int configChanges,
        boolean notResumed, Configuration config) throws RemoteException;
void scheduleNewIntent(List<Intent> intent, IBinder token) throws
RemoteException;
void scheduleDestroyActivity(IBinder token, boolean finished,
        int configChanges) throws RemoteException;
void scheduleReceiver(Intent intent, ActivityInfo info, Compatibility-
Info compatInfo,
        int resultCode, String data, Bundle extras, boolean sync,
        int sendingUser, int processState) throws RemoteException;
static final int BACKUP_MODE_INCREMENTAL = 0;
static final int BACKUP_MODE_FULL = 1;
static final int BACKUP_MODE_RESTORE = 2;
static final int BACKUP_MODE_RESTORE_FULL = 3;
void scheduleCreateBackupAgent(ApplicationInfo app, CompatibilityInfo
compatInfo,
        int backupMode) throws RemoteException;
void scheduleDestroyBackupAgent(ApplicationInfo app, CompatibilityInfo
compatInfo)
        throws RemoteException;
void scheduleCreateService(IBinder token, ServiceInfo info,
        CompatibilityInfo compatInfo, int processState) throws
        RemoteException;
void scheduleBindService(IBinder token,
        Intent intent, boolean rebind, int processState) throws
        RemoteException;
void scheduleUnbindService(IBinder token,
```

```
        Intent intent) throws RemoteException;
    void scheduleServiceArgs(IBinder token, boolean taskRemoved, int
startId,
        int flags, Intent args) throws RemoteException;
    void scheduleStopService(IBinder token) throws RemoteException;
    ...
}
```

因为它继承了 IInterface 接口，所以它是一个 Binder 类型的接口。从 IApplicationThread 声明的接口方法可以看出，其内部包含了大量启动、停止 Activity 的接口，此外还包含了启动和停止服务的接口。从接口方法的命名可以猜测，IApplicationThread 这个 Binder 接口的实现者完成了大量和 Activity 以及 Service 启动/停止相关的功能，事实证明的确是这样的。

那么 IApplicationThread 的实现者到底是什么呢？答案就是 ActivityThread 中的内部类 ApplicationThread，下面来看一下 ApplicationThread 的定义，如下所示。

```
private class ApplicationThread extends ApplicationThreadNative
public abstract class ApplicationThreadNative extends Binder
    implements IApplicationThread
```

可以看出，ApplicationThread 继承了 ApplicationThreadNative，而 ApplicationThreadNative 则继承了 Binder 并实现了 IApplicationThread 接口。如果读者还记得系统为 AIDL 文件自动生成的代码，就会发现 ApplicationThreadNative 的作用其实和系统为 AIDL 文件生成的类是一样的，这方面的知识在第 2 章已经做了介绍，读者可以查看第 2 章的相关内容。

在 ApplicationThreadNative 的内部，还有一个 ApplicationThreadProxy 类，这个类的实现如下所示。相信读者有一种似曾相识的感觉，其实这个内部类也是系统为 AIDL 文件自动生成的代理类。种种迹象表明，ApplicationThreadNative 就是 IApplicationThread 的实现者，由于 ApplicationThreadNative 被系统定义为抽象类，所以 ApplicationThread 就成了 IApplicationThread 最终的实现者。

```
class ApplicationThreadProxy implements IApplicationThread {
    private final IBinder mRemote;

    public ApplicationThreadProxy(IBinder remote) {
        mRemote = remote;
    }
```

```
    public final IBinder asBinder() {
        return mRemote;
    }

    public final void schedulePauseActivity(IBinder token, boolean finished,
            boolean userLeaving, int configChanges, boolean dontReport)
            throws RemoteException {
        Parcel data = Parcel.obtain();
        data.writeInterfaceToken(IApplicationThread.descriptor);
        data.writeStrongBinder(token);
        data.writeInt(finished ? 1 : 0);
        data.writeInt(userLeaving ? 1 :0);
        data.writeInt(configChanges);
        data.writeInt(dontReport ? 1 : 0);
        mRemote.transact(SCHEDULE_PAUSE_ACTIVITY_TRANSACTION, data, null,
                IBinder.FLAG_ONEWAY);
        data.recycle();
    }

    public final void scheduleStopActivity(IBinder token, boolean showWindow,
            int configChanges) throws RemoteException {
        Parcel data = Parcel.obtain();
        data.writeInterfaceToken(IApplicationThread.descriptor);
        data.writeStrongBinder(token);
        data.writeInt(showWindow ? 1 : 0);
        data.writeInt(configChanges);
        mRemote.transact(SCHEDULE_STOP_ACTIVITY_TRANSACTION, data, null,
                IBinder.FLAG_ONEWAY);
        data.recycle();
    }
    ...
}
```

绕了一大圈，Activity 的启动过程最终回到了 ApplicationThread 中，ApplicationThread 通过 scheduleLaunchActivity 方法来启动 Activity，代码如下所示。

```
// we use token to identify this activity without having to send the
// activity itself back to the activity manager. (matters more with ipc)
public final void scheduleLaunchActivity(Intent intent, IBinder token, int ident,
        ActivityInfo info, Configuration curConfig, CompatibilityInfo
        compatInfo,
```

```
            IVoiceInteractor voiceInteractor, int procState, Bundle state,
            PersistableBundle persistentState, List<ResultInfo> pendingResults,
            List<Intent> pendingNewIntents, boolean notResumed, boolean isForward,
            ProfilerInfo profilerInfo) {

    updateProcessState(procState, false);

    ActivityClientRecord r = new ActivityClientRecord();

    r.token = token;
    r.ident = ident;
    r.intent = intent;
    r.voiceInteractor = voiceInteractor;
    r.activityInfo = info;
    r.compatInfo = compatInfo;
    r.state = state;
    r.persistentState = persistentState;

    r.pendingResults = pendingResults;
    r.pendingIntents = pendingNewIntents;

    r.startsNotResumed = notResumed;
    r.isForward = isForward;

    r.profilerInfo = profilerInfo;

    updatePendingConfiguration(curConfig);

    sendMessage(H.LAUNCH_ACTIVITY, r);
}
```

在 ApplicationThread 中，scheduleLaunchActivity 的实现很简单，就是发送一个启动 Activity 的消息交由 Handler 处理，这个 Handler 有着一个很简洁的名字：H。sendMessage 的作用是发送一个消息给 H 处理，它的实现如下所示。

```
private void sendMessage(int what, Object obj, int arg1, int arg2, boolean
async) {
    if (DEBUG_MESSAGES) Slog.v(
        TAG, "SCHEDULE " + what + " " + mH.codeToString(what)
        + ": " + arg1 + " / " + obj);
```

```
    Message msg = Message.obtain();
    msg.what = what;
    msg.obj = obj;
    msg.arg1 = arg1;
    msg.arg2 = arg2;
    if (async) {
        msg.setAsynchronous(true);
    }
    mH.sendMessage(msg);
}
```

接着来看一下 Handler H 对消息的处理,如下所示。

```
private class H extends Handler {
    public static final int LAUNCH_ACTIVITY         = 100;
    public static final int PAUSE_ACTIVITY          = 101;
    public static final int PAUSE_ACTIVITY_FINISHING = 102;
    public static final int STOP_ACTIVITY_SHOW      = 103;
    public static final int STOP_ACTIVITY_HIDE      = 104;
    public static final int SHOW_WINDOW             = 105;
    public static final int HIDE_WINDOW             = 106;
    public static final int RESUME_ACTIVITY         = 107;
    ...

    public void handleMessage(Message msg) {
        if (DEBUG_MESSAGES) Slog.v(TAG, ">>> handling: " + codeToString
        (msg.what));
        switch (msg.what) {
            case LAUNCH_ACTIVITY: {
                Trace.traceBegin(Trace.TRACE_TAG_ACTIVITY_MANAGER,
                "activityStart");
                final ActivityClientRecord r = (ActivityClientRecord) msg.obj;

                r.packageInfo = getPackageInfoNoCheck(
                        r.activityInfo.applicationInfo, r.compatInfo);
                handleLaunchActivity(r, null);
                Trace.traceEnd(Trace.TRACE_TAG_ACTIVITY_MANAGER);
            } break;
            case RELAUNCH_ACTIVITY: {
                Trace.traceBegin(Trace.TRACE_TAG_ACTIVITY_MANAGER,
                "activityRestart");
```

```
            ActivityClientRecord r = (ActivityClientRecord)msg.obj;
            handleRelaunchActivity(r);
            Trace.traceEnd(Trace.TRACE_TAG_ACTIVITY_MANAGER);
        } break;
        case PAUSE_ACTIVITY:
            Trace.traceBegin(Trace.TRACE_TAG_ACTIVITY_MANAGER,
            "activityPause");
            handlePauseActivity((IBinder)msg.obj, false, (msg.arg1&1) !=
        0, msg.arg2,
                    (msg.arg1&2) != 0);
            maybeSnapshot();
            Trace.traceEnd(Trace.TRACE_TAG_ACTIVITY_MANAGER);
            break;
        ...
        }
        if (DEBUG_MESSAGES) Slog.v(TAG, "<<< done: " + codeToString(msg.what));
    }
}
```

从 Handler H 对 "LAUNCH_ACTIVITY" 这个消息的处理可以知道，Activity 的启动过程由 ActivityThread 的 handleLaunchActivity 方法来实现，它的源码如下所示。

```
private void handleLaunchActivity(ActivityClientRecord r, Intent customIntent) {
    ...
    if (localLOGV) Slog.v(
        TAG, "Handling launch of " + r);

    Activity a = performLaunchActivity(r, customIntent);

    if (a != null) {
        r.createdConfig = new Configuration(mConfiguration);
        Bundle oldState = r.state;
        handleResumeActivity(r.token, false, r.isForward,
                !r.activity.mFinished && !r.startsNotResumed);
        ...
    }
    ...
}
```

从上面的源码可以看出，performLaunchActivity 方法最终完成了 Activity 对象的创建和启动过程，并且 ActivityThread 通过 handleResumeActivity 方法来调用被启动 Activity 的

onResume 这一生命周期方法。

performLaunchActivity 这个方法主要完成了如下几件事。

1. 从 ActivityClientRecord 中获取待启动的 Activity 的组件信息

```
ActivityInfo aInfo = r.activityInfo;
if (r.packageInfo == null) {
    r.packageInfo = getPackageInfo(aInfo.applicationInfo, r.compatInfo,
            Context.CONTEXT_INCLUDE_CODE);
}

ComponentName component = r.intent.getComponent();
if (component == null) {
    component = r.intent.resolveActivity(
        mInitialApplication.getPackageManager());
    r.intent.setComponent(component);
}

if (r.activityInfo.targetActivity != null) {
    component = new ComponentName(r.activityInfo.packageName,
            r.activityInfo.targetActivity);
}
```

2. 通过 Instrumentation 的 newActivity 方法使用类加载器创建 Activity 对象

```
Activity activity = null;
try {
    java.lang.ClassLoader cl = r.packageInfo.getClassLoader();
    activity = mInstrumentation.newActivity(
            cl, component.getClassName(), r.intent);
    StrictMode.incrementExpectedActivityCount(activity.getClass());
    r.intent.setExtrasClassLoader(cl);
    r.intent.prepareToEnterProcess();
    if (r.state != null) {
        r.state.setClassLoader(cl);
    }
} catch (Exception e) {
    if (!mInstrumentation.onException(activity, e)) {
        throw new RuntimeException(
            "Unable to instantiate activity " + component
```

```
        + ": " + e.toString(), e);
    }
}
```

至于 Instrumentation 的 newActivity，它的实现比较简单，就是通过类加载器来创建 Activity 对象：

```
public Activity newActivity(ClassLoader cl, String className,
        Intent intent)
        throws InstantiationException, IllegalAccessException,
        ClassNotFoundException {
    return (Activity)cl.loadClass(className).newInstance();
}
```

3. 通过 LoadedApk 的 makeApplication 方法来尝试创建 Application 对象

```
public Application makeApplication(boolean forceDefaultAppClass,
        Instrumentation instrumentation) {
    if (mApplication != null) {
        return mApplication;
    }

    Application app = null;
    String appClass = mApplicationInfo.className;
    if (forceDefaultAppClass || (appClass == null)) {
        appClass = "android.app.Application";
    }

    try {
        java.lang.ClassLoader cl = getClassLoader();
        if (!mPackageName.equals("android")) {
            initializeJavaContextClassLoader();
        }
        ContextImpl appContext = ContextImpl.createAppContext(mActivityThread, this);
        app = mActivityThread.mInstrumentation.newApplication(
                cl, appClass, appContext);
        appContext.setOuterContext(app);
    } catch (Exception e) {
        if (!mActivityThread.mInstrumentation.onException(app, e)) {
            throw new RuntimeException(
```

```
            "Unable to instantiate application " + appClass
            + ": " + e.toString(), e);
        }
    }
    mActivityThread.mAllApplications.add(app);
    mApplication = app;

    if (instrumentation != null) {
        try {
            instrumentation.callApplicationOnCreate(app);
        } catch (Exception e) {
            if (!instrumentation.onException(app, e)) {
                throw new RuntimeException(
                    "Unable to create application " + app.getClass().getName()
                    + ": " + e.toString(), e);
            }
        }
    }
    ...
    return app;
}
```

从 makeApplication 的实现可以看出，如果 Application 已经被创建过了，那么就不会再重复创建了，这也意味着一个应用只有一个 Application 对象。Application 对象的创建也是通过 Instrumentation 来完成的，这个过程和 Activity 对象的创建一样，都是通过类加载器来实现的。Application 创建完毕后，系统会通过 Instrumentation 的 callApplicationOnCreate 来调用 Application 的 onCreate 方法。

4. 创建 ContextImpl 对象并通过 Activity 的 attach 方法来完成一些重要数据的初始化

```
Context appContext = createBaseContextForActivity(r, activity);
CharSequence   title   =   r.activityInfo.loadLabel(appContext.getPackage-
Manager());
Configuration config = new Configuration(mCompatConfiguration);
if (DEBUG_CONFIGURATION) Slog.v(TAG, "Launching activity "
    + r.activityInfo.name + " with config " + config);
activity.attach(appContext, this, getInstrumentation(), r.token,
        r.ident, app, r.intent, r.activityInfo, title, r.parent,
        r.embeddedID, r.lastNonConfigurationInstances, config,
        r.voiceInteractor);
```

ContextImpl 是一个很重要的数据结构，它是 Context 的具体实现，Context 中的大部分逻辑都是由 ContextImpl 来完成的。ContextImpl 是通过 Activity 的 attach 方法来和 Activity 建立关联的，除此以外，在 attach 方法中 Activity 还会完成 Window 的创建并建立自己和 Window 的关联，这样当 Window 接收到外部输入事件后就可以将事件传递给 Activity。

5. 调用 Activity 的 onCreate 方法

mInstrumentation.callActivityOnCreate(activity, r.state)，由于 Activity 的 onCreate 已经被调用，这也意味着 Activity 已经完成了整个启动过程。

9.3　Service 的工作过程

在 9.2 节中介绍了 Activity 的工作过程，本节将介绍 Service 的工作过程，通过本节的分析，读者将会对 Service 的一些工作原理有更进一步的认识，比如 Service 的启动过程和绑定过程。在分析 Service 的工作过程之前，先看一下如何使用一个 Service。Service 分为两种工作状态，一种是启动状态，主要用于执行后台计算；另一种是绑定状态，主要用于其他组件和 Service 的交互。需要注意的是，Service 的这两种状态是可以共存的，即 Service 既可以处于启动状态也可以同时处于绑定状态。通过 Context 的 startService 方法即可启动一个 Service，如下所示。

```
Intent intentService = new Intent(this, MyService.class);
startService(intentService);
```

通过 Context 的 bindService 方法即可以绑定的方式启动一个 Service，如下所示。

```
Intent intentService = new Intent(this, MyService.class);
bindService(intentService, mServiceConnection, BIND_AUTO_CREATE);
```

9.3.1　Service 的启动过程

Service 的启动过程从 ContextWrapper 的 startService 开始，如下所示。

```
public ComponentName startService(Intent service) {
    return mBase.startService(service);
}
```

上面代码的 mBase 的类型是 ContextImpl，在 9.2 节中我们知道，Activity 被创建时会通过 attach 方法将一个 ContextImpl 对象关联起来,这个 ContextImpl 对象就是上述代码中的 mBase。从 ContextWrapper 的实现可以看出,其大部分操作都是通过 mBase 来实现的,在设计模式中这是一种典型的桥接模式。下面继续看 ContextImpl 的 startService 的实现,如下所示。

```
public ComponentName startService(Intent service) {
    warnIfCallingFromSystemProcess();
    return startServiceCommon(service, mUser);
}

private ComponentName startServiceCommon(Intent service, UserHandle user) {
    try {
        validateServiceIntent(service);
        service.prepareToLeaveProcess();
        ComponentName cn = ActivityManagerNative.getDefault().startService(
            mMainThread.getApplicationThread(), service,
            service.resolveTypeIfNeeded(getContentResolver()), user.get-
            Identifier());
        if (cn != null) {
            if (cn.getPackageName().equals("!")) {
                throw new SecurityException(
                    "Not allowed to start service " + service
                    + " without permission " + cn.getClassName());
            } else if (cn.getPackageName().equals("!!")) {
                throw new SecurityException(
                    "Unable to start service " + service
                    + ": " + cn.getClassName());
            }
        }
        return cn;
    } catch (RemoteException e) {
        return null;
    }
}
```

在 ContextImpl 中,startService 方法会调用 startServiceCommon 方法,而 startService-Common 方法又会通过 ActivityManagerNative.getDefault()这个对象来启动一个服务。对于 ActivityManagerNative.getDefault()这个对象,我们应该有点印象,在 9.2 节中进行了详细的分析,它实际上就是 AMS(ActivityManagerService),这里就不再重复说明了。需要注意

的是，在上述代码中通过 AMS 来启动服务的行为是一个远程过程调用。AMS 的 startService 方法的实现如下所示。

```
public ComponentName startService(IApplicationThread caller, Intent service,
        String resolvedType, int userId) {
    enforceNotIsolatedCaller("startService");
    // Refuse possible leaked file descriptors
    if (service != null && service.hasFileDescriptors() == true) {
        throw new IllegalArgumentException("File descriptors passed in
        Intent");
    }

    if (DEBUG_SERVICE)
        Slog.v(TAG, "startService: " + service + " type=" + resolvedType);
    synchronized(this) {
        final int callingPid = Binder.getCallingPid();
        final int callingUid = Binder.getCallingUid();
        final long origId = Binder.clearCallingIdentity();
        ComponentName res = mServices.startServiceLocked(caller, service,
                resolvedType, callingPid, callingUid, userId);
        Binder.restoreCallingIdentity(origId);
        return res;
    }
}
```

在上面的代码中，AMS 会通过 mServices 这个对象来完成 Service 后续的启动过程，mServices 对象的类型是 ActiveServices，ActiveServices 是一个辅助 AMS 进行 Service 管理的类，包括 Service 的启动、绑定和停止等。在 ActiveServices 的 startServiceLocked 方法的尾部会调用 startServiceInnerLocked 方法，startServiceInnerLocked 的实现如下所示。

```
ComponentName startServiceInnerLocked(ServiceMap smap, Intent service,
        ServiceRecord r, boolean callerFg, boolean addToStarting) {
    ProcessStats.ServiceState stracker = r.getTracker();
    if (stracker != null) {
        stracker.setStarted(true, mAm.mProcessStats.getMemFactorLocked(),
        r.lastActivity);
    }
    r.callStart = false;
    synchronized (r.stats.getBatteryStats()) {
```

```
            r.stats.startRunningLocked();
    }
    String error = bringUpServiceLocked(r, service.getFlags(), callerFg,
    false);
    if (error != null) {
        return new ComponentName("!!", error);
    }

    if (r.startRequested && addToStarting) {
        boolean first = smap.mStartingBackground.size() == 0;
        smap.mStartingBackground.add(r);
        r.startingBgTimeout = SystemClock.uptimeMillis() + BG_START_TIMEOUT;
        if (DEBUG_DELAYED_SERVICE) {
            RuntimeException here = new RuntimeException("here");
            here.fillInStackTrace();
            Slog.v(TAG, "Starting background (first=" + first + "): " + r, here);
        } else if (DEBUG_DELAYED_STARTS) {
            Slog.v(TAG, "Starting background (first=" + first + "): " + r);
        }
        if (first) {
            smap.rescheduleDelayedStarts();
        }
    } else if (callerFg) {
        smap.ensureNotStartingBackground(r);
    }

    return r.name;
}
```

在上述代码中，ServiceRecord 描述的是一个 Service 记录，ServiceRecord 一直贯穿着整个 Service 的启动过程。startServiceInnerLocked 方法并没有完成具体的启动工作，而是把后续的工作交给了 bringUpServiceLocked 方法来处理，在 bringUpServiceLocked 方法中又调用了 realStartServiceLocked 方法。从名字上来看，这个方法应该是真正地启动一个 Service，它的实现如下所示。

```
private final void realStartServiceLocked(ServiceRecord r,
        ProcessRecord app, boolean execInFg) throws RemoteException {
    ...
    boolean created = false;
```

```java
    try {
        String nameTerm;
        int lastPeriod = r.shortName.lastIndexOf('.');
        nameTerm = lastPeriod >= 0 ? r.shortName.substring(lastPeriod) :
        r.shortName;
        if (LOG_SERVICE_START_STOP) {
            EventLogTags.writeAmCreateService(
                    r.userId, System.identityHashCode(r), nameTerm, r.app.
                    uid, r.app.pid);
        }
        synchronized (r.stats.getBatteryStats()) {
            r.stats.startLaunchedLocked();
        }
        mAm.ensurePackageDexOpt(r.serviceInfo.packageName);
        app.forceProcessStateUpTo(ActivityManager.PROCESS_STATE_SERVICE);
        app.thread.scheduleCreateService(r, r.serviceInfo,
                mAm.compatibilityInfoForPackageLocked(r.serviceInfo.
                applicationInfo),
                app.repProcState);
        r.postNotification();
        created = true;
    } catch (DeadObjectException e) {
        Slog.w(TAG, "Application dead when creating service " + r);
        mAm.appDiedLocked(app);
    } finally {
        if (!created) {
            app.services.remove(r);
            r.app = null;
            scheduleServiceRestartLocked(r, false);
            return;
        }
    }

    requestServiceBindingsLocked(r, execInFg);

    updateServiceClientActivitiesLocked(app, null, true);

    // If the service is in the started state, and there are no
    // pending arguments, then fake up one so its onStartCommand() will
    // be called.
```

```
    if (r.startRequested && r.callStart && r.pendingStarts.size() == 0) {
        r.pendingStarts.add(new ServiceRecord.StartItem(r, false, r.make-
        NextStartId(),
            null, null));
    }

    sendServiceArgsLocked(r, execInFg, true);
    ...
}
```

在 realStartServiceLocked 方法中，首先通过 app.thread 的 scheduleCreateService 方法来创建 Service 对象并调用其 onCreate，接着再通过 sendServiceArgsLocked 方法来调用 Service 的其他方法，比如 onStartCommand，这两个过程均是进程间通信。app.thread 对象是 IApplicationThread 类型，它实际上是一个 Binder，它的具体实现是 ApplicationThread 和 ApplicationThreadNative，在 9.2 节已经对这个问题做了说明。由于 ApplicationThread 继承了 ApplicationThreadNative，因此只需要看 ApplicationThread 对 Service 启动过程的处理即可，这对应着它的 scheduleCreateService 方法，如下所示。

```
public final void scheduleCreateService(IBinder token,
        ServiceInfo info,CompatibilityInfo compatInfo, int processState) {
    updateProcessState(processState, false);
    CreateServiceData s = new CreateServiceData();
    s.token = token;
    s.info = info;
    s.compatInfo = compatInfo;

    sendMessage(H.CREATE_SERVICE, s);
}
```

很显然，这个过程和 Activity 的启动过程是类似的，都是通过发送消息给 Handler H 来完成的。H 会接收这个 CREATE_SERVICE 消息并通过 ActivityThread 的 handleCreateService 方法来完成 Service 的最终启动，handleCreateService 的源码如下所示。

```
private void handleCreateService(CreateServiceData data) {
    // If we are getting ready to gc after going to the background, well
    // we are back active so skip it.
    unscheduleGcIdler();

    LoadedApk packageInfo = getPackageInfoNoCheck(
```

```
            data.info.applicationInfo, data.compatInfo);
    Service service = null;
    try {
        java.lang.ClassLoader cl = packageInfo.getClassLoader();
        service = (Service) cl.loadClass(data.info.name).newInstance();
    } catch (Exception e) {
        if (!mInstrumentation.onException(service, e)) {
            throw new RuntimeException(
                "Unable to instantiate service " + data.info.name
                + ": " + e.toString(), e);
        }
    }

    try {
        if (localLOGV) Slog.v(TAG, "Creating service " + data.info.name);

        ContextImpl context = ContextImpl.createAppContext(this, packageInfo);
        context.setOuterContext(service);

        Application app = packageInfo.makeApplication(false, mInstrumen-
        tation);
        service.attach(context, this, data.info.name, data.token, app,
                ActivityManagerNative.getDefault());
        service.onCreate();
        mServices.put(data.token, service);
        try {
            ActivityManagerNative.getDefault().serviceDoneExecuting(
                    data.token, 0, 0, 0);
        } catch (RemoteException e) {
            // nothing to do.
        }
    } catch (Exception e) {
        if (!mInstrumentation.onException(service, e)) {
            throw new RuntimeException(
                "Unable to create service " + data.info.name
                + ": " + e.toString(), e);
        }
    }
}
```

handleCreateService 主要完成了如下几件事。

首先通过类加载器创建 Service 的实例。

然后创建 Application 对象并调用其 onCreate，当然 Application 的创建过程只会有一次。

接着创建 ConTextImpl 对象并通过 Service 的 attach 方法建立二者之间的关系，这个过程和 Activity 实际上是类似的，毕竟 Service 和 Activity 都是一个 Context。

最后调用 Service 的 onCreate 方法并将 Service 对象存储到 ActivityThread 中的一个列表中。这个列表的定义如下所示。

```
final ArrayMap<IBinder, Service> mServices = new ArrayMap<IBinder, Service>()
```

由于 Service 的 onCreate 方法被执行了，这也意味着 Service 已经启动了。除此之外，ActivityThread 中还会通过 handleServiceArgs 方法调用 Service 的 onStartCommand 方法，如下所示。

```
private void handleServiceArgs(ServiceArgsData data) {
    Service s = mServices.get(data.token);
    if (s != null) {
        try {
            if (data.args != null) {
                data.args.setExtrasClassLoader(s.getClassLoader());
                data.args.prepareToEnterProcess();
            }
            int res;
            if (!data.taskRemoved) {
                res = s.onStartCommand(data.args, data.flags, data.startId);
            } else {
                s.onTaskRemoved(data.args);
                res = Service.START_TASK_REMOVED_COMPLETE;
            }

            QueuedWork.waitToFinish();

            try {
                ActivityManagerNative.getDefault().serviceDoneExecuting(
                        data.token, 1, data.startId, res);
            } catch (RemoteException e) {
                // nothing to do.
```

```
            }
            ensureJitEnabled();
        } catch (Exception e) {
            if (!mInstrumentation.onException(s, e)) {
                throw new RuntimeException(
                        "Unable to start service " + s
                        + " with " + data.args + ": " + e.toString(), e);
            }
        }
    }
}
```

到这里，Service 的启动过程已经分析完了，下面分析 Service 的绑定过程。

9.3.2 Service 的绑定过程

和 Service 的启动过程一样，Service 的绑定过程也是从 ContextWrapper 开始的，如下所示。

```
public boolean bindService(Intent service, ServiceConnection conn,
        int flags) {
    return mBase.bindService(service, conn, flags);
}
```

这个过程和 Service 的启动过程是类似的，mBase 同样是 ContextImpl 类型的对象。ContextImpl 的 bindService 方法最终会调用自己的 bindServiceCommon 方法，如下所示。

```
private boolean bindServiceCommon(Intent service, ServiceConnection conn,
int flags,
        UserHandle user) {
    IServiceConnection sd;
    if (conn == null) {
        throw new IllegalArgumentException("connection is null");
    }
    if (mPackageInfo != null) {
        sd = mPackageInfo.getServiceDispatcher(conn, getOuterContext(),
                mMainThread.getHandler(), flags);
    } else {
        throw new RuntimeException("Not supported in system context");
    }
```

```
        validateServiceIntent(service);
        try {
            IBinder token = getActivityToken();
            if (token == null && (flags&BIND_AUTO_CREATE) == 0 && mPackageInfo !=
null
                    && mPackageInfo.getApplicationInfo().targetSdkVersion
                    < android.os.Build.VERSION_CODES.ICE_CREAM_SANDWICH) {
                flags |= BIND_WAIVE_PRIORITY;
            }
            service.prepareToLeaveProcess();
            int res = ActivityManagerNative.getDefault().bindService(
                mMainThread.getApplicationThread(), getActivityToken(),
                service, service.resolveTypeIfNeeded(getContentResolver()),
                sd, flags, user.getIdentifier());
            if (res < 0) {
                throw new SecurityException(
                        "Not allowed to bind to service " + service);
            }
            return res != 0;
        } catch (RemoteException e) {
            return false;
        }
    }
```

bindServiceCommon 方法主要完成如下两件事情。

首先将客户端的 ServiceConnection 对象转化为 ServiceDispatcher.InnerConnection 对象。之所以不能直接使用 ServiceConnection 对象，这是因为服务的绑定有可能是跨进程的，因此 ServiceConnection 对象必须借助于 Binder 才能让远程服务端回调自己的方法，而 ServiceDispatcher 的内部类 InnerConnection 刚好充当了 Binder 这个角色。那么 ServiceDispatcher 的作用是什么呢？其实 ServiceDispatcher 起着连接 ServiceConnection 和 InnerConnection 的作用。这个过程由 LoadedApk 的 getServiceDispatcher 方法来完成，它的实现如下：

```
    public final IServiceConnection getServiceDispatcher(ServiceConnection c,
            Context context, Handler handler, int flags) {
        synchronized (mServices) {
            LoadedApk.ServiceDispatcher sd = null;
            ArrayMap<ServiceConnection, LoadedApk.ServiceDispatcher> map =
```

```
            mServices.get(context);
        if (map != null) {
            sd = map.get(c);
        }
        if (sd == null) {
            sd = new ServiceDispatcher(c, context, handler, flags);
            if (map == null) {
                map = new ArrayMap<ServiceConnection, LoadedApk.Service-
                    Dispatcher>();
                mServices.put(context, map);
            }
            map.put(c, sd);
        } else {
            sd.validate(context, handler);
        }
        return sd.getIServiceConnection();
    }
}
```

在上面的代码中，mServices 是一个 ArrayMap，它存储了一个应用当前活动的 ServiceConnection 和 ServiceDispatcher 的映射关系，它的定义如下所示。

```
private final ArrayMap<Context, ArrayMap<ServiceConnection, LoadedApk.
ServiceDispatcher>> mServices = new ArrayMap<Context, ArrayMap
<ServiceConnection, LoadedApk.ServiceDispatcher>>();
```

系统首先会查找是否存在相同的 ServiceConnection，如果不存在就重新创建一个 ServiceDispatcher 对象并将其存储在 mServices 中，其中映射关系的 key 是 ServiceConnection，value 是 ServiceDispatcher，在 ServiceDispatcher 的内部又保存了 ServiceConnection 和 InnerConnection 对象。当 Service 和客户端建立连接后，系统会通过 InnerConnection 来调用 ServiceConnection 中的 onServiceConnected 方法，这个过程有可能是跨进程的。当 ServiceDispatcher 创建好了以后，getServiceDispatcher 会返回其保存的 InnerConnection 对象。

接着 bindServiceCommon 方法会通过 AMS 来完成 Service 的具体的绑定过程，这对应于 AMS 的 bindService 方法，如下所示。

```
public int bindService(IApplicationThread caller, IBinder token,
        Intent service, String resolvedType,
        IServiceConnection connection, int flags, int userId) {
```

```
    enforceNotIsolatedCaller("bindService");

    // Refuse possible leaked file descriptors
    if (service != null && service.hasFileDescriptors() == true) {
        throw new IllegalArgumentException("File descriptors passed in Intent");
    }

    synchronized(this) {
        return mServices.bindServiceLocked(caller, token, service,
        resolvedType,
            connection, flags, userId);
    }
}
```

接下来，AMS 会调用 ActiveServices 的 bindServiceLocked 方法，bindServiceLocked 再调用 bringUpServiceLocked，bringUpServiceLocked 又会调用 realStartServiceLocked 方法，realStartServiceLocked 方法的执行逻辑和 9.3.1 节中的逻辑类似，最终都是通过 ApplicationThread 来完成 Service 实例的创建并执行其 onCreate 方法，这里不再重复讲解了。和启动 Service 不同的是，Service 的绑定过程会调用 app.thread 的 scheduleBindService 方法，这个过程的实现在 ActiveServices 的 requestServiceBindingLocked 方法中，如下所示。

```
private final boolean requestServiceBindingLocked(ServiceRecord r,
        IntentBindRecord i, boolean execInFg, boolean rebind) {
    if (r.app == null || r.app.thread == null) {
        // If service is not currently running, can't yet bind.
        return false;
    }
    if ((!i.requested || rebind) && i.apps.size() > 0) {
        try {
            bumpServiceExecutingLocked(r, execInFg, "bind");
            r.app.forceProcessStateUpTo(ActivityManager.PROCESS_STATE_
            SERVICE);
            r.app.thread.scheduleBindService(r, i.intent.getIntent(), rebind,
                    r.app.repProcState);
            if (!rebind) {
                i.requested = true;
            }
            i.hasBound = true;
            i.doRebind = false;
```

```
            } catch (RemoteException e) {
                if (DEBUG_SERVICE) Slog.v(TAG, "Crashed while binding " + r);
                return false;
            }
        }
    }
    return true;
}
```

在上述代码中，app.thread 这个对象多次出现过，对于它我们应该再熟悉不过了，它实际上就是 ApplicationThread。ApplicationThread 的一系列以 schedule 开头的方法，其内部都是通过 Handler H 来中转的，对于 scheduleBindService 方法来说也是如此，它的实现如下所示。

```
public final void scheduleBindService(IBinder token, Intent intent,
        boolean rebind, int processState) {
    updateProcessState(processState, false);
    BindServiceData s = new BindServiceData();
    s.token = token;
    s.intent = intent;
    s.rebind = rebind;

    if (DEBUG_SERVICE)
        Slog.v(TAG, "scheduleBindService token=" + token + " intent=" +
        intent + " uid="
            + Binder.getCallingUid() + " pid=" + Binder.getCallingPid());
    sendMessage(H.BIND_SERVICE, s);
}
```

在 H 内部，接收到 BIND_SERVICE 这类消息时，会交给 ActivityThread 的 handleBindService 方法来处理。在 handleBindService 中，首先根据 Service 的 token 取出 Service 对象，然后调用 Service 的 onBind 方法，Service 的 onBind 方法会返回一个 Binder 对象给客户端使用，这个过程我们在 Service 的开发过程中应该都比较熟悉了。原则上来说，Service 的 onBind 方法被调用以后，Service 就处于绑定状态了，但是 onBind 方法是 Service 的方法，这个时候客户端并不知道已经成功连接 Service 了，所以还必须调用客户端的 ServiceConnection 中的 onServiceConnected,这个过程是由 ActivityManagerNative.getDefault() 的 publishService 方法来完成的，而前面多次提到，ActivityManagerNative.getDefault()就是 AMS。handleBindService 的实现过程如下所示。

```
private void handleBindService(BindServiceData data) {
```

```
        Service s = mServices.get(data.token);
        if (DEBUG_SERVICE)
            Slog.v(TAG, "handleBindService s=" + s + " rebind=" + data.rebind);
        if (s != null) {
            try {
                data.intent.setExtrasClassLoader(s.getClassLoader());
                data.intent.prepareToEnterProcess();
                try {
                    if (!data.rebind) {
                        IBinder binder = s.onBind(data.intent);
                        ActivityManagerNative.getDefault().publishService(
                                data.token, data.intent, binder);
                    } else {
                        s.onRebind(data.intent);
                        ActivityManagerNative.getDefault().serviceDoneExecuting(
                                data.token, 0, 0, 0);
                    }
                    ensureJitEnabled();
                } catch (RemoteException ex) {
                }
            } catch (Exception e) {
                if (!mInstrumentation.onException(s, e)) {
                    throw new RuntimeException(
                            "Unable to bind to service " + s
                            + " with " + data.intent + ": " + e.toString(), e);
                }
            }
        }
    }
```

Service 有一个特性，当多次绑定同一个 Service 时，Service 的 onBind 方法只会执行一次，除非 Service 被终止了。当 Service 的 onBind 执行以后，系统还需要告知客户端已经成功连接 Service 了。根据上面的分析，这个过程由 AMS 的 publishService 方法来实现，它的源码如下所示。

```
public void publishService(IBinder token,Intent intent,IBinder service) {
    // Refuse possible leaked file descriptors
    if (intent != null && intent.hasFileDescriptors() == true) {
        throw new IllegalArgumentException("File descriptors passed in
        Intent");
```

```
        }

        synchronized(this) {
            if (!(token instanceof ServiceRecord)) {
                throw new IllegalArgumentException("Invalid service token");
            }
            mServices.publishServiceLocked((ServiceRecord)token, intent,
            service);
        }
    }
```

从上面代码可以看出，AMS 的 publishService 方法将具体的工作交给了 ActiveServices 类型的 mServices 对象来处理。ActiveServices 的 publishServiceLocked 方法看起来很复杂，但其实核心代码就只有一句话：c.conn.connected(r.name, service)，其中 c 的类型是 ConnectionRecord，c.conn 的类型是 ServiceDispatcher.InnerConnection，service 就是 Service 的 onBind 方法返回的 Binder 对象。为了分析具体的逻辑，下面看一下 ServiceDispatcher. InnerConnection 的定义，如下所示。

```
private static class InnerConnection extends IServiceConnection.Stub {
    final WeakReference<LoadedApk.ServiceDispatcher> mDispatcher;

    InnerConnection(LoadedApk.ServiceDispatcher sd) {
        mDispatcher = new WeakReference<LoadedApk.ServiceDispatcher>(sd);
    }

    public void connected(ComponentName name, IBinder service) throws
    RemoteException {
        LoadedApk.ServiceDispatcher sd = mDispatcher.get();
        if (sd != null) {
            sd.connected(name, service);
        }
    }
}
```

从 InnerConnection 的定义可以看出，它的 connected 方法又调用了 ServiceDispatcher 的 connected 方法，ServiceDispatcher 的 connected 方法的实现如下所示。

```
public void connected(ComponentName name, IBinder service) {
    if (mActivityThread != null) {
```

```
            mActivityThread.post(new RunConnection(name, service, 0));
        } else {
            doConnected(name, service);
        }
    }
```

对于 Service 的绑定过程来说，ServiceDispatcher 的 mActivityThread 是一个 Handler，其实它就是 ActivityThread 中的 H，从前面 ServiceDispatcher 的创建过程来说，mActivityThread 不会为 null，这样一来，RunConnection 就可以经由 H 的 post 方法从而运行在主线程中，因此，客户端 ServiceConnection 中的方法是在主线程被回调的。RunConnection 的定义如下所示。

```
private final class RunConnection implements Runnable {
    RunConnection(ComponentName name, IBinder service, int command) {
        mName = name;
        mService = service;
        mCommand = command;
    }

    public void run() {
        if (mCommand == 0) {
            doConnected(mName, mService);
        } else if (mCommand == 1) {
            doDeath(mName, mService);
        }
    }

    final ComponentName mName;
    final IBinder mService;
    final int mCommand;
}
```

很显然，RunConnection 的 run 方法也是简单调用了 ServiceDispatcher 的 doConnected 方法，由于 ServiceDispatcher 内部保存了客户端的 ServiceConnection 对象，因此它可以很方便地调用 ServiceConnection 对象的 onServiceConnected 方法，如下所示。

```
// If there is a new service, it is now connected.
if (service != null) {
    mConnection.onServiceConnected(name, service);
}
```

客户端的 onServiceConnected 方法执行后，Service 的绑定过程也就分析完成了，至于 Service 的停止过程和解除绑定的过程，系统的执行过程是类似的，读者可以自行分析源码，这里就不再分析了。

9.4 BroadcastReceiver 的工作过程

本节将介绍 BroadcastReceiver 的工作过程，主要包含两方面的内容，一个是广播的注册过程，另一个是广播的发送和接收过程。这里先简单回顾一下广播的使用方法，首先要定义广播接收者，只需要继承 BroadcastReceiver 并重写 onReceive 方法即可，下面是一个典型的广播接收者的实现：

```
public class MyReceiver extends BroadcastReceiver {

    @Override
    public void onReceive(Context context, Intent intent) {
        // onReceive 函数不能做耗时的事情，参考值：10s 以内
        Log.d("scott", "on receive action=" + intent.getAction());
        String action = intent.getAction();
        // do some works
    }

}
```

定义好了广播接收者，接着还需要注册广播接收者，注册分为两种方式，既可以在 AndroidManifest 文件中静态注册，也可以通过代码动态注册。

静态注册的示例如下：

```
<receiver android:name=".MyReceiver" >
    <intent-filter>
        <action android:name="com.ryg.receiver.LAUNCH" />
    </intent-filter>
</receiver>
```

通过代码来动态注册广播也是很简单的，如下所示。需要注意的是，动态注册的广播需要在合适的时机进行解注册，解注册采用 unregisterReceiver 方法。

```
IntentFilter filter = new IntentFilter();
filter.addAction("com.ryg.receiver.LAUNCH");
registerReceiver(new MyReceiver(), filter);
```

前面两步都完成了以后，就可以通过 send 方法来发送广播了，如下所示。

```
Intent intent = new Intent();
intent.setAction("com.ryg.receiver.LAUNCH");
sendBroadcast(intent);
```

上面简单回顾了广播的使用方法，下面就开始分析广播的工作过程，首先分析广播的注册过程，接着再分析广播的发送和接收过程。

9.4.1 广播的注册过程

广播的注册分为静态注册和动态注册，其中静态注册的广播在应用安装时由系统自动完成注册，具体来说是由 PMS（PackageManagerService）来完成整个注册过程的，除了广播以外，其他三大组件也都是在应用安装时由 PMS 解析并注册的。这里只分析广播的动态注册的过程，动态注册的过程是从 ContextWrapper 的 registerReceiver 方法开始的，和 Activity 以及 Service 一样。ContextWrapper 并没有做实际的工作，而是将注册过程直接交给了 ContextImpl 来完成，如下所示。

```
public Intent registerReceiver(
    BroadcastReceiver receiver, IntentFilter filter) {
    return mBase.registerReceiver(receiver, filter);
}
```

ContextImpl 的 registerReceiver 方法调用了自己的 registerReceiverInternal 方法，它的实现如下所示。

```
private Intent registerReceiverInternal(BroadcastReceiver receiver, int userId,
        IntentFilter filter, String broadcastPermission,
        Handler scheduler, Context context) {
    IIntentReceiver rd = null;
    if (receiver != null) {
        if (mPackageInfo != null && context != null) {
            if (scheduler == null) {
                scheduler = mMainThread.getHandler();
            }
```

```
            rd = mPackageInfo.getReceiverDispatcher(
                receiver, context, scheduler,
                mMainThread.getInstrumentation(), true);
        } else {
            if (scheduler == null) {
                scheduler = mMainThread.getHandler();
            }
            rd = new LoadedApk.ReceiverDispatcher(
                receiver, context, scheduler, null, true).getIIntent-
                Receiver();
        }
    }
    try {
        return ActivityManagerNative.getDefault().registerReceiver(
            mMainThread.getApplicationThread(), mBasePackageName,
            rd, filter, broadcastPermission, userId);
    } catch (RemoteException e) {
        return null;
    }
}
```

在上面的代码中，系统首先从 mPackageInfo 获取 IIntentReceiver 对象，然后再采用跨进程的方式向 AMS 发送广播注册的请求。之所以采用 IIntentReceiver 而不是直接采用 BroadcastReceiver，这是因为上述注册过程是一个进程间通信的过程，而 BroadcastReceiver 作为 Android 的一个组件是不能直接跨进程传递的，所以需要通过 IIntentReceiver 来中转一下。毫无疑问，IIntentReceiver 必须是一个 Binder 接口，它的具体实现是 LoadedApk.Receiver-Dispatcher.InnerReceiver，ReceiverDispatcher 的内部同时保存了 BroadcastReceiver 和 InnerReceiver，这样当接收到广播时，ReceiverDispatcher 可以很方便地调用 BroadcastReceiver 的 onReceive 方法，具体会在 9.4.2 节中说明。可以发现，BroadcastReceiver 的这个过程和 Service 的实现原理类似，Service 也有一个叫 ServiceDispatcher 的类，并且其内部类 InnerConnection 也是一个 Binder 接口，作用同样也是为了进程间通信，这一点在 9.3.2 节中已经描述过了，这里不再重复说明。

关于 ActivityManagerNative.getDefault()，这里就不用再做说明了，它就是 AMS，在前面的章节中已经多次提到它。下面看一下 ReceiverDispatcher 的 getIIntentReceiver 的实现，如下所示。很显然，getReceiverDispatcher 方法重新创建了一个 ReceiverDispatcher 对象并将其保存的 InnerReceiver 对象作为返回值返回，其中 InnerReceiver 对象和 BroadcastReceiver

都是在 ReceiverDispatcher 的构造方法中被保存起来的。

```java
public IIntentReceiver getReceiverDispatcher(BroadcastReceiver r,
        Context context, Handler handler,
        Instrumentation instrumentation, boolean registered) {
    synchronized (mReceivers) {
        LoadedApk.ReceiverDispatcher rd = null;
        ArrayMap<BroadcastReceiver, LoadedApk.ReceiverDispatcher> map = null;
        if (registered) {
            map = mReceivers.get(context);
            if (map != null) {
                rd = map.get(r);
            }
        }
        if (rd == null) {
            rd = new ReceiverDispatcher(r, context, handler,
                    instrumentation, registered);
            if (registered) {
                if (map == null) {
                    map = new ArrayMap<BroadcastReceiver, LoadedApk.
                    ReceiverDispatcher>();
                    mReceivers.put(context, map);
                }
                map.put(r, rd);
            }
        } else {
            rd.validate(context, handler);
        }
        rd.mForgotten = false;
        return rd.getIIntentReceiver();
    }
}
```

由于注册广播的真正实现过程是在 AMS 中，因此我们需要看一下 AMS 的具体实现。AMS 的 registerReceiver 方法看起来很长，其实关键点就只有下面一部分，最终会把远程的 InnerReceiver 对象以及 IntentFilter 对象存储起来，这样整个广播的注册过程就完成了，代码如下所示。

```java
public Intent registerReceiver(IApplicationThread caller, String callerPackage,
    IIntentReceiver receiver, IntentFilter filter, String permission, int
    userId) {
    ...
```

```
    mRegisteredReceivers.put(receiver.asBinder(), rl);

    BroadcastFilter bf = new BroadcastFilter(filter, rl, callerPackage,
            permission, callingUid, userId);
    rl.add(bf);
    if (!bf.debugCheck()) {
        Slog.w(TAG, "==> For Dynamic broadast");
    }
    mReceiverResolver.addFilter(bf);
    ...
}
```

9.4.2 广播的发送和接收过程

上面分析了广播的注册过程，可以发现注册过程的逻辑还是比较简单的，下面来分析广播的发送和接收过程。当通过 send 方法来发送广播时，AMS 会查找出匹配的广播接收者并将广播发送给它们处理。广播的发送有几种类型：普通广播、有序广播和粘性广播，有序广播和粘性广播与普通广播相比具有不同的特性，但是它们的发送/接收过程的流程是类似的，因此这里只分析普通广播的实现。

广播的发送和接收，其本质是一个过程的两个阶段。这里从广播的发送可以说起，广播的发送仍然开始于 ContextWrapper 的 sendBroadcast 方法，之所以不是 Context，那是因为 Context 的 sendBroadcast 是一个抽象方法。和广播的注册过程一样，ContextWrapper 的 sendBroadcast 方法仍然什么都不做，只是把事情交给 ContextImpl 去处理，ContextImpl 的 sendBroadcast 方法的源码如下所示。

```
public void sendBroadcast(Intent intent) {
    warnIfCallingFromSystemProcess();
    String resolvedType = intent.resolveTypeIfNeeded(getContentResolver());
    try {
        intent.prepareToLeaveProcess();
        ActivityManagerNative.getDefault().broadcastIntent(
            mMainThread.getApplicationThread(), intent, resolvedType, null,
            Activity.RESULT_OK, null, null, null, AppOpsManager.OP_NONE,
            false, false,
            getUserId());
    } catch (RemoteException e) {
    }
}
```

从上面的代码来看，ContextImpl 也是几乎什么事都没干，它直接向 AMS 发起了一个异步请求用于发送广播。因此，下面直接看 AMS 对广播发送过程的处理，AMS 的 broadcastIntent 方法的源码如下所示。

```java
public final int broadcastIntent(IApplicationThread caller,
        Intent intent, String resolvedType, IIntentReceiver resultTo,
        int resultCode, String resultData, Bundle map,
        String requiredPermission, int appOp, boolean serialized, boolean
        sticky, int userId) {
    enforceNotIsolatedCaller("broadcastIntent");
    synchronized(this) {
        intent = verifyBroadcastLocked(intent);

        final ProcessRecord callerApp = getRecordForAppLocked(caller);
        final int callingPid = Binder.getCallingPid();
        final int callingUid = Binder.getCallingUid();
        final long origId = Binder.clearCallingIdentity();
        int res = broadcastIntentLocked(callerApp,
                callerApp != null ? callerApp.info.packageName : null,
                intent, resolvedType, resultTo,
                resultCode, resultData, map, requiredPermission, appOp,
                serialized, sticky,
                callingPid, callingUid, userId);
        Binder.restoreCallingIdentity(origId);
        return res;
    }
}
```

从上面代码来看，broadcastIntent 调用了 broadcastIntentLocked 方法，AMS 的 broadcastIntentLocked 方法有 436 行代码，看起来比较复杂。在代码最开始有如下一行：

```java
// By default broadcasts do not go to stopped apps.
intent.addFlags(Intent.FLAG_EXCLUDE_STOPPED_PACKAGES);
```

这表示在 Android 5.0 中，默认情况下广播不会发送给已经停止的应用，其实不仅仅是 Android 5.0，从 Android 3.1 开始广播已经具有这种特性了。这是因为系统在 Android 3.1 中为 Intent 添加了两个标记位，分别是 FLAG_INCLUDE_STOPPED_PACKAGES 和 FLAG_EXCLUDE_STOPPED_PACKAGES，用来控制广播是否要对处于停止状态的应用起作用，它们的含义如下所示。

FLAG_INCLUDE_STOPPED_PACKAGES

表示包含已经停止的应用，这个时候广播会发送给已经停止的应用。

FLAG_EXCLUDE_STOPPED_PACKAGES

表示不包含已经停止的应用，这个时候广播不会发送给已经停止的应用。

从 Android 3.1 开始，系统为所有广播默认添加了 FLAG_EXCLUDE_STOPPED_PACKAGES 标志，这样做是为了防止广播无意间或者在不必要的时候调起已经停止运行的应用。如果的确需要调起未启动的应用，那么只需要为广播的 Intent 添加 FLAG_INCLUDE_STOPPED_PACKAGES 标记即可。当 FLAG_EXCLUDE_STOPPED_PACKAGES 和 FLAG_INCLUDE_STOPPED_PACKAGES 两种标记位共存时，以 FLAG_INCLUDE_STOPPED_PACKAGES 为准。这里需要补充一下，一个应用处于停止状态分为两种情形：第一种是应用安装后未运行，第二种是应用被手动或者其他应用强停了。Android 3.1 中广播的这个特性同样会影响开机广播，从 Android 3.1 开始，处于停止状态的应用同样无法接收到开机广播，而在 Android 3.1 之前，处于停止状态的应用是可以收到开机广播的。

在 broadcastIntentLocked 的内部，会根据 intent-filter 查找出匹配的广播接收者并经过一系列的条件过滤，最终会将满足条件的广播接收者添加到 BroadcastQueue 中，接着 BroadcastQueue 就会将广播发送给相应的广播接收者，这个过程的源码如下所示。

```
if ((receivers != null && receivers.size() > 0)
        || resultTo != null) {
    BroadcastQueue queue = broadcastQueueForIntent(intent);
    BroadcastRecord r = new BroadcastRecord(queue, intent, callerApp,
            callerPackage, callingPid, callingUid, resolvedType,
            requiredPermission, appOp, receivers, resultTo, resultCode,
            resultData, map, ordered, sticky, false, userId);
    if (DEBUG_BROADCAST) Slog.v(
            TAG, "Enqueueing ordered broadcast " + r
            + ": prev had " + queue.mOrderedBroadcasts.size());
    if (DEBUG_BROADCAST) {
        int seq = r.intent.getIntExtra("seq", -1);
        Slog.i(TAG, "Enqueueing broadcast " + r.intent.getAction() + " seq="
            + seq);
    }
```

```
        boolean replaced = replacePending && queue.replaceOrderedBroadcast-
        Locked(r);
        if (!replaced) {
            queue.enqueueOrderedBroadcastLocked(r);
            queue.scheduleBroadcastsLocked();
        }
}
```

下面看一下 BroadcastQueue 中广播的发送过程的实现，如下所示。

```
public void scheduleBroadcastsLocked() {
    if (DEBUG_BROADCAST) Slog.v(TAG, "Schedule broadcasts ["
            + mQueueName + "]: current="
            + mBroadcastsScheduled);

    if (mBroadcastsScheduled) {
        return;
    }
    mHandler.sendMessage(mHandler.obtainMessage(BROADCAST_INTENT_MSG,
    this));
    mBroadcastsScheduled = true;
}
```

BroadcastQueue 的 scheduleBroadcastsLocked 方法并没有立即发送广播，而是发送了一个 BROADCAST_INTENT_MSG 类型的消息，BroadcastQueue 收到消息后会调用 processNextBroadcast 方法，BroadcastQueue 的 processNextBroadcast 方法对普通广播的处理如下所示。

```
// First, deliver any non-serialized broadcasts right away.
while (mParallelBroadcasts.size() > 0) {
    r = mParallelBroadcasts.remove(0);
    r.dispatchTime = SystemClock.uptimeMillis();
    r.dispatchClockTime = System.currentTimeMillis();
    final int N = r.receivers.size();
    if (DEBUG_BROADCAST_LIGHT) Slog.v(TAG, "Processing parallel broadcast ["
            + mQueueName + "] " + r);
    for (int i=0; i<N; i++) {
        Object target = r.receivers.get(i);
        if (DEBUG_BROADCAST) Slog.v(TAG,
                "Delivering non-ordered on["+mQueueName + "] to registered "
```

```
            + target + ": " + r);
        deliverToRegisteredReceiverLocked(r, (BroadcastFilter)target, false);
    }
    addBroadcastToHistoryLocked(r);
    if (DEBUG_BROADCAST_LIGHT) Slog.v(TAG, "Done with parallel broadcast ["
            + mQueueName + "] " + r);
}
```

可以看到，无序广播存储在 mParallelBroadcasts 中，系统会遍历 mParallelBroadcasts 并将其中的广播发送给它们所有的接收者，具体的发送过程是通过 deliverToRegisteredReceiverLocked 方法来实现的。deliverToRegisteredReceiverLocked 方法负责将一个广播发送给一个特定的接收者，它内部调用了 performReceiveLocked 方法来完成具体的发送过程：

```
performReceiveLocked(filter.receiverList.app, filter.receiverList.receiver,
    new Intent(r.intent), r.resultCode, r.resultData,
    r.resultExtras, r.ordered, r.initialSticky, r.userId);
```

performReceiveLocked 方法的实现如下所示。由于接收广播会调起应用程序，因此 app.thread 不为 null，根据前面的分析我们知道这里的 app.thread 仍然指 ApplicationThread。

```
private static void performReceiveLocked(ProcessRecord app, IIntentReceiver receiver,
        Intent intent, int resultCode, String data, Bundle extras,
        boolean ordered, boolean sticky, int sendingUser) throws RemoteException {
    // Send the intent to the receiver asynchronously using one-way binder calls.
    if (app != null) {
        if (app.thread != null) {
            // If we have an app thread, do the call through that so it is
            // correctly ordered with other one-way calls.
            app.thread.scheduleRegisteredReceiver(receiver, intent,
                resultCode,
                data, extras, ordered, sticky, sendingUser, app.repProcState);
        } else {
            // Application has died. Receiver doesn't exist.
            throw new RemoteException("app.thread must not be null");
        }
    } else {
        receiver.performReceive(intent, resultCode, data, extras, ordered,
```

```
        sticky, sendingUser);
    }
}
```

ApplicationThread 的 scheduleRegisteredReceiver 的实现比较简单，它通过 InnerReceiver 来实现广播的接收，如下所示。

```
public void scheduleRegisteredReceiver(IIntentReceiver receiver, Intent intent,
        int resultCode, String dataStr, Bundle extras, boolean ordered,
        boolean sticky, int sendingUser, int processState) throws
        RemoteException {
    updateProcessState(processState, false);
    receiver.performReceive(intent, resultCode, dataStr, extras, ordered,
            sticky, sendingUser);
}
```

InnerReceiver 的 performReceive 方法会调用 LoadedApk.ReceiverDispatcher 的 performReceive 方法，LoadedApk.ReceiverDispatcher 的 performReceive 方法的实现如下所示。

```
public void performReceive(Intent intent, int resultCode, String data,
        Bundle extras, boolean ordered, boolean sticky, int sendingUser) {
    if (ActivityThread.DEBUG_BROADCAST) {
        int seq = intent.getIntExtra("seq", -1);
        Slog.i(ActivityThread.TAG, "Enqueueing broadcast " + intent.
        getAction() + " seq=" + seq + " to " + mReceiver);
    }
    Args args = new Args(intent, resultCode, data, extras, ordered,
            sticky, sendingUser);
    if (!mActivityThread.post(args)) {
        if (mRegistered && ordered) {
            IActivityManager mgr = ActivityManagerNative.getDefault();
            if (ActivityThread.DEBUG_BROADCAST) Slog.i(ActivityThread.TAG,
                    "Finishing sync broadcast to " + mReceiver);
            args.sendFinished(mgr);
        }
    }
}
```

在上面的代码中，会创建一个 Args 对象并通过 mActivityThread 的 post 方法来执行 Args

中的逻辑，而 Args 实现了 Runnable 接口。mActivityThread 是一个 Handler，它其实就是 ActivityThread 中的 mH，mH 的类型是 ActivityThread 的内部类 H，关于 H 这个类前面已经介绍过了，这里就不再多说了。在 Args 的 run 方法中有如下几行代码：

```
final BroadcastReceiver receiver = mReceiver;
receiver.setPendingResult(this);
receiver.onReceive(mContext, intent);
```

很显然，这个时候 BroadcastReceiver 的 onReceive 方法被执行了，也就是说应用已经接收到广播了，同时 onReceive 方法是在广播接收者的主线程中被调用的。到这里，整个广播的注册、发送和接收过程已经分析完了，读者应该对广播的整个工作过程有了一定的理解。

9.5　ContentProvider 的工作过程

ContentProvider 的使用方法在第 2 章已经做了介绍，这里再简单说明一下。ContentProvider 是一种内容共享型组件，它通过 Binder 向其他组件乃至其他应用提供数据。当 ContentProvider 所在的进程启动时，ContentProvider 会同时启动并被发布到 AMS 中。需要注意的是，这个时候 ContentProvider 的 onCreate 要先于 Application 的 onCreate 而执行，这在四大组件中是一个少有的现象。

当一个应用启动时，入口方法为 ActivityThread 的 main 方法，main 方法是一个静态方法，在 main 方法中会创建 ActivityThread 的实例并创建主线程的消息队列，然后在 ActivityThread 的 attach 方法中会远程调用 AMS 的 attachApplication 方法并将 ApplicationThread 对象提供给 AMS。ApplicationThread 是一个 Binder 对象，它的 Binder 接口是 IApplicationThread，它主要用于 ActivityThread 和 AMS 之间的通信，这一点在前面多次提到。在 AMS 的 attachApplication 方法中，会调用 ApplicationThread 的 bindApplication 方法，注意这个过程同样是跨进程完成的，bindApplication 的逻辑会经过 ActivityThread 中的 mH Handler 切换到 ActivityThread 中去执行，具体的方法是 handleBindApplication。在 handleBindApplication 方法中，ActivityThread 会创建 Application 对象并加载 ContentProvider。需要注意的是，ActivityThread 会先加载 ContentProvider，然后再调用 Application 的 onCreate 方法，整个流程可以参看图 9-2。

这就是 ContentProvider 的启动过程，ContentProvider 启动后，外界就可以通过它所提供的增删改查这四个接口来操作 ContentProvider 中的数据源，即 insert、delete、update 和 query 四个方法。这四个方法都是通过 Binder 来调用的，外界无法直接访问 ContentProvider，它只能通过 AMS 根据 Uri 来获取对应的 ContentProvider 的 Binder 接口 IConentProvider，然后再通过 IConentProvider 来访问 ContentProvider 中的数据源。

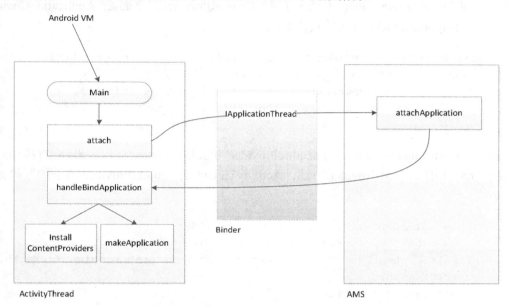

图 9-2　ContentProvider 的启动过程

　　一般来说，ContentProvider 都应该是单实例的。ContentProvider 到底是不是单实例，这是由它的 android:multiprocess 属性来决定的，当 android:multiprocess 为 false 时，ContentProvider 是单实例，这也是默认值；当 android:multiprocess 为 true 时，ContentProvider 为多实例，这个时候在每个调用者的进程中都存在一个 ContentProvider 对象。由于在实际的开发中，并未发现多实例的 ContentProvider 的具体使用场景，官方文档中的解释是这样可以避免进程间通信的开销，但是这在实际开发中仍然缺少使用价值。因此，我们可以简单认为 ContentProvider 都是单实例的。下面分析单实例的 ContentProvider 的启动过程。

　　访问 ContentProvider 需要通过 ContentResolver，ContentResolver 是一个抽象类，通过 Context 的 getContentResolver 方法获取的实际上是 ApplicationContentResolver 对象，ApplicationContentResolver 类继承了 ContentResolver 并实现了 ContentResolver 中的抽象方法。当 ContentProvider 所在的进程未启动时，第一次访问它时就会触发 ContentProvider 的创建，当然这也伴随着 ContentProvider 所在进程的启动。通过 ContentProvider 的四个方法

的任何一个都可以触发 ContentProvider 的启动过程，这里选择 query 方法。

ContentProvider 的 query 方法中，首先会获取 IContentProvider 对象，不管是通过 acquireUnstableProvider 方法还是直接通过 acquireProvider 方法，它们的本质都是一样的，最终都是通过 acquireProvider 方法来获取 ContentProvider。下面是 ApplicationContentResolver 的 acquireProvider 方法的具体实现：

```
protected IContentProvider acquireProvider(Context context, String auth) {
    return mMainThread.acquireProvider(context,
            ContentProvider.getAuthorityWithoutUserId(auth),
            resolveUserIdFromAuthority(auth), true);
}
```

ApplicationContentResolver 的 acquireProvider 方法并没有处理任何逻辑，它直接调用了 ActivityThread 的 acquireProvider 方法，ActivityThread 的 acquireProvider 方法的源码如下所示。

```
public final IContentProvider acquireProvider(
        Context c, String auth, int userId, boolean stable) {
    final IContentProvider provider = acquireExistingProvider(c, auth,
    userId, stable);
    if (provider != null) {
        return provider;
    }

    // There is a possible race here. Another thread may try to acquire
    // the same provider at the same time. When this happens, we want to
      ensure
    // that the first one wins.
    // Note that we cannot hold the lock while acquiring and installing the
    // provider since it might take a long time to run and it could also
      potentially
    // be re-entrant in the case where the provider is in the same process.
    IActivityManager.ContentProviderHolder holder = null;
    try {
        holder = ActivityManagerNative.getDefault().getContentProvider(
                getApplicationThread(), auth, userId, stable);
    } catch (RemoteException ex) {
    }
```

```
    if (holder == null) {
        Slog.e(TAG, "Failed to find provider info for " + auth);
        return null;
    }

    // Install provider will increment the reference count for us, and break
    // any ties in the race.
    holder = installProvider(c, holder, holder.info,
            true /*noisy*/, holder.noReleaseNeeded, stable);
    return holder.provider;
}
```

上面的代码首先会从 ActivityThread 中查找是否已经存在目标 ContentProvider 了，如果存在就直接返回。ActivityThread 中通过 mProviderMap 来存储已经启动的 ContentProvider 对象，mProviderMap 的声明如下所示。

```
final ArrayMap<ProviderKey, ProviderClientRecord> mProviderMap
    = new ArrayMap<ProviderKey, ProviderClientRecord>();
```

如果目前 ContentProvider 没有启动，那么就发送一个进程间请求给 AMS 让其启动目标 ContentProvider，最后再通过 installProvider 方法来修改引用计数。那么 AMS 是如何启动 ContentProvider 的呢？我们知道，ContentProvider 被启动时会伴随着进程的启动，在 AMS 中，首先会启动 ContentProvider 所在的进程，然后再启动 ContentProvider。启动进程是由 AMS 的 startProcessLocked 方法来完成的，其内部主要是通过 Process 的 start 方法来完成一个新进程的启动，新进程启动后其入口方法为 ActivityThread 的 main 方法，如下所示。

```
public static void main(String[] args) {
    SamplingProfilerIntegration.start();

    // CloseGuard defaults to true and can be quite spammy. We
    // disable it here, but selectively enable it later (via
    // StrictMode) on debug builds, but using DropBox, not logs.
    CloseGuard.setEnabled(false);

    Environment.initForCurrentUser();

    // Set the reporter for event logging in libcore
    EventLogger.setReporter(new EventLoggingReporter());
```

```
        Security.addProvider(new AndroidKeyStoreProvider());

        // Make sure TrustedCertificateStore looks in the right place for CA
            certificates
        final File configDir = Environment.getUserConfigDirectory(User-
        Handle.myUserId());
        TrustedCertificateStore.setDefaultUserDirectory(configDir);

        Process.setArgV0("<pre-initialized>");

        Looper.prepareMainLooper();

        ActivityThread thread = new ActivityThread();
        thread.attach(false);

        if (sMainThreadHandler == null) {
            sMainThreadHandler = thread.getHandler();
        }

        AsyncTask.init();

        if (false) {
            Looper.myLooper().setMessageLogging(new
                    LogPrinter(Log.DEBUG, "ActivityThread"));
        }

        Looper.loop();

        throw new RuntimeException("Main thread loop unexpectedly exited");
    }
```

可以看到，ActivityThread 的 main 方法是一个静态方法，在它内部首先会创建 Activity-Thread 的实例并调用 attach 方法来进行一系列初始化，接着就开始进行消息循环了。ActivityThread 的 attach 方法会将 ApplicationThread 对象通过 AMS 的 attachApplication 方法跨进程传递给 AMS，最终 AMS 会完成 ContentProvider 的创建过程，源码如下所示。

```
    try {
        mgr.attachApplication(mAppThread);
    } catch (RemoteException ex) {
        // Ignore
```

}

AMS 的 attachApplication 方法调用了 attachApplicationLocked 方法，attachApplication-Locked 中又调用了 ApplicationThread 的 bindApplication，注意这个过程也是进程间调用，如下所示。

```
thread.bindApplication(processName, appInfo, providers, app.instrumen-
tationClass,
        profilerInfo, app.instrumentationArguments, app.instrumentation-
        Watcher,
        app.instrumentationUiAutomationConnection, testMode, enableOpen-
        GlTrace,
        isRestrictedBackupMode || !normalMode, app.persistent,
        new Configuration(mConfiguration), app.compat, getCommonServices-
        Locked(),
        mCoreSettingsObserver.getCoreSettingsLocked());
```

ActivityThread 的 bindApplication 会发送一个 BIND_APPLICATION 类型的消息给 mH，mH 是一个 Handler，它收到消息后会调用 ActivityThread 的 handleBindApplication 方法，bindApplication 发送消息的过程如下所示。

```
AppBindData data = new AppBindData();
data.processName = processName;
data.appInfo = appInfo;
data.providers = providers;
data.instrumentationName = instrumentationName;
data.instrumentationArgs = instrumentationArgs;
data.instrumentationWatcher = instrumentationWatcher;
data.instrumentationUiAutomationConnection = instrumentationUiConnection;
data.debugMode = debugMode;
data.enableOpenGlTrace = enableOpenGlTrace;
data.restrictedBackupMode = isRestrictedBackupMode;
data.persistent = persistent;
data.config = config;
data.compatInfo = compatInfo;
data.initProfilerInfo = profilerInfo;
sendMessage(H.BIND_APPLICATION, data);
```

ActivityThread 的 handleBindApplication 则完成了 Application 的创建以及 Content-

Provider 的创建，可以分为如下四个步骤。

1. 创建 ContextImpl 和 Instrumentation

```
ContextImpl instrContext = ContextImpl.createAppContext(this, pi);

try {
    java.lang.ClassLoader cl = instrContext.getClassLoader();
    mInstrumentation = (Instrumentation)
        cl.loadClass(data.instrumentationName.getClassName()).newInstance();
} catch (Exception e) {
    throw new RuntimeException(
        "Unable to instantiate instrumentation "
        + data.instrumentationName + ": " + e.toString(), e);
}

mInstrumentation.init(this, instrContext, appContext,
    new ComponentName(ii.packageName, ii.name), data.instrumentation-
    Watcher,
    data.instrumentationUiAutomationConnection);
```

2. 创建 Application 对象

```
Application app = data.info.makeApplication(data.restrictedBackupMode, null);
mInitialApplication = app;
```

3. 启动当前进程的 ContentProvider 并调用其 onCreate 方法

```
List<ProviderInfo> providers = data.providers;
if (providers != null) {
    installContentProviders(app, providers);
    // For process that contains content providers, we want to
    // ensure that the JIT is enabled "at some point".
    mH.sendEmptyMessageDelayed(H.ENABLE_JIT, 10*1000);
}
```

installContentProviders 完成了 ContentProvider 的启动工作，它的实现如下所示。首先会遍历当前进程的 ProviderInfo 的列表并一一调用调用 installProvider 方法来启动它们，接着将已经启动的 ContentProvider 发布到 AMS 中，AMS 会把它们存储在 ProviderMap 中，这样一来外部调用者就可以直接从 AMS 中获取 ContentProvider 了。

```
private void installContentProviders(
        Context context, List<ProviderInfo> providers) {
    final ArrayList<IActivityManager.ContentProviderHolder> results =
        new ArrayList<IActivityManager.ContentProviderHolder>();

    for (ProviderInfo cpi : providers) {
        if (DEBUG_PROVIDER) {
            StringBuilder buf = new StringBuilder(128);
            buf.append("Pub ");
            buf.append(cpi.authority);
            buf.append(": ");
            buf.append(cpi.name);
            Log.i(TAG, buf.toString());
        }
        IActivityManager.ContentProviderHolder cph = installProvider
        (context, null, cpi,
                false /*noisy*/, true /*noReleaseNeeded*/, true /*stable*/);
        if (cph != null) {
            cph.noReleaseNeeded = true;
            results.add(cph);
        }
    }

    try {
        ActivityManagerNative.getDefault().publishContentProviders(
            getApplicationThread(), results);
    } catch (RemoteException ex) {
    }
}
```

下面看一下ContentProvider对象的创建过程,在installProvider方法中有下面一段代码,其通过类加载器完成了ContentProvider对象的创建:

```
final java.lang.ClassLoader cl = c.getClassLoader();
localProvider = (ContentProvider)cl.
    loadClass(info.name).newInstance();
provider = localProvider.getIContentProvider();
if (provider == null) {
    Slog.e(TAG, "Failed to instantiate class " +
        info.name + " from sourceDir " +
```

```
        info.applicationInfo.sourceDir);
    return null;
}
if (DEBUG_PROVIDER) Slog.v(
    TAG, "Instantiating local provider " + info.name);
// XXX Need to create the correct context for this provider.
localProvider.attachInfo(c, info);
```

在上述代码中，除了完成 ContentProvider 对象的创建，还会通过 ContentProvider 的 attachInfo 方法来调用它的 onCreate 方法，如下所示。

```
private void attachInfo(Context context,ProviderInfo info,boolean testing) {
    ...
    if (mContext == null) {
        mContext = context;
        if (context != null) {
            mTransport.mAppOpsManager = (AppOpsManager) context.getSystem-
            Service(
                Context.APP_OPS_SERVICE);
        }
        mMyUid = Process.myUid();
        ...
        ContentProvider.this.onCreate();
    }
}
```

到此为止，ContentProvider 已经被创建并且其 onCreate 方法也已经被调用，这意味着 ContentProvider 已经启动完成了。

4．调用 Application 的 onCreate 方法

```
try {
    mInstrumentation.callApplicationOnCreate(app);
} catch (Exception e) {
    if (!mInstrumentation.onException(app, e)) {
        throw new RuntimeException(
            "Unable to create application " + app.getClass().getName()
            + ": " + e.toString(), e);
    }
}
```

经过上面的四个步骤，ContentProvider 已经成功启动，并且其所在进程的 Application 也已经启动，这意味着 ContentProvider 所在的进程已经完成了整个的启动过程，然后其他应用就可以通过 AMS 来访问这个 ContentProvider 了。拿到了 ContentProvider 以后，就可以通过它所提供的接口方法来访问它了。需要注意的是，这里的 ContentProvider 并不是原始的 ContentProvider，而是 ContentProvider 的 Binder 类型的对象 IContentProvider，IContentProvider 的具体实现是 ContentProviderNative 和 ContentProvider.Transport，其中 ContentProvider.Transport 继承了 ContentProviderNative。这里仍然选择 query 方法，首先其他应用会通过 AMS 获取到 ContentProvider 的 Binder 对象即 IContentProvider，而 IContentProvider 的实现者实际上是 ContentProvider.Transport。因此其他应用调用 IContentProvider 的 query 方法时最终会以进程间通信的方式调用到 ContentProvider.Transport 的 query 方法，它的实现如下所示。

```
public Cursor query(String callingPkg, Uri uri, String[] projection,
        String selection, String[] selectionArgs, String sortOrder,
        ICancellationSignal cancellationSignal) {
    validateIncomingUri(uri);
    uri = getUriWithoutUserId(uri);
    if (enforceReadPermission(callingPkg, uri) != AppOpsManager.MODE_
ALLOWED) {
        return rejectQuery(uri, projection, selection, selectionArgs,
        sortOrder,
                CancellationSignal.fromTransport(cancellationSignal));
    }
    final String original = setCallingPackage(callingPkg);
    try {
        return ContentProvider.this.query(
                uri, projection, selection, selectionArgs, sortOrder,
                CancellationSignal.fromTransport(cancellationSignal));
    } finally {
        setCallingPackage(original);
    }
}
```

很显然，ContentProvider.Transport 的 query 方法调用了 ContentProvider 的 query 方法，query 方法的执行结果再通过 Binder 返回给调用者，这样一来整个调用过程就完成了。除了 query 方法，insert、delete 和 update 方法也是类似的，这里就不再分析了。

第 10 章　Android 的消息机制

本章所要讲述的内容是 Android 的消息机制。提到消息机制读者应该都不陌生，在日常开发中不可避免地要涉及这方面的内容。从开发的角度来说，Handler 是 Android 消息机制的上层接口，这使得在开发过程中只需要和 Handler 交互即可。Handler 的使用过程很简单，通过它可以轻松地将一个任务切换到 Handler 所在的线程中去执行。很多人认为 Handler 的作用是更新 UI，这的确没错，但是更新 UI 仅仅是 Handler 的一个特殊的使用场景。具体来说是这样的：有时候需要在子线程中进行耗时的 I/O 操作，可能是读取文件或者访问网络等，当耗时操作完成以后可能需要在 UI 上做一些改变，由于 Android 开发规范的限制，我们并不能在子线程中访问 UI 控件，否则就会触发程序异常，这个时候通过 Handler 就可以将更新 UI 的操作切换到主线程中执行。因此，本质上来说，Handler 并不是专门用于更新 UI 的，它只是常被开发者用来更新 UI。

Android 的消息机制主要是指 Handler 的运行机制，Handler 的运行需要底层的 MessageQueue 和 Looper 的支撑。MessageQueue 的中文翻译是消息队列，顾名思义，它的内部存储了一组消息，以队列的形式对外提供插入和删除的工作。虽然叫消息队列，但是它的内部存储结构并不是真正的队列，而是采用单链表的数据结构来存储消息列表。Looper 的中文翻译为循环，在这里可以理解为消息循环。由于 MessageQueue 只是一个消息的存储单元，它不能去处理消息，而 Looper 就填补了这个功能，Looper 会以无限循环的形式去查找是否有新消息，如果有的话就处理消息，否则就一直等待着。Looper 中还有一个特殊的概念，那就是 ThreadLocal，ThreadLocal 并不是线程，它的作用是可以在每个线程中存储数据。我们知道，Handler 创建的时候会采用当前线程的 Looper 来构造消息循环系统，那么 Handler 内部如何获取到当前线程的 Looper 呢？这就要使用 ThreadLocal 了，ThreadLocal 可以在不同的线程中互不干扰地存储并提供数据，通过 ThreadLocal 可以轻松获取每个线程的 Looper。当然需要注意的是，线程是默认没有 Looper 的，如果需要使用 Handler 就必须为线程创建 Looper。

我们经常提到的主线程，也叫 UI 线程，它就是 ActivityThread，ActivityThread 被创建时就会初始化 Looper，这也是在主线程中默认可以使用 Handler 的原因。

10.1 Android 的消息机制概述

前面提到，Android 的消息机制主要是指 Handler 的运行机制以及 Handler 所附带的 MessageQueue 和 Looper 的工作过程，这三者实际上是一个整体，只不过我们在开发过程中比较多地接触到 Handler 而已。Handler 的主要作用是将一个任务切换到某个指定的线程中去执行，那么 Android 为什么要提供这个功能呢？或者说 Android 为什么需要提供在某个具体的线程中执行任务这种功能呢？这是因为 Android 规定访问 UI 只能在主线程中进行，如果在子线程中访问 UI，那么程序就会抛出异常。ViewRootImpl 对 UI 操作做了验证，这个验证工作是由 ViewRootImpl 的 checkThread 方法来完成的，如下所示。

```
void checkThread() {
    if (mThread != Thread.currentThread()) {
        throw new CalledFromWrongThreadException(
            "Only the original thread that created a view hierarchy can touch its views.");
    }
}
```

针对 checkThread 方法中抛出的异常信息，相信读者在开发中都曾经遇到过。由于这一点的限制，导致必须在主线程中访问 UI，但是 Android 又建议不要在主线程中进行耗时操作，否则会导致程序无法响应即 ANR。考虑一种情况，假如我们需要从服务端拉取一些信息并将其显示在 UI 上，这个时候必须在子线程中进行拉取工作，拉取完毕后又不能在子线程中直接访问 UI，如果没有 Handler，那么我们的确没有办法将访问 UI 的工作切换到主线程中去执行。因此，系统之所以提供 Handler，主要原因就是为了解决在子线程中无法访问 UI 的矛盾。

这里再延伸一点，系统为什么不允许在子线程中访问 UI 呢？这是因为 Android 的 UI 控件不是线程安全的，如果在多线程中并发访问可能会导致 UI 控件处于不可预期的状态，那为什么系统不对 UI 控件的访问加上锁机制呢？缺点有两个：首先加上锁机制会让 UI 访问的逻辑变得复杂；其次锁机制会降低 UI 访问的效率，因为锁机制会阻塞某些线程的执行。鉴于这两个缺点，最简单且高效的方法就是采用单线程模型来处理 UI 操作，对于开发者来说也不是很麻烦，只是需要通过 Handler 切换一下 UI 访问的执行线程即可。

Handler 的使用方法这里就不做介绍了，这里描述一下 Handler 的工作原理。Handler 创建时会采用当前线程的 Looper 来构建内部的消息循环系统，如果当前线程没有 Looper，那么就会报错，如下所示。

```
E/AndroidRuntime(27568): FATAL EXCEPTION: Thread-43484
E/AndroidRuntime(27568): java.lang.RuntimeException: Can't create handler
inside thread that has not called Looper.prepare()
E/AndroidRuntime(27568): at android.os.Handler.<init>(Handler.java:121)
E/AndroidRuntime(27568):                         at            com.ryg.chapter_
10.TestActivity$3.run(TestActivity.java:57)
```

如何解决上述问题呢？其实很简单，只需要为当前线程创建 Looper 即可，或者在一个有 Looper 的线程中创建 Handler 也行，具体会在 10.2.3 节中进行介绍。

Handler 创建完毕后，这个时候其内部的 Looper 以及 MessageQueue 就可以和 Handler 一起协同工作了，然后通过 Handler 的 post 方法将一个 Runnable 投递到 Handler 内部的 Looper 中去处理，也可以通过 Handler 的 send 方法发送一个消息，这个消息同样会在 Looper 中去处理。其实 post 方法最终也是通过 send 方法来完成的，接下来主要来看一下 send 方法的工作过程。当 Handler 的 send 方法被调用时，它会调用 MessageQueue 的 enqueueMessage 方法将这个消息放入消息队列中，然后 Looper 发现有新消息到来时，就会处理这个消息，最终消息中的 Runnable 或者 Handler 的 handleMessage 方法就会被调用。注意 Looper 是运行在创建 Handler 所在的线程中的，这样一来 Handler 中的业务逻辑就被切换到创建 Handler 所在的线程中去执行了，这个过程可以用图 10-1 来表示。

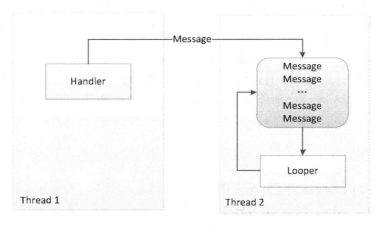

图 10-1　Handler 的工作过程

10.2　Android 的消息机制分析

在 10.1 节中对 Android 的消息机制已经做了一个概括性的描述，通过图 10-1 也能够比较好地理解 Handler 的工作过程。本节将对 Android 消息机制的实现原理做一个全面的分析。由于 Android 的消息机制实际上就是 Handler 的运行机制，因此本节主要围绕着 Handler 的工作过程来分析 Android 的消息机制，主要包括 Handler、MessageQueue 和 Looper。同时为了更好地理解 Looper 的工作原理，本节还会介绍 ThreadLocal，通过本节的介绍可以让读者对 Android 的消息机制有一个深入的理解。

10.2.1　ThreadLocal 的工作原理

ThreadLocal 是一个线程内部的数据存储类，通过它可以在指定的线程中存储数据，数据存储以后，只有在指定线程中可以获取到存储的数据，对于其他线程来说则无法获取到数据。在日常开发中用到 ThreadLocal 的地方较少，但是在某些特殊的场景下，通过 ThreadLocal 可以轻松地实现一些看起来很复杂的功能，这一点在 Android 的源码中也有所体现，比如 Looper、ActivityThread 以及 AMS 中都用到了 ThreadLocal。具体到 ThreadLocal 的使用场景，这个不好统一来描述，一般来说，当某些数据是以线程为作用域并且不同线程具有不同的数据副本的时候，就可以考虑采用 ThreadLocal。比如对于 Handler 来说，它需要获取当前线程的 Looper，很显然 Looper 的作用域就是线程并且不同线程具有不同的 Looper，这个时候通过 ThreadLocal 就可以轻松实现 Looper 在线程中的存取。如果不采用 ThreadLocal，那么系统就必须提供一个全局的哈希表供 Handler 查找指定线程的 Looper，这样一来就必须提供一个类似于 LooperManager 的类了，但是系统并没有这么做而是选择了 ThreadLocal，这就是 ThreadLocal 的好处。

ThreadLocal 另一个使用场景是复杂逻辑下的对象传递，比如监听器的传递，有些时候一个线程中的任务过于复杂，这可能表现为函数调用栈比较深以及代码入口的多样性，在这种情况下，我们又需要监听器能够贯穿整个线程的执行过程，这个时候可以怎么做呢？其实这时就可以采用 ThreadLocal，采用 ThreadLocal 可以让监听器作为线程内的全局对象而存在，在线程内部只要通过 get 方法就可以获取到监听器。如果不采用 ThreadLocal，那么我们能想到的可能是如下两种方法：第一种方法是将监听器通过参数的形式在函数调用栈中进行传递，第二种方法就是将监听器作为静态变量供线程访问。上述这两种方法都是有局限性的。第一种方法的问题是当函数调用栈很深的时候，通过函数参数来传递监听器

对象这几乎是不可接受的，这会让程序的设计看起来很糟糕。第二种方法是可以接受的，但是这种状态是不具有可扩充性的，比如同时有两个线程在执行，那么就需要提供两个静态的监听器对象，如果有 10 个线程在并发执行呢？提供 10 个静态的监听器对象？这显然是不可思议的，而采用 ThreadLocal，每个监听器对象都在自己的线程内部存储，根本就不会有方法 2 的这种问题。

介绍了那么多 ThreadLocal 的知识，可能还是有点抽象，下面通过实际的例子来演示 ThreadLocal 的真正含义。首先定义一个 ThreadLocal 对象，这里选择 Boolean 类型的，如下所示。

```
private ThreadLocal<Boolean> mBooleanThreadLocal = new ThreadLocal<Boolean>();
```

然后分别在主线程、子线程 1 和子线程 2 中设置和访问它的值，代码如下所示。

```
mBooleanThreadLocal.set(true);
Log.d(TAG, "[Thread#main]mBooleanThreadLocal=" + mBooleanThreadLocal.get());

new Thread("Thread#1") {
    @Override
    public void run() {
        mBooleanThreadLocal.set(false);
        Log.d(TAG, "[Thread#1]mBooleanThreadLocal=" + mBooleanThreadLocal.
        get());
    };
}.start();

new Thread("Thread#2") {
    @Override
    public void run() {
        Log.d(TAG, "[Thread#2]mBooleanThreadLocal=" + mBooleanThreadLocal.
        get());
    };
}.start();
```

在上面的代码中，在主线程中设置 mBooleanThreadLocal 的值为 true，在子线程 1 中设置 mBooleanThreadLocal 的值为 false，在子线程 2 中不设置 mBooleanThreadLocal 的值。然后分别在 3 个线程中通过 get 方法获取 mBooleanThreadLocal 的值，根据前面对 ThreadLocal 的描述，这个时候，主线程中应该是 true，子线程 1 中应该是 false，而子线程 2 中由于没有设置值，所以应该是 null。安装并运行程序，日志如下所示。

```
D/TestActivity(8676): [Thread#main]mBooleanThreadLocal=true
D/TestActivity(8676): [Thread#1]mBooleanThreadLocal=false
D/TestActivity(8676): [Thread#2]mBooleanThreadLocal=null
```

从上面日志可以看出，虽然在不同线程中访问的是同一个 ThreadLocal 对象，但是它们通过 ThreadLocal 获取到的值却是不一样的，这就是 ThreadLocal 的奇妙之处。结合这个例子然后再看一遍前面对 ThreadLocal 的两个使用场景的理论分析，我们应该就能比较好地理解 ThreadLocal 的使用方法了。ThreadLocal 之所以有这么奇妙的效果，是因为不同线程访问同一个 ThreadLocal 的 get 方法，ThreadLocal 内部会从各自的线程中取出一个数组，然后再从数组中根据当前 ThreadLocal 的索引去查找出对应的 value 值。很显然，不同线程中的数组是不同的，这就是为什么通过 ThreadLocal 可以在不同的线程中维护一套数据的副本并且彼此互不干扰。

对 ThreadLocal 的使用方法和工作过程做了介绍后，下面分析 ThreadLocal 的内部实现，ThreadLocal 是一个泛型类，它的定义为 public class ThreadLocal<T>，只要弄清楚 ThreadLocal 的 get 和 set 方法就可以明白它的工作原理。

首先看 ThreadLocal 的 set 方法，如下所示。

```
public void set(T value) {
    Thread currentThread = Thread.currentThread();
    Values values = values(currentThread);
    if (values == null) {
        values = initializeValues(currentThread);
    }
    values.put(this, value);
}
```

在上面的 set 方法中，首先会通过 values 方法来获取当前线程中的 ThreadLocal 数据，如何获取呢？其实获取的方式也是很简单的，在 Thread 类的内部有一个成员专门用于存储线程的 ThreadLocal 的数据：ThreadLocal.Values localValues，因此获取当前线程的 ThreadLocal 数据就变得异常简单了。如果 localValues 的值为 null，那么就需要对其进行初始化，初始化后再将 ThreadLocal 的值进行存储。下面看一下 ThreadLocal 的值到底是如何在 localValues 中进行存储的。在 localValues 内部有一个数组：private Object[] table，ThreadLocal 的值就存在在这个 table 数组中。下面看一下 localValues 是如何使用 put 方法将 ThreadLocal 的值存储到 table 数组中的，如下所示。

```
void put(ThreadLocal<?> key, Object value) {
    cleanUp();

    // Keep track of first tombstone. That's where we want to go back
    // and add an entry if necessary.
    int firstTombstone = -1;

    for (int index = key.hash & mask;; index = next(index)) {
        Object k = table[index];

        if (k == key.reference) {
            // Replace existing entry.
            table[index + 1] = value;
            return;
        }

        if (k == null) {
            if (firstTombstone == -1) {
                // Fill in null slot.
                table[index] = key.reference;
                table[index + 1] = value;
                size++;
                return;
            }

            // Go back and replace first tombstone.
            table[firstTombstone] = key.reference;
            table[firstTombstone + 1] = value;
            tombstones--;
            size++;
            return;
        }

        // Remember first tombstone.
        if (firstTombstone == -1 && k == TOMBSTONE) {
            firstTombstone = index;
        }
    }
}
```

上面的代码实现了数据的存储过程，这里不去分析它的具体算法，但是我们可以得出一个存储规则，那就是 ThreadLocal 的值在 table 数组中的存储位置总是为 ThreadLocal 的 reference 字段所标识的对象的下一个位置，比如 ThreadLocal 的 reference 对象在 table 数组中的索引为 index，那么 ThreadLocal 的值在 table 数组中的索引就是 index+1。最终 ThreadLocal 的值将会被存储在 table 数组中：table[index + 1] = value。

上面分析了 ThreadLocal 的 set 方法，这里分析它的 get 方法，如下所示。

```java
public T get() {
    // Optimized for the fast path.
    Thread currentThread = Thread.currentThread();
    Values values = values(currentThread);
    if (values != null) {
        Object[] table = values.table;
        int index = hash & values.mask;
        if (this.reference == table[index]) {
            return (T) table[index + 1];
        }
    } else {
        values = initializeValues(currentThread);
    }

    return (T) values.getAfterMiss(this);
}
```

可以发现，ThreadLocal 的 get 方法的逻辑也比较清晰，它同样是取出当前线程的 localValues 对象，如果这个对象为 null 那么就返回初始值，初始值由 ThreadLocal 的 initialValue 方法来描述，默认情况下为 null，当然也可以重写这个方法，它的默认实现如下所示。

```java
/**
 * Provides the initial value of this variable for the current thread.
 * The default implementation returns {@code null}.
 *
 * @return the initial value of the variable.
 */
protected T initialValue() {
    return null;
}
```

如果 localValues 对象不为 null，那就取出它的 table 数组并找出 ThreadLocal 的 reference 对象在 table 数组中的位置，然后 table 数组中的下一个位置所存储的数据就是 ThreadLocal

的值。

从 ThreadLocal 的 set 和 get 方法可以看出，它们所操作的对象都是当前线程的 localValues 对象的 table 数组，因此在不同线程中访问同一个 ThreadLocal 的 set 和 get 方法，它们对 ThreadLocal 所做的读/写操作仅限于各自线程的内部，这就是为什么 ThreadLocal 可以在多个线程中互不干扰地存储和修改数据，理解 ThreadLocal 的实现方式有助于理解 Looper 的工作原理。

10.2.2 消息队列的工作原理

消息队列在 Android 中指的是 MessageQueue，MessageQueue 主要包含两个操作：插入和读取。读取操作本身会伴随着删除操作，插入和读取对应的方法分别为 enqueueMessage 和 next，其中 enqueueMessage 的作用是往消息队列中插入一条消息，而 next 的作用是从消息队列中取出一条消息并将其从消息队列中移除。尽管 MessageQueue 叫消息队列，但是它的内部实现并不是用的队列，实际上它是通过一个单链表的数据结构来维护消息列表，单链表在插入和删除上比较有优势。下面主要看一下它的 enqueueMessage 和 next 方法的实现，enqueueMessage 的源码如下所示。

```
boolean enqueueMessage(Message msg, long when) {
    ...
    synchronized (this) {
        ...
        msg.markInUse();
        msg.when = when;
        Message p = mMessages;
        boolean needWake;
        if (p == null || when == 0 || when < p.when) {
            // New head, wake up the event queue if blocked.
            msg.next = p;
            mMessages = msg;
            needWake = mBlocked;
        } else {
            // Inserted within the middle of the queue.  Usually we don't
            have to wake
            // up the event queue unless there is a barrier at the head of
            the queue
            // and the message is the earliest asynchronous message in the
            queue.
```

```
            needWake = mBlocked && p.target == null && msg.isAsynchronous();
            Message prev;
            for (;;) {
                prev = p;
                p = p.next;
                if (p == null || when < p.when) {
                    break;
                }
                if (needWake && p.isAsynchronous()) {
                    needWake = false;
                }
            }
            msg.next = p; // invariant: p == prev.next
            prev.next = msg;
        }

        // We can assume mPtr != 0 because mQuitting is false.
        if (needWake) {
            nativeWake(mPtr);
        }
    }
    return true;
}
```

从 enqueueMessage 的实现来看，它的主要操作其实就是单链表的插入操作，这里就不再过多解释了，下面看一下 next 方法的实现，next 的主要逻辑如下所示。

```
Message next() {
    ...
    int pendingIdleHandlerCount = -1; // -1 only during first iteration
    int nextPollTimeoutMillis = 0;
    for (;;) {
        if (nextPollTimeoutMillis != 0) {
            Binder.flushPendingCommands();
        }

        nativePollOnce(ptr, nextPollTimeoutMillis);

        synchronized (this) {
            // Try to retrieve the next message. Return if found.
```

```java
                final long now = SystemClock.uptimeMillis();
            Message prevMsg = null;
            Message msg = mMessages;
            if (msg != null && msg.target == null) {
                // Stalled by a barrier.  Find the next asynchronous message
                in the queue.
                do {
                    prevMsg = msg;
                    msg = msg.next;
                } while (msg != null && !msg.isAsynchronous());
            }
            if (msg != null) {
                if (now < msg.when) {
                    // Next message is not ready.  Set a timeout to wake up
                    when it is ready.
                    nextPollTimeoutMillis = (int) Math.min(msg.when - now,
                    Integer.MAX_VALUE);
                } else {
                    // Got a message.
                    mBlocked = false;
                    if (prevMsg != null) {
                        prevMsg.next = msg.next;
                    } else {
                        mMessages = msg.next;
                    }
                    msg.next = null;
                    if (false) Log.v("MessageQueue", "Returning message: "
                    + msg);
                    return msg;
                }
            } else {
                // No more messages.
                nextPollTimeoutMillis = -1;
            }
            
            ...
        }
        ...
    }
}
```

可以发现 next 方法是一个无限循环的方法，如果消息队列中没有消息，那么 next 方法会一直阻塞在这里。当有新消息到来时，next 方法会返回这条消息并将其从单链表中移除。

10.2.3　Looper 的工作原理

在 10.2.2 节中介绍了消息队列的主要实现，本节将分析 Looper 的具体实现。Looper 在 Android 的消息机制中扮演着消息循环的角色，具体来说就是它会不停地从 MessageQueue 中查看是否有新消息，如果有新消息就会立刻处理，否则就一直阻塞在那里。首先看一下它的构造方法，在构造方法中它会创建一个 MessageQueue 即消息队列，然后将当前线程的对象保存起来，如下所示。

```
private Looper(boolean quitAllowed) {
    mQueue = new MessageQueue(quitAllowed);
    mThread = Thread.currentThread();
}
```

我们知道，Handler 的工作需要 Looper，没有 Looper 的线程就会报错，那么如何为一个线程创建 Looper 呢？其实很简单，通过 Looper.prepare() 即可为当前线程创建一个 Looper，接着通过 Looper.loop() 来开启消息循环，如下所示。

```
new Thread("Thread#2") {
    @Override
    public void run() {
        Looper.prepare();
        Handler handler = new Handler();
        Looper.loop();
    };
}.start();
```

Looper 除了 prepare 方法外，还提供了 prepareMainLooper 方法，这个方法主要是给主线程也就是 ActivityThread 创建 Looper 使用的，其本质也是通过 prepare 方法来实现的。由于主线程的 Looper 比较特殊，所以 Looper 提供了一个 getMainLooper 方法，通过它可以在任何地方获取到主线程的 Looper。Looper 也是可以退出的，Looper 提供了 quit 和 quitSafely 来退出一个 Looper，二者的区别是：quit 会直接退出 Looper，而 quitSafely 只是设定一个退出标记，然后把消息队列中的已有消息处理完毕后才安全地退出。Looper 退出后，通过 Handler 发送的消息会失败，这个时候 Handler 的 send 方法会返回 false。在子线程中，如果手动为其创建了 Looper，那么在所有的事情完成以后应该调用 quit 方法来终止消息循环，否则这个子线程就会一直处于等待的状态，而如果退出 Looper 以后，这个线程就会立刻终

止，因此建议不需要的时候终止 Looper。

Looper 最重要的一个方法是 loop 方法，只有调用了 loop 后，消息循环系统才会真正地起作用，它的实现如下所示。

```java
/**
 * Run the message queue in this thread. Be sure to call
 * {@link #quit()} to end the loop.
 */
public static void loop() {
    final Looper me = myLooper();
    if (me == null) {
        throw new RuntimeException("No Looper; Looper.prepare() wasn't
        called on this thread.");
    }
    final MessageQueue queue = me.mQueue;

    // Make sure the identity of this thread is that of the local process,
    // and keep track of what that identity token actually is.
    Binder.clearCallingIdentity();
    final long ident = Binder.clearCallingIdentity();

    for (;;) {
        Message msg = queue.next(); // might block
        if (msg == null) {
            // No message indicates that the message queue is quitting.
            return;
        }

        // This must be in a local variable, in case a UI event sets the logger
        Printer logging = me.mLogging;
        if (logging != null) {
            logging.println(">>>>> Dispatching to " + msg.target + " " +
                    msg.callback + ": " + msg.what);
        }

        msg.target.dispatchMessage(msg);

        if (logging != null) {
            logging.println("<<<<< Finished to " + msg.target + " " +
            msg.callback);
```

```
        }

        // Make sure that during the course of dispatching the
        // identity of the thread wasn't corrupted.
        final long newIdent = Binder.clearCallingIdentity();
        if (ident != newIdent) {
            Log.wtf(TAG, "Thread identity changed from 0x"
                    + Long.toHexString(ident) + " to 0x"
                    + Long.toHexString(newIdent)+" while dispatching to "
                    + msg.target.getClass().getName() + " "
                    + msg.callback + " what=" + msg.what);
        }

        msg.recycleUnchecked();
    }
}
```

Looper 的 loop 方法的工作过程也比较好理解，loop 方法是一个死循环，唯一跳出循环的方式是 MessageQueue 的 next 方法返回了 null。当 Looper 的 quit 方法被调用时，Looper 就会调用 MessageQueue 的 quit 或者 quitSafely 方法来通知消息队列退出，当消息队列被标记为退出状态时，它的 next 方法就会返回 null。也就是说，Looper 必须退出，否则 loop 方法就会无限循环下去。loop 方法会调用 MessageQueue 的 next 方法来获取新消息，而 next 是一个阻塞操作，当没有消息时，next 方法会一直阻塞在那里，这也导致 loop 方法一直阻塞在那里。如果 MessageQueue 的 next 方法返回了新消息，Looper 就会处理这条消息：msg.target.dispatchMessage(msg)，这里的 msg.target 是发送这条消息的 Handler 对象，这样 Handler 发送的消息最终又交给它的 dispatchMessage 方法来处理了。但是这里不同的是，Handler 的 dispatchMessage 方法是在创建 Handler 时所使用的 Looper 中执行的，这样就成功地将代码逻辑切换到指定的线程中去执行了。

10.2.4 Handler 的工作原理

Handler 的工作主要包含消息的发送和接收过程。消息的发送可以通过 post 的一系列方法以及 send 的一系列方法来实现，post 的一系列方法最终是通过 send 的一系列方法来实现的。发送一条消息的典型过程如下所示。

```
public final boolean sendMessage(Message msg)
{
    return sendMessageDelayed(msg, 0);
```

```
}

public final boolean sendMessageDelayed(Message msg, long delayMillis)
{
    if (delayMillis < 0) {
        delayMillis = 0;
    }
    return sendMessageAtTime(msg, SystemClock.uptimeMillis() + delayMillis);
}

public boolean sendMessageAtTime(Message msg, long uptimeMillis) {
    MessageQueue queue = mQueue;
    if (queue == null) {
        RuntimeException e = new RuntimeException(
                this + " sendMessageAtTime() called with no mQueue");
        Log.w("Looper", e.getMessage(), e);
        return false;
    }
    return enqueueMessage(queue, msg, uptimeMillis);
}

private boolean enqueueMessage(MessageQueue queue, Message msg, long uptimeMillis) {
    msg.target = this;
    if (mAsynchronous) {
        msg.setAsynchronous(true);
    }
    return queue.enqueueMessage(msg, uptimeMillis);
}
```

可以发现，Handler 发送消息的过程仅仅是向消息队列中插入了一条消息，MessageQueue 的 next 方法就会返回这条消息给 Looper，Looper 收到消息后就开始处理了，最终消息由 Looper 交由 Handler 处理，即 Handler 的 dispatchMessage 方法会被调用，这时 Handler 就进入了处理消息的阶段。dispatchMessage 的实现如下所示。

```
public void dispatchMessage(Message msg) {
    if (msg.callback != null) {
        handleCallback(msg);
    } else {
        if (mCallback != null) {
```

```
            if (mCallback.handleMessage(msg)) {
                return;
            }
        }
        handleMessage(msg);
    }
}
```

Handler 处理消息的过程如下：

首先，检查 Message 的 callback 是否为 null，不为 null 就通过 handleCallback 来处理消息。Message 的 callback 是一个 Runnable 对象，实际上就是 Handler 的 post 方法所传递的 Runnable 参数。handleCallback 的逻辑也是很简单，如下所示。

```
private static void handleCallback(Message message) {
    message.callback.run();
}
```

其次，检查 mCallback 是否为 null，不为 null 就调用 mCallback 的 handleMessage 方法来处理消息。Callback 是个接口，它的定义如下：

```
/**
 * Callback interface you can use when instantiating a Handler to avoid
 * having to implement your own subclass of Handler.
 *
 * @param msg A {@link android.os.Message Message} object
 * @return True if no further handling is desired
 */
public interface Callback {
    public boolean handleMessage(Message msg);
}
```

通过 Callback 可以采用如下方式来创建 Handler 对象：Handler handler = new Handler(callback)。那么 Callback 的意义是什么呢？源码里面的注释已经做了说明：可以用来创建一个 Handler 的实例但并不需要派生 Handler 的子类。在日常开发中，创建 Handler 最常见的方式就是派生一个 Handler 的子类并重写其 handleMessage 方法来处理具体的消息，而 Callback 给我们提供了另外一种使用 Handler 的方式，当我们不想派生子类时，就可以通过 Callback 来实现。

最后，调用 Handler 的 handleMessage 方法来处理消息。Handler 处理消息的过程可以归纳为一个流程图，如图 10-2 所示。

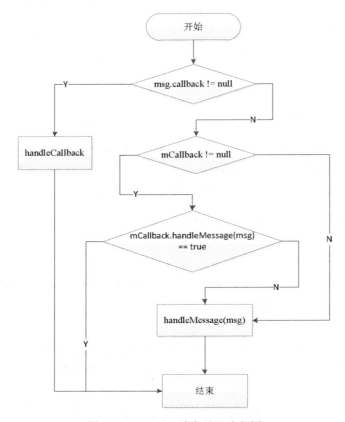

图 10-2　Handler 消息处理流程图

Handler 还有一个特殊的构造方法，那就是通过一个特定的 Looper 来构造 Handler，它的实现如下所示。通过这个构造方法可以实现一些特殊的功能。

```
public Handler(Looper looper) {
    this(looper, null, false);
}
```

下面看一下 Handler 的一个默认构造方法 public Handler()，这个构造方法会调用下面的构造方法。很明显，如果当前线程没有 Looper 的话，就会抛出"Can't create handler inside thread that has not called Looper.prepare()"这个异常，这也解释了在没有 Looper 的子线程中创建 Handler 会引发程序异常的原因。

```
public Handler(Callback callback, boolean async) {
    ...
    mLooper = Looper.myLooper();
    if (mLooper == null) {
        throw new RuntimeException(
            "Can't create handler inside thread that has not called
            Looper.prepare()");
    }
    mQueue = mLooper.mQueue;
    mCallback = callback;
    mAsynchronous = async;
}
```

10.3 主线程的消息循环

Android 的主线程就是 ActivityThread，主线程的入口方法为 main，在 main 方法中系统会通过 Looper.prepareMainLooper()来创建主线程的 Looper 以及 MessageQueue，并通过 Looper.loop()来开启主线程的消息循环，这个过程如下所示。

```
public static void main(String[] args) {
    ...
    Process.setArgV0("<pre-initialized>");

    Looper.prepareMainLooper();

    ActivityThread thread = new ActivityThread();
    thread.attach(false);

    if (sMainThreadHandler == null) {
        sMainThreadHandler = thread.getHandler();
    }

    AsyncTask.init();

    if (false) {
        Looper.myLooper().setMessageLogging(new
            LogPrinter(Log.DEBUG, "ActivityThread"));
```

```
    }

    Looper.loop();

    throw new RuntimeException("Main thread loop unexpectedly exited");
}
```

主线程的消息循环开始了以后，ActivityThread 还需要一个 Handler 来和消息队列进行交互，这个 Handler 就是 ActivityThread.H，它内部定义了一组消息类型，主要包含了四大组件的启动和停止等过程，如下所示。

```
private class H extends Handler {
    public static final int LAUNCH_ACTIVITY         = 100;
    public static final int PAUSE_ACTIVITY          = 101;
    public static final int PAUSE_ACTIVITY_FINISHING = 102;
    public static final int STOP_ACTIVITY_SHOW      = 103;
    public static final int STOP_ACTIVITY_HIDE      = 104;
    public static final int SHOW_WINDOW             = 105;
    public static final int HIDE_WINDOW             = 106;
    public static final int RESUME_ACTIVITY         = 107;
    public static final int SEND_RESULT             = 108;
    public static final int DESTROY_ACTIVITY        = 109;
    public static final int BIND_APPLICATION        = 110;
    public static final int EXIT_APPLICATION        = 111;
    public static final int NEW_INTENT              = 112;
    public static final int RECEIVER                = 113;
    public static final int CREATE_SERVICE          = 114;
    public static final int SERVICE_ARGS            = 115;
    public static final int STOP_SERVICE            = 116;
    ...
}
```

ActivityThread 通过 ApplicationThread 和 AMS 进行进程间通信，AMS 以进程间通信的方式完成 ActivityThread 的请求后会回调 ApplicationThread 中的 Binder 方法，然后 ApplicationThread 会向 H 发送消息，H 收到消息后会将 ApplicationThread 中的逻辑切换到 ActivityThread 中去执行，即切换到主线程中去执行，这个过程就是主线程的消息循环模型。

第 11 章 Android 的线程和线程池

本章的主题是 Android 中的线程和线程池。线程在 Android 中是一个很重要的概念，从用途上来说，线程分为主线程和子线程，主线程主要处理和界面相关的事情，而子线程则往往用于执行耗时操作。由于 Android 的特性，如果在主线程中执行耗时操作那么就会导致程序无法及时地响应，因此耗时操作必须放在子线程中去执行。除了 Thread 本身以外，在 Android 中可以扮演线程角色的还有很多，比如 AsyncTask 和 IntentService，同时 HandlerThread 也是一种特殊的线程。尽管 AsyncTask、IntentService 以及 HandlerThread 的表现形式都有别于传统的线程，但是它们的本质仍然是传统的线程。对于 AsyncTask 来说，它的底层用到了线程池，对于 IntentService 和 HandlerThread 来说，它们的底层则直接使用了线程。

不同形式的线程虽然都是线程，但是它们仍然具有不同的特性和使用场景。AsyncTask 封装了线程池和 Handler，它主要是为了方便开发者在子线程中更新 UI。HandlerThread 是一种具有消息循环的线程，在它的内部可以使用 Handler。IntentService 是一个服务，系统对其进行了封装使其可以更方便地执行后台任务，IntentService 内部采用 HandlerThread 来执行任务，当任务执行完毕后 IntentService 会自动退出。从任务执行的角度来看，IntentService 的作用很像一个后台线程，但是 IntentService 是一种服务，它不容易被系统杀死从而可以尽量保证任务的执行，而如果是一个后台线程，由于这个时候进程中没有活动的四大组件，那么这个进程的优先级就会非常低，会很容易被系统杀死，这就是 IntentService 的优点。

在操作系统中，线程是操作系统调度的最小单元，同时线程又是一种受限的系统资源，即线程不可能无限制地产生，并且线程的创建和销毁都会有相应的开销。当系统中存在大量的线程时，系统会通过时间片轮转的方式调度每个线程，因此线程不可能做到绝对的并行，除非线程数量小于等于 CPU 的核心数，一般来说这是不可能的。试想一下，如果在一个进程中频繁地创建和销毁线程，这显然不是高效的做法。正确的做法是采用线程池，一个线程池中会缓存一定数量的线程，通过线程池就可以避免因为频繁创建和销毁线程所带来的系统开销。Android 中的线程池来源于 Java，主要是通过 Executor 来派生特定类型的线程池，不同种类的线程池又具有各自的特性，详细内容会在 11.3 节中进行介绍。

11.1 主线程和子线程

主线程是指进程所拥有的线程，在 Java 中默认情况下一个进程只有一个线程，这个线程就是主线程。主线程主要处理界面交互相关的逻辑，因为用户随时会和界面发生交互，因此主线程在任何时候都必须有较高的响应速度，否则就会产生一种界面卡顿的感觉。为了保持较高的响应速度，这就要求主线程中不能执行耗时的任务，这个时候子线程就派上用场了。子线程也叫工作线程，除了主线程以外的线程都是子线程。

Android 沿用了 Java 的线程模型，其中的线程也分为主线程和子线程，其中主线程也叫 UI 线程。主线程的作用是运行四大组件以及处理它们和用户的交互，而子线程的作用则是执行耗时任务，比如网络请求、I/O 操作等。从 Android 3.0 开始系统要求网络访问必须在子线程中进行，否则网络访问将会失败并抛出 NetworkOnMainThreadException 这个异常，这样做是为了避免主线程由于被耗时操作所阻塞从而出现 ANR 现象。

11.2 Android 中的线程形态

本节将对 Android 中的线程形态做一个全面的介绍，除了传统的 Thread 以外，还包含 AsyncTask、HandlerThread 以及 IntentService，这三者的底层实现也是线程，但是它们具有特殊的表现形式，同时在使用上也各有优缺点。为了简化在子线程中访问 UI 的过程，系统提供了 AsyncTask，AsyncTask 经过几次修改，导致了对于不同的 API 版本 AsyncTask 具有不同的表现，尤其是多任务的并发执行上。由于这个原因，很多开发者对 AsyncTask 的使用上存在误区，本节将详细介绍使用 AsyncTask 时的注意事项，并从源码的角度来分析 AsyncTask 的执行过程。

11.2.1 AsyncTask

AsyncTask 是一种轻量级的异步任务类，它可以在线程池中执行后台任务，然后把执行的进度和最终结果传递给主线程并在主线程中更新 UI。从实现上来说，AsyncTask 封装了 Thread 和 Handler，通过 AsyncTask 可以更加方便地执行后台任务以及在主线程中访问 UI，但是 AsyncTask 并不适合进行特别耗时的后台任务，对于特别耗时的任务来说，建议使用线程池。

AsyncTask 是一个抽象的泛型类，它提供了 Params、Progress 和 Result 这三个泛型参数，其中 Params 表示参数的类型，Progress 表示后台任务的执行进度的类型，而 Result 则表示后台任务的返回结果的类型，如果 AsyncTask 确实不需要传递具体的参数，那么这三个泛型参数可以用 Void 来代替。AsyncTask 这个类的声明如下所示。

```
public abstract class AsyncTask<Params, Progress, Result>.
```

AsyncTask 提供了 4 个核心方法，它们的含义如下所示。

（1）onPreExecute()，在主线程中执行，在异步任务执行之前，此方法会被调用，一般可以用于做一些准备工作。

（2）doInBackground(Params...params)，在线程池中执行，此方法用于执行异步任务，params 参数表示异步任务的输入参数。在此方法中可以通过 publishProgress 方法来更新任务的进度，publishProgress 方法会调用 onProgressUpdate 方法。另外此方法需要返回计算结果给 onPostExecute 方法。

（3）onProgressUpdate(Progress...values)，在主线程中执行，当后台任务的执行进度发生改变时此方法会被调用。

（4）onPostExecute(Result result)，在主线程中执行，在异步任务执行之后，此方法会被调用，其中 result 参数是后台任务的返回值，即 doInBackground 的返回值。

上面这几个方法，onPreExecute 先执行，接着是 doInBackground，最后才是 onPostExecute。除了上述四个方法以外，AsyncTask 还提供了 onCancelled()方法，它同样在主线程中执行，当异步任务被取消时，onCancelled()方法会被调用，这个时候 onPostExecute 则不会被调用。下面提供一个典型的示例，如下所示。

```
private class DownloadFilesTask extends AsyncTask<URL, Integer, Long> {
    protected Long doInBackground(URL... urls) {
        int count = urls.length;
        long totalSize = 0;
```

```
        for (int i = 0; i < count; i++) {
            totalSize += Downloader.downloadFile(urls[i]);
            publishProgress((int) ((i / (float) count) * 100));
            // Escape early if cancel() is called
            if (isCancelled())
                break;
        }
        return totalSize;
    }

    protected void onProgressUpdate(Integer... progress) {
        setProgressPercent(progress[0]);
    }

    protected void onPostExecute(Long result) {
        showDialog("Downloaded " + result + " bytes");
    }
}
```

在上面的代码中，实现了一个具体的 AsyncTask 类，这个类主要用于模拟文件的下载过程，它的输入参数类型为 URL，后台任务的进程参数为 Integer，而后台任务的返回结果为 Long 类型。注意到 doInBackground 和 onProgressUpdate 方法它们的参数中均包含...的字样，在 Java 中...表示参数的数量不定，它是一种数组型参数，...的概念和 C 语言中的...是一致的。当要执行上述下载任务时，可以通过如下方式来完成：

```
new DownloadFilesTask().execute(url1, url2, url3);
```

在 DownloadFilesTask 中，doInBackground 用来执行具体的下载任务并通过 publishProgress 方法来更新下载的进度，同时还要判断下载任务是否被外界取消了。当下载任务完成后，doInBackground 会返回结果，即下载的总字节数。需要注意的是，doInBackground 是在线程池中执行的。onProgressUpdate 用于更新界面中下载的进度，它运行在主线程，当 publishProgress 被调用时，此方法就会被调用。当下载任务完成后，onPostExecute 方法就会被调用，它也是运行在主线程中，这个时候我们就可以在界面上做出一些提示，比如弹出一个对话框告知用户下载已经完成。

AsyncTask 在具体的使用过程中也是有一些条件限制的，主要有如下几点：

（1）AsyncTask 的类必须在主线程中加载，这就意味着第一次访问 AsyncTask 必须发生在主线程，当然这个过程在 Android 4.1 及以上版本中已经被系统自动完成。在 Android 5.0

的源码中，可以查看 ActivityThread 的 main 方法，它会调用 AsyncTask 的 init 方法，这就满足了 AsyncTask 的类必须在主线程中进行加载这个条件了。至于为什么必须要满足这个条件，在 11.2.2 节中会结合 AsyncTask 的源码再次分析这个问题。

（2）AsyncTask 的对象必须在主线程中创建。

（3）execute 方法必须在 UI 线程调用。

（4）不要在程序中直接调用 onPreExecute()、onPostExecute、doInBackground 和 onProgressUpdate 方法。

（5）一个 AsyncTask 对象只能执行一次，即只能调用一次 execute 方法，否则会报运行时异常。

（6）在 Android 1.6 之前，AsyncTask 是串行执行任务的，Android 1.6 的时候 AsyncTask 开始采用线程池里处理并行任务，但是从 Android 3.0 开始，为了避免 AsyncTask 所带来的并发错误，AsyncTask 又采用一个线程来串行执行任务。尽管如此，在 Android 3.0 以及后续的版本中，我们仍然可以通过 AsyncTask 的 executeOnExecutor 方法来并行地执行任务。

11.2.2 AsyncTask 的工作原理

为了分析 AsyncTask 的工作原理，我们从它的 execute 方法开始分析，execute 方法又会调用 executeOnExecutor 方法，它们的实现如下所示。

```
public final AsyncTask<Params, Progress, Result> execute(Params... params) {
    return executeOnExecutor(sDefaultExecutor, params);
}

public final AsyncTask<Params, Progress, Result> executeOnExecutor(Executor exec,
        Params... params) {
    if (mStatus != Status.PENDING) {
        switch (mStatus) {
            case RUNNING:
                throw new IllegalStateException("Cannot execute task:"
                        + " the task is already running.");
            case FINISHED:
                throw new IllegalStateException("Cannot execute task:"
```

```
                + " the task has already been executed "
                + "(a task can be executed only once)");
        }
    }
    mStatus = Status.RUNNING;
    onPreExecute();
    mWorker.mParams = params;
    exec.execute(mFuture);
    return this;
}
```

在上面的代码中，sDefaultExecutor 实际上是一个串行的线程池，一个进程中所有的 AsyncTask 全部在这个串行的线程池中排队执行，这个排队执行的过程后面会再进行分析。在 executeOnExecutor 方法中，AsyncTask 的 onPreExecute 方法最先执行，然后线程池开始执行。下面分析线程池的执行过程，如下所示。

```
public static final Executor SERIAL_EXECUTOR = new SerialExecutor();
private static volatile Executor sDefaultExecutor = SERIAL_EXECUTOR;

private static class SerialExecutor implements Executor {
    final ArrayDeque<Runnable> mTasks = new ArrayDeque<Runnable>();
    Runnable mActive;

    public synchronized void execute(final Runnable r) {
        mTasks.offer(new Runnable() {
            public void run() {
                try {
                    r.run();
                } finally {
                    scheduleNext();
                }
            }
        });
        if (mActive == null) {
            scheduleNext();
        }
    }

    protected synchronized void scheduleNext() {
        if ((mActive = mTasks.poll()) != null) {
```

```
            THREAD_POOL_EXECUTOR.execute(mActive);
        }
    }
}
```

从 SerialExecutor 的实现可以分析 AsyncTask 的排队执行的过程。首先系统会把 AsyncTask 的 Params 参数封装为 FutureTask 对象，FutureTask 是一个并发类，在这里它充当了 Runnable 的作用。接着这个 FutureTask 会交给 SerialExecutor 的 execute 方法去处理，SerialExecutor 的 execute 方法首先会把 FutureTask 对象插入到任务队列 mTasks 中，如果这个时候没有正在活动的 AsyncTask 任务，那么就会调用 SerialExecutor 的 scheduleNext 方法来执行下一个 AsyncTask 任务。同时当一个 AsyncTask 任务执行完后，AsyncTask 会继续执行其他任务直到所有的任务都被执行为止，从这一点可以看出，在默认情况下，AsyncTask 是串行执行的。

AsyncTask 中有两个线程池（SerialExecutor 和 THREAD_POOL_EXECUTOR）和一个 Handler（InternalHandler），其中线程池 SerialExecutor 用于任务的排队，而线程池 THREAD_POOL_EXECUTOR 用于真正地执行任务，InternalHandler 用于将执行环境从线程池切换到主线程，关于线程池的概念将在第 11.3 节中详细介绍，其本质仍然是线程的调用过程。在 AsyncTask 的构造方法中有如下这么一段代码，由于 FutureTask 的 run 方法会调用 mWorker 的 call 方法，因此 mWorker 的 call 方法最终会在线程池中执行。

```
mWorker = new WorkerRunnable<Params, Result>() {
    public Result call() throws Exception {
        mTaskInvoked.set(true);

        Process.setThreadPriority(Process.THREAD_PRIORITY_BACKGROUND);
        //noinspection unchecked
        return postResult(doInBackground(mParams));
    }
};
```

在 mWorker 的 call 方法中，首先将 mTaskInvoked 设为 true，表示当前任务已经被调用过了，然后执行 AsyncTask 的 doInBackground 方法，接着将其返回值传递给 postResult 方法，它的实现如下所示。

```
private Result postResult(Result result) {
    @SuppressWarnings("unchecked")
    Message message = sHandler.obtainMessage(MESSAGE_POST_RESULT,
```

```
            new AsyncTaskResult<Result>(this, result));
    message.sendToTarget();
    return result;
}
```

在上面的代码中，postResult 方法会通过 sHandler 发送一个 MESSAGE_POST_RESULT 的消息，这个 sHandler 的定义如下所示。

```
private static final InternalHandler sHandler = new InternalHandler();

private static class InternalHandler extends Handler {
    @SuppressWarnings({"unchecked", "RawUseOfParameterizedType"})
    @Override
    public void handleMessage(Message msg) {
        AsyncTaskResult result = (AsyncTaskResult) msg.obj;
        switch (msg.what) {
            case MESSAGE_POST_RESULT:
                // There is only one result
                result.mTask.finish(result.mData[0]);
                break;
            case MESSAGE_POST_PROGRESS:
                result.mTask.onProgressUpdate(result.mData);
                break;
        }
    }
}
```

可以发现，sHandler 是一个静态的 Handler 对象，为了能够将执行环境切换到主线程，这就要求 sHandler 这个对象必须在主线程中创建。由于静态成员会在加载类的时候进行初始化，因此这就变相要求 AsyncTask 的类必须在主线程中加载，否则同一个进程中的 AsyncTask 都将无法正常工作。sHandler 收到 MESSAGE_POST_RESULT 这个消息后会调用 AsyncTask 的 finish 方法，如下所示。

```
private void finish(Result result) {
    if (isCancelled()) {
        onCancelled(result);
    } else {
        onPostExecute(result);
    }
}
```

```
    mStatus = Status.FINISHED;
}
```

AsyncTask 的 finish 方法的逻辑比较简单,如果 AsyncTask 被取消执行了,那么就调用 onCancelled 方法,否则就会调用 onPostExecute 方法,可以看到 doInBackground 的返回结果会传递给 onPostExecute 方法,到这里 AsyncTask 的整个工作过程就分析完毕了。

通过分析 AsyncTask 的源码,可以进一步确定,从 Android 3.0 开始,默认情况下 AsyncTask 的确是串行执行的,在这里通过一系列实验来证实这个判断。

请看如下实验代码,代码很简单,就是单击按钮的时候同时执行 5 个 AsyncTask 任务,每个 AsyncTask 会休眠 3s 来模拟耗时操作,同时把每个 AsyncTask 执行结束的时间打印出来,这样我们就能观察出 AsyncTask 到底是串行执行还是并行执行。

```
@Override
public void onClick(View v) {
    if (v == mButton) {
        new MyAsyncTask("AsyncTask#1").execute("");
        new MyAsyncTask("AsyncTask#2").execute("");
        new MyAsyncTask("AsyncTask#3").execute("");
        new MyAsyncTask("AsyncTask#4").execute("");
        new MyAsyncTask("AsyncTask#5").execute("");
    }

}

private static class MyAsyncTask extends AsyncTask<String, Integer, String> {

    private String mName = "AsyncTask";

    public MyAsyncTask(String name) {
        super();
        mName = name;
    }

    @Override
    protected String doInBackground(String... params) {
        try {
            Thread.sleep(3000);
        } catch (InterruptedException e) {
```

```
            e.printStackTrace();
        }
        return mName;
    }

    @Override
    protected void onPostExecute(String result) {
        super.onPostExecute(result);
        SimpleDateFormat df = new SimpleDateFormat("yyyy-MM-dd HH:mm:ss");
        Log.e(TAG, result + "execute finish at " + df.format(new Date()));
    }
}
```

分别在 Android 4.1.1 和 Android 2.3.3 的设备上运行程序，按照本节前面的描述，AsyncTask 在 4.1.1 上应该是串行的，在 2.3.3 上应该是并行的，到底是不是这样呢？请看下面的运行结果。

Android 4.1.1 上执行：如图 11-1 所示，5 个 AsyncTask 共耗时 15s 且时间间隔为 3s，很显然是串行执行的。

Application	Tag	Text
com.example.test	AsyncTaskTest	AsyncTask#1execute finish at 2013-12-27 01:44:47
com.example.test	AsyncTaskTest	AsyncTask#2execute finish at 2013-12-27 01:44:50
com.example.test	AsyncTaskTest	AsyncTask#3execute finish at 2013-12-27 01:44:53
com.example.test	AsyncTaskTest	AsyncTask#4execute finish at 2013-12-27 01:44:56
com.example.test	AsyncTaskTest	AsyncTask#5execute finish at 2013-12-27 01:44:59

图 11-1　AsyncTask 在 Android 4.1.1 上的执行顺序

Android 2.3.3 上执行：如图 11-2 所示，5 个 AsyncTask 的结束时间是一样的，很显然是并行执行的。

Application	Tag	Text
com.example.test	AsyncTaskTest	AsyncTask#1execute finish at 2013-12-27 01:45:39
com.example.test	AsyncTaskTest	AsyncTask#2execute finish at 2013-12-27 01:45:39
com.example.test	AsyncTaskTest	AsyncTask#3execute finish at 2013-12-27 01:45:39
com.example.test	AsyncTaskTest	AsyncTask#4execute finish at 2013-12-27 01:45:39
com.example.test	AsyncTaskTest	AsyncTask#5execute finish at 2013-12-27 01:45:39

图 11-2　AsyncTask 在 Android 2.3.3 上的执行顺序

为了让 AsyncTask 可以在 Android 3.0 及以上的版本上并行，可以采用 AsyncTask 的

executeOnExecutor 方法，需要注意的是这个方法是 Android 3.0 新添加的方法，并不能在低版本上使用，如下所示。

```
@TargetApi(Build.VERSION_CODES.HONEYCOMB)
@Override
public void onClick(View v) {
    if (v == mButton) {
        if (Build.VERSION.SDK_INT >= Build.VERSION_CODES.HONEYCOMB) {
            new MyAsyncTask("AsyncTask#1").executeOnExecutor(AsyncTask.THREAD_POOL_EXECUTOR,"");
            new MyAsyncTask("AsyncTask#2").executeOnExecutor(AsyncTask.THREAD_POOL_EXECUTOR,"");
            new MyAsyncTask("AsyncTask#3").executeOnExecutor(AsyncTask.THREAD_POOL_EXECUTOR,"");
            new MyAsyncTask("AsyncTask#4").executeOnExecutor(AsyncTask.THREAD_POOL_EXECUTOR,"");
            new MyAsyncTask("AsyncTask#5").executeOnExecutor(AsyncTask.THREAD_POOL_EXECUTOR,"");
        }
    }
}

private static class MyAsyncTask extends AsyncTask<String, Integer, String> {

    private String mName = "AsyncTask";

    public MyAsyncTask(String name) {
        super();
        mName = name;
    }

    @Override
    protected String doInBackground(String... params) {
        try {
            Thread.sleep(3000);
        } catch (InterruptedException e) {
            e.printStackTrace();
        }
        return mName;
    }
```

```
@Override
protected void onPostExecute(String result) {
    super.onPostExecute(result);
    SimpleDateFormat df = new SimpleDateFormat("yyyy-MM-dd HH:mm:ss");
    Log.e(TAG, result + "execute finish at " + df.format(new Date()));
}
}
```

在 Android 4.1.1 的设备上运行上述程序，日志如图 11-3 所示，很显然，我们的目的达到了，成功地让 AsyncTask 在 4.1.1 的手机上并行起来了。

Application	Tag	Text
com.example.test	AsyncTaskTest	AsyncTask#1execute finish at 2013-12-27 01:52:40
com.example.test	AsyncTaskTest	AsyncTask#2execute finish at 2013-12-27 01:52:40
com.example.test	AsyncTaskTest	AsyncTask#4execute finish at 2013-12-27 01:52:40
com.example.test	AsyncTaskTest	AsyncTask#3execute finish at 2013-12-27 01:52:40
com.example.test	AsyncTaskTest	AsyncTask#5execute finish at 2013-12-27 01:52:40

图 11-3　AsyncTask 的 executeOnExecutor 方法的作用

11.2.3　HandlerThread

HandlerThread 继承了 Thread，它是一种可以使用 Handler 的 Thread，它的实现也很简单，就是在 run 方法中通过 Looper.prepare() 来创建消息队列，并通过 Looper.loop() 来开启消息循环，这样在实际的使用中就允许在 HandlerThread 中创建 Handler 了。HandlerThread 的 run 方法如下所示。

```
public void run() {
    mTid = Process.myTid();
    Looper.prepare();
    synchronized (this) {
        mLooper = Looper.myLooper();
        notifyAll();
    }
    Process.setThreadPriority(mPriority);
    onLooperPrepared();
    Looper.loop();
    mTid = -1;
}
```

从 HandlerThread 的实现来看，它和普通的 Thread 有显著的不同之处。普通 Thread 主要用于在 run 方法中执行一个耗时任务，而 HandlerThread 在内部创建了消息队列，外界需要通过 Handler 的消息方式来通知 HandlerThread 执行一个具体的任务。HandlerThread 是一个很有用的类，它在 Android 中的一个具体的使用场景是 IntentService，IntentService 将在 11.2.4 节中进行介绍。由于 HandlerThread 的 run 方法是一个无限循环，因此当明确不需要再使用 HandlerThread 时，可以通过它的 quit 或者 quitSafely 方法来终止线程的执行，这是一个良好的编程习惯。

11.2.4　IntentService

IntentService 是一种特殊的 Service，它继承了 Service 并且它是一个抽象类，因此必须创建它的子类才能使用 IntentService。IntentService 可用于执行后台耗时的任务，当任务执行后它会自动停止，同时由于 IntentService 是服务的原因，这导致它的优先级比单纯的线程要高很多，所以 IntentService 比较适合执行一些高优先级的后台任务，因为它优先级高不容易被系统杀死。在实现上，IntentService 封装了 HandlerThread 和 Handler，这一点可以从它的 onCreate 方法中看出来，如下所示。

```
public void onCreate() {
    // TODO: It would be nice to have an option to hold a partial wakelock
    // during processing, and to have a static startService(Context, Intent)
    // method that would launch the service & hand off a wakelock.

    super.onCreate();
    HandlerThread thread = new HandlerThread("IntentService[" + mName + "]");
    thread.start();

    mServiceLooper = thread.getLooper();
    mServiceHandler = new ServiceHandler(mServiceLooper);
}
```

当 IntentService 被第一次启动时，它的 onCreate 方法会被调用，onCreate 方法会创建一个 HandlerThread，然后使用它的 Looper 来构造一个 Handler 对象 mServiceHandler，这样通过 mServiceHandler 发送的消息最终都会在 HandlerThread 中执行，从这个角度来看，IntentService 也可以用于执行后台任务。每次启动 IntentService，它的 onStartCommand 方法就会调用一次，IntentService 在 onStartCommand 中处理每个后台任务的 Intent。下面看一下 onStartCommand 方法是如何处理外界的 Intent 的，onStartCommand 调用了 onStart，

onStart 方法的实现如下所示。

```
public void onStart(Intent intent, int startId) {
    Message msg = mServiceHandler.obtainMessage();
    msg.arg1 = startId;
    msg.obj = intent;
    mServiceHandler.sendMessage(msg);
}
```

可以看出，IntentService 仅仅是通过 mServiceHandler 发送了一个消息，这个消息会在 HandlerThread 中被处理。mServiceHandler 收到消息后，会将 Intent 对象传递给 onHandleIntent 方法去处理。注意这个 Intent 对象的内容和外界的 startService(intent)中的 intent 的内容是完全一致的，通过这个 Intent 对象即可解析出外界启动 IntentService 时所传递的参数，通过这些参数就可以区分具体的后台任务，这样在 onHandleIntent 方法中就可以对不同的后台任务做处理了。当 onHandleIntent 方法执行结束后，IntentService 会通过 stopSelf(int startId)方法来尝试停止服务。这里之所以采用 stopSelf(int startId)而不是 stopSelf() 来停止服务，那是因为 stopSelf()会立刻停止服务，而这个时候可能还有其他消息未处理，stopSelf(int startId)则会等待所有的消息都处理完毕后才终止服务。一般来说，stopSelf(int startId)在尝试停止服务之前会判断最近启动服务的次数是否和 startId 相等，如果相等就立刻停止服务，不相等则不停止服务，这个策略可以从 AMS 的 stopServiceToken 方法的实现中找到依据，读者感兴趣的话可以自行查看源码实现。ServiceHandler 的实现如下所示。

```
private final class ServiceHandler extends Handler {
    public ServiceHandler(Looper looper) {
        super(looper);
    }

    @Override
    public void handleMessage(Message msg) {
        onHandleIntent((Intent)msg.obj);
        stopSelf(msg.arg1);
    }
}
```

IntentService 的 onHandleIntent 方法是一个抽象方法，它需要我们在子类中实现，它的作用是从 Intent 参数中区分具体的任务并执行这些任务。如果目前只存在一个后台任务，那么 onHandleIntent 方法执行完这个任务后，stopSelf(int startId)就会直接停止服务；如果目前存在多个后台任务，那么当 onHandleIntent 方法执行完最后一个任务时，stopSelf(int startId)

才会直接停止服务。另外，由于每执行一个后台任务就必须启动一次 IntentService，而 IntentService 内部则通过消息的方式向 HandlerThread 请求执行任务，Handler 中的 Looper 是顺序处理消息的，这就意味着 IntentService 也是顺序执行后台任务的，当有多个后台任务同时存在时，这些后台任务会按照外界发起的顺序排队执行。

下面通过一个示例来进一步说明 IntentService 的工作方式，首先派生一个 IntentService 的子类，比如 LocalIntentService，它的实现如下所示。

```java
public class LocalIntentService extends IntentService {
    private static final String TAG = "LocalIntentService";

    public LocalIntentService() {
        super(TAG);
    }

    @Override
    protected void onHandleIntent(Intent intent) {
        String action = intent.getStringExtra("task_action");
        Log.d(TAG, "receive task :" + action);
        SystemClock.sleep(3000);
        if ("com.ryg.action.TASK1".equals(action)) {
            Log.d(TAG, "handle task: " + action);
        }
    }

    @Override
    public void onDestroy() {
        Log.d(TAG, "service destroyed.");
        super.onDestroy();
    }
}
```

这里对 LocalIntentService 的实现做一下简单的说明。在 onHandleIntent 方法中会从参数中解析出后台任务的标识，即 task_action 字段所代表的内容，然后根据不同的任务标识来执行具体的后台任务。这里为了简单起见，直接通过 SystemClock.sleep(3000)来休眠 3000 毫秒从而模拟一种耗时的后台任务，另外为了验证 IntentService 的停止时机，这里在 onDestroy()中打印了一句日志。LocalIntentService 实现完成了以后，就可以在外界请求执行后台任务了，在下面的代码中先后发起了 3 个后台任务的请求：

```java
Intent service = new Intent(this, LocalIntentService.class);
service.putExtra("task_action", "com.ryg.action.TASK1");
```

```
startService(service);
service.putExtra("task_action", "com.ryg.action.TASK2");
startService(service);
service.putExtra("task_action", "com.ryg.action.TASK3");
startService(service);
```

运行程序，观察日志，如下所示。

```
05-17 17:08:23.186 E/dalvikvm(25793): threadid=11: calling run(),name=
IntentService[LocalIntentService]
05-17 17:08:23.196 D/LocalIntentService(25793): receive task :com.ryg.
action.TASK1
05-17 17:08:26.199 D/LocalIntentService(25793): handle task: com.ryg.
action.TASK1
05-17 17:08:26.199 D/LocalIntentService(25793): receive task :com.ryg.
action.TASK2
05-17 17:08:29.192 D/LocalIntentService(25793): receive task :com.ryg.
action.TASK3
05-17 17:08:32.205 D/LocalIntentService(25793): service destroyed.
05-17 17:08:32.205 E/dalvikvm(25793): threadid=11: exiting,name=
IntentService[LocalIntentService]
```

从上面的日志可以看出，三个后台任务是排队执行的，它们的执行顺序就是它们发起请求对的顺序，即 TASK1、TASK2、TASK3。另外一点就是当 TASK3 执行完毕后，LocalIntentService 才真正地停止，从日志中可以看出 LocalIntentService 执行了 onDestroy()，这也意味着服务正在停止。

11.3　Android 中的线程池

提到线程池就必须先说一下线程池的好处，相信读者都有所体会，线程池的优点可以概括为以下三点：

（1）重用线程池中的线程，避免因为线程的创建和销毁所带来的性能开销。

（2）能有效控制线程池的最大并发数，避免大量的线程之间因互相抢占系统资源而导致的阻塞现象。

(3)能够对线程进行简单的管理,并提供定时执行以及指定间隔循环执行等功能。

Android 中的线程池的概念来源于 Java 中的 Executor,Executor 是一个接口,真正的线程池的实现为 ThreadPoolExecutor。ThreadPoolExecutor 提供了一系列参数来配置线程池,通过不同的参数可以创建不同的线程池,从线程池的功能特性上来说,Android 的线程池主要分为 4 类,这 4 类线程池可以通过 Executors 所提供的工厂方法来得到,具体会在 11.3.2 节中进行详细介绍。由于 Android 中的线程池都是直接或者间接通过配置 ThreadPoolExecutor 来实现的,因此在介绍它们之前需要先介绍 ThreadPoolExecutor。

11.3.1　ThreadPoolExecutor

ThreadPoolExecutor 是线程池的真正实现,它的构造方法提供了一系列参数来配置线程池,下面介绍 ThreadPoolExecutor 的构造方法中各个参数的含义,这些参数将会直接影响到线程池的功能特性,下面是 ThreadPoolExecutor 的一个比较常用的构造方法。

```
public ThreadPoolExecutor(int corePoolSize,
                          int maximumPoolSize,
                          long keepAliveTime,
                          TimeUnit unit,
                          BlockingQueue<Runnable> workQueue,
                          ThreadFactory threadFactory)
```

corePoolSize

线程池的核心线程数,默认情况下,核心线程会在线程池中一直存活,即使它们处于闲置状态。如果将 ThreadPoolExecutor 的 allowCoreThreadTimeOut 属性设置为 true,那么闲置的核心线程在等待新任务到来时会有超时策略,这个时间间隔由 keepAliveTime 所指定,当等待时间超过 keepAliveTime 所指定的时长后,核心线程就会被终止。

maximumPoolSize

线程池所能容纳的最大线程数,当活动线程数达到这个数值后,后续的新任务将会被阻塞。

keepAliveTime

非核心线程闲置时的超时时长,超过这个时长,非核心线程就会被回收。当 ThreadPool-Executor 的 allowCoreThreadTimeOut 属性设置为 true 时,keepAliveTime 同样会作用于核心线程。

unit

用于指定 keepAliveTime 参数的时间单位,这是一个枚举,常用的有 TimeUnit.MILLISECONDS(毫秒)、TimeUnit.SECONDS(秒)以及 TimeUnit.MINUTES(分钟)等。

workQueue

线程池中的任务队列,通过线程池的 execute 方法提交的 Runnable 对象会存储在这个参数中。

threadFactory

线程工厂,为线程池提供创建新线程的功能。ThreadFactory 是一个接口,它只有一个方法:Thread newThread(Runnable r)。

除了上面的这些主要参数外,ThreadPoolExecutor 还有一个不常用的参数 RejectedExecutionHandler handler。当线程池无法执行新任务时,这可能是由于任务队列已满或者是无法成功执行任务,这个时候 ThreadPoolExecutor 会调用 handler 的 rejectedExecution 方法来通知调用者,默认情况下 rejectedExecution 方法会直接抛出一个 RejectedExecutionException。ThreadPoolExecutor 为 RejectedExecutionHandler 提供了几个可选值:CallerRunsPolicy、AbortPolicy、DiscardPolicy 和 DiscardOldestPolicy,其中 AbortPolicy 是默认值,它会直接抛出 RejectedExecutionException,由于 handler 这个参数不常用,这里就不再具体介绍了。

ThreadPoolExecutor 执行任务时大致遵循如下规则:

(1)如果线程池中的线程数量未达到核心线程的数量,那么会直接启动一个核心线程来执行任务。

(2)如果线程池中的线程数量已经达到或者超过核心线程的数量,那么任务会被插入到任务队列中排队等待执行。

(3)如果在步骤 2 中无法将任务插入到任务队列中,这往往是由于任务队列已满,这个时候如果线程数量未达到线程池规定的最大值,那么会立刻启动一个非核心线程来执行任务。

（4）如果步骤 3 中线程数量已经达到线程池规定的最大值，那么就拒绝执行此任务，ThreadPoolExecutor 会调用 RejectedExecutionHandler 的 rejectedExecution 方法来通知调用者。

ThreadPoolExecutor 的参数配置在 AsyncTask 中有明显的体现，下面是 AsyncTask 中的线程池的配置情况：

```
private static final int CPU_COUNT = Runtime.getRuntime().availableProcessors();
private static final int CORE_POOL_SIZE = CPU_COUNT + 1;
private static final int MAXIMUM_POOL_SIZE = CPU_COUNT * 2 + 1;
private static final int KEEP_ALIVE = 1;

private static final ThreadFactory sThreadFactory = new ThreadFactory() {
    private final AtomicInteger mCount = new AtomicInteger(1);

    public Thread newThread(Runnable r) {
        return new Thread(r, "AsyncTask #" + mCount.getAndIncrement());
    }
};

private static final BlockingQueue<Runnable> sPoolWorkQueue =
        new LinkedBlockingQueue<Runnable>(128);

/**
 * An {@link Executor} that can be used to execute tasks in parallel.
 */
public static final Executor THREAD_POOL_EXECUTOR
        = new ThreadPoolExecutor(CORE_POOL_SIZE, MAXIMUM_POOL_SIZE, KEEP_ALIVE, TimeUnit.SECONDS, sPoolWorkQueue, sThreadFactory);
```

从上面的代码可以知道，AsyncTask 对 THREAD_POOL_EXECUTOR 这个线程池进行了配置，配置后的线程池规格如下：

- 核心线程数等于 CPU 核心数 + 1；

- 线程池的最大线程数为 CPU 核心数的 2 倍 + 1；

- 核心线程无超时机制，非核心线程在闲置时的超时时间为 1 秒；

- 任务队列的容量为 128。

11.3.2 线程池的分类

在 11.3.1 节中对 ThreadPoolExecutor 的配置细节进行了详细的介绍，本节将接着介绍 Android 中最常见的四类具有不同功能特性的线程池，它们都直接或间接地通过配置 ThreadPoolExecutor 来实现自己的功能特性，这四类线程池分别是 FixedThreadPool、CachedThreadPool、ScheduledThreadPool 以及 SingleThreadExecutor。

1. FixedThreadPool

通过 Executors 的 newFixedThreadPool 方法来创建。它是一种线程数量固定的线程池，当线程处于空闲状态时，它们并不会被回收，除非线程池被关闭了。当所有的线程都处于活动状态时，新任务都会处于等待状态，直到有线程空闲出来。由于 FixedThreadPool 只有核心线程并且这些核心线程不会被回收，这意味着它能够更加快速地响应外界的请求。newFixedThreadPool 方法的实现如下，可以发现 FixedThreadPool 中只有核心线程并且这些核心线程没有超时机制，另外任务队列也是没有大小限制的。

```
public static ExecutorService newFixedThreadPool(int nThreads) {
    return new ThreadPoolExecutor(nThreads, nThreads,
                                  0L, TimeUnit.MILLISECONDS,
                                  new LinkedBlockingQueue<Runnable>());
}
```

2. CachedThreadPool

通过 Executors 的 newCachedThreadPool 方法来创建。它是一种线程数量不定的线程池，它只有非核心线程，并且其最大线程数为 Integer.MAX_VALUE。由于 Integer.MAX_VALUE 是一个很大的数，实际上就相当于最大线程数可以任意大。当线程池中的线程都处于活动状态时，线程池会创建新的线程来处理新任务，否则就会利用空闲的线程来处理新任务。线程池中的空闲线程都有超时机制，这个超时时长为 60 秒，超过 60 秒闲置线程就会被回收。和 FixedThreadPool 不同的是，CachedThreadPool 的任务队列其实相当于一个空集合，这将导致任何任务都会立即被执行，因为在这种场景下 SynchronousQueue 是无法插入任务的。SynchronousQueue 是一个非常特殊的队列，在很多情况下可以把它简单理解为一个无法存储元素的队列，由于它在实际中较少使用，这里就不深入探讨它了。从 CachedThreadPool 的特性来看，这类线程池比较适合执行大量的耗时较少的任务。当整个线程池都处于闲置状态时，线程池中的线程都会超时而被停止，这个时候 CachedThreadPool 之中实际上是没有任何线程的，它几乎是不占用任何系统资源的。newCachedThreadPool

方法的实现如下所示。

```
public static ExecutorService newCachedThreadPool() {
    return new ThreadPoolExecutor(0, Integer.MAX_VALUE,
                                  60L, TimeUnit.SECONDS,
                                  new SynchronousQueue<Runnable>());
}
```

3. ScheduledThreadPool

通过 Executors 的 newScheduledThreadPool 方法来创建。它的核心线程数量是固定的，而非核心线程数是没有限制的，并且当非核心线程闲置时会被立即回收。ScheduledThreadPool 这类线程池主要用于执行定时任务和具有固定周期的重复任务，newScheduledThreadPool 方法的实现如下所示。

```
public static ScheduledExecutorService newScheduledThreadPool(int corePoolSize) {
    return new ScheduledThreadPoolExecutor(corePoolSize);
}

public ScheduledThreadPoolExecutor(int corePoolSize) {
    super(corePoolSize, Integer.MAX_VALUE, 0, NANOSECONDS,
          new DelayedWorkQueue());
}
```

4. SingleThreadExecutor

通过 Executors 的 newSingleThreadExecutor 方法来创建。这类线程池内部只有一个核心线程，它确保所有的任务都在同一个线程中按顺序执行。SingleThreadExecutor 的意义在于统一所有的外界任务到一个线程中，这使得在这些任务之间不需要处理线程同步的问题。newSingleThreadExecutor 方法的实现如下所示。

```
public static ExecutorService newSingleThreadExecutor() {
    return new FinalizableDelegatedExecutorService
        (new ThreadPoolExecutor(1, 1,
                                0L, TimeUnit.MILLISECONDS,
                                new LinkedBlockingQueue<Runnable>()));
}
```

上面对 Android 中常见的 4 种线程池进行了详细的介绍，除了上面系统提供的 4 类线程池以外，也可以根据实际需要灵活地配置线程池。下面的代码演示了系统预置的 4 种线

程池的典型使用方法。

```
Runnable command = new Runnable() {
    @Override
    public void run() {
        SystemClock.sleep(2000);
    }
};

ExecutorService fixedThreadPool = Executors.newFixedThreadPool(4);
fixedThreadPool.execute(command);

ExecutorService cachedThreadPool = Executors.newCachedThreadPool();
cachedThreadPool.execute(command);

ScheduledExecutorService scheduledThreadPool = Executors.newScheduled-
ThreadPool(4);
// 2000ms 后执行 command
scheduledThreadPool.schedule(command, 2000, TimeUnit.MILLISECONDS);
// 延迟 10ms 后，每隔 1000ms 执行一次 command
scheduledThreadPool.scheduleAtFixedRate(command, 10, 1000, TimeUnit.
MILLISECONDS);

ExecutorService singleThreadExecutor = Executors.newSingleThread-
Executor();
singleThreadExecutor.execute(command);
```

第 12 章 Bitmap 的加载和 Cache

本章的主题是 Bitmap 的加载和 Cache，主要包含三个方面的内容。首先讲述如何有效地加载一个 Bitmap，这是一个很有意义的话题，由于 Bitmap 的特殊性以及 Android 对单个应用所施加的内存限制，比如 16MB，这导致加载 Bitmap 的时候很容易出现内存溢出。下面这个异常信息在开发中应该时常遇到：

```
java.lang.OutofMemoryError: bitmap size exceeds VM budget
```

因此如何高效地加载 Bitmap 是一个很重要也很容易被开发者忽视的问题。

接着介绍 Android 中常用的缓存策略，缓存策略是一个通用的思想，可以用在很多场景中，但是实际开发中经常需要用 Bitmap 做缓存。通过缓存策略，我们不需要每次都从网络上请求图片或者从存储设备中加载图片，这样就极大地提高了图片的加载效率以及产品的用户体验。目前比较常用的缓存策略是 LruCache 和 DiskLruCache，其中 LruCache 常被用做内存缓存，而 DiskLruCache 常被用做存储缓存。Lru 是 Least Recently Used 的缩写，即最近最少使用算法，这种算法的核心思想为：当缓存快满时，会淘汰近期最少使用的缓存目标，很显然 Lru 算法的思想是很容易被接受的。

最后本章会介绍如何优化列表的卡顿现象，ListView 和 GridView 由于要加载大量的子视图，当用户快速滑动时就容易出现卡顿的现象，因此本章最后针对这个问题将会给出一些优化建议。

为了更好地介绍上述三个主题，本章提供了一个示例程序，该程序会尝试从网络加载大量图片并在 GridView 中显示，可以发现这个程序具有很强的实用性，并且其技术细节完全覆盖了本章的三个主题：图片加载、缓存策略、列表的滑动流畅性，通过这个示例程序读者可以很好地理解本章的全部内容并能够在实际中灵活应用。

12.1 Bitmap 的高效加载

在介绍 Bitmap 的高效加载之前，先说一下如何加载一个 Bitmap，Bitmap 在 Android 中指的是一张图片，可以是 png 格式也可以是 jpg 等其他常见的图片格式。那么如何加载一个图片呢？BitmapFactory 类提供了四类方法：decodeFile、decodeResource、decodeStream 和 decodeByteArray，分别用于支持从文件系统、资源、输入流以及字节数组中加载出一个 Bitmap 对象，其中 decodeFile 和 decodeResource 又间接调用了 decodeStream 方法，这四类方法最终是在 Android 的底层实现的，对应着 BitmapFactory 类的几个 native 方法。

如何高效地加载 Bitmap 呢？其实核心思想也很简单，那就是采用 BitmapFactory.Options 来加载所需尺寸的图片。这里假设通过 ImageView 来显示图片，很多时候 ImageView 并没有图片的原始尺寸那么大，这个时候把整个图片加载进来后再设给 ImageView，这显然是没必要的，因为 ImageView 并没有办法显示原始的图片。通过 BitmapFactory.Options 就可以按一定的采样率来加载缩小后的图片，将缩小后的图片在 ImageView 中显示，这样就会降低内存占用从而在一定程度上避免 OOM，提高了 Bitmap 加载时的性能。BitmapFactory 提供的加载图片的四类方法都支持 BitmapFactory.Options 参数，通过它们就可以很方便地对一个图片进行采样缩放。

通过 BitmapFactory.Options 来缩放图片，主要是用到了它的 inSampleSize 参数，即采样率。当 inSampleSize 为 1 时，采样后的图片大小为图片的原始大小；当 inSampleSize 大于 1 时，比如为 2，那么采样后的图片其宽/高均为原图大小的 1/2，而像素数为原图的 1/4，其占有的内存大小也为原图的 1/4。拿一张 1024×1024 像素的图片来说，假定采用 ARGB8888 格式存储，那么它占有的内存为 1024×1024×4，即 4MB，如果 inSampleSize 为 2，那么采样后的图片其内存占用只有 512×512×4，即 1MB。可以发现采样率 inSampleSize 必须是大于 1 的整数图片才会有缩小的效果，并且采样率同时作用于宽/高，这将导致缩放后的图片大小以采样率的 2 次方形式递减，即缩放比例为 1/（inSampleSize 的 2 次方），比如 inSampleSize 为 4，那么缩放比例就是 1/16。有一种特殊情况，那就是当 inSampleSize 小于 1 时，其作用相当于 1，即无缩放效果。另外最新的官方文档中指出，inSampleSize 的取值应该总是为 2 的指数，比如 1、2、4、8、16，等等。如果外界传递给系统的 inSampleSize 不为 2 的指数，那么系统会向下取整并选择一个最接近的 2 的指数来代替，比如 3，系统会选择 2 来代替，但是经过验证发现这个结论并非在所有的 Android 版本上都成立，因此把它当成一个开发建议即可。

考虑以下实际的情况，比如 ImageView 的大小是 100×100 像素，而图片的原始大小为 200×200，那么只需将采样率 inSampleSize 设为 2 即可。但是如果图片大小为 200×300 呢？这个时候采样率还应该选择 2，这样缩放后的图片大小为 100×150 像素，仍然是适合 ImageView 的，如果采样率为 3，那么缩放后的图片大小就会小于 ImageView 所期望的大小，这样图片就会被拉伸从而导致模糊。

通过采样率即可有效地加载图片，那么到底如何获取采样率呢？获取采样率也很简单，遵循如下流程：

（1）将 BitmapFactory.Options 的 inJustDecodeBounds 参数设为 true 并加载图片。

（2）从 BitmapFactory.Options 中取出图片的原始宽高信息，它们对应于 outWidth 和 outHeight 参数。

（3）根据采样率的规则并结合目标 View 的所需大小计算出采样率 inSampleSize。

（4）将 BitmapFactory.Options 的 inJustDecodeBounds 参数设为 false，然后重新加载图片。

经过上面 4 个步骤，加载出的图片就是最终缩放后的图片，当然也有可能不需要缩放。这里说明一下 inJustDecodeBounds 参数，当此参数设为 true 时，BitmapFactory 只会解析图片的原始宽/高信息，并不会去真正地加载图片，所以这个操作是轻量级的。另外需要注意的是，这个时候 BitmapFactory 获取的图片宽/高信息和图片的位置以及程序运行的设备有关，比如同一张图片放在不同的 drawable 目录下或者程序运行在不同屏幕密度的设备上，这都可能导致 BitmapFactory 获取到不同的结果，之所以会出现这个现象，这和 Android 的资源加载机制有关，相信读者平日里肯定有所体会，这里就不再详细说明了。

将上面的 4 个流程用程序来实现，就产生了下面的代码：

```
public static Bitmap decodeSampledBitmapFromResource(Resources res, int resId,
        int reqWidth, int reqHeight) {
    // First decode with inJustDecodeBounds=true to check dimensions
    final BitmapFactory.Options options = new BitmapFactory.Options();
    options.inJustDecodeBounds = true;
```

```
        BitmapFactory.decodeResource(res, resId, options);

        // Calculate inSampleSize
        options.inSampleSize = calculateInSampleSize(options, reqWidth, reqHeight);

        // Decode bitmap with inSampleSize set
        options.inJustDecodeBounds = false;
        return BitmapFactory.decodeResource(res, resId, options);
    }

    public static int calculateInSampleSize(
            BitmapFactory.Options options, int reqWidth, int reqHeight) {
        // Raw height and width of image
        final int height = options.outHeight;
        final int width = options.outWidth;
        int inSampleSize = 1;

        if (height > reqHeight || width > reqWidth) {
            final int halfHeight = height / 2;
            final int halfWidth = width / 2;
            // Calculate the largest inSampleSize value that is a power of 2 and
              keeps both
            // height and width larger than the requested height and width.
            while ((halfHeight / inSampleSize) >= reqHeight
                    && (halfWidth / inSampleSize) >= reqWidth) {
                inSampleSize *= 2;
            }
        }

        return inSampleSize;
    }
```

有了上面的两个方法，实际使用的时候就很简单了，比如 ImageView 所期望的图片大小为 100×100 像素，这个时候就可以通过如下方式高效地加载并显示图片：

```
mImageView.setImageBitmap(
    decodeSampledBitmapFromResource(getResources(), R.id.myimage, 100, 100));
```

除了 BitmapFactory 的 decodeResource 方法，其他三个 decode 系列的方法也是支持采样加载的，并且处理方式也是类似的，但是 decodeStream 方法稍微有点特殊，这个会在后续内容中详细介绍。通过本节的介绍，读者应该能很好地掌握这种高效地加载图片的方法了。

12.2 Android 中的缓存策略

缓存策略在 Android 中有着广泛的使用场景，尤其在图片加载这个场景下，缓存策略就变得更为重要。考虑一种场景：有一批网络图片，需要下载后在用户界面上予以显示，这个场景在 PC 环境下是很简单的，直接把所有的图片下载到本地再显示即可，但是放到移动设备上就不一样了。不管是 Android 还是 iOS 设备，流量对于用户来说都是一种宝贵的资源，由于流量是收费的，所以在应用开发中并不能过多地消耗用户的流量，否则这个应用肯定不能被用户所接受。再加上目前国内公共场所的 WiFi 普及率并不算太高，因此用户在很多情况下手机上都是用的移动网络而非 WiFi，因此必须提供一种解决方案来解决流量的消耗问题。

如何避免过多的流量消耗呢？那就是本节所要讨论的主题：缓存。当程序第一次从网络加载图片后，就将其缓存到存储设备上，这样下次使用这张图片就不用再从网络上获取了，这样就为用户节省了流量。很多时候为了提高应用的用户体验，往往还会把图片在内存中再缓存一份，这样当应用打算从网络上请求一张图片时，程序会首先从内存中去获取，如果内存中没有那就从存储设备中去获取，如果存储设备中也没有，那就从网络上下载这张图片。因为从内存中加载图片比从存储设备中加载图片要快，所以这样既提高了程序的效率又为用户节约了不必要的流量开销。上述的缓存策略不仅仅适用于图片，也适用于其他文件类型。

说到缓存策略，其实并没有统一的标准。一般来说，缓存策略主要包含缓存的添加、获取和删除这三类操作。如何添加和获取缓存这个比较好理解，那么为什么还要删除缓存呢？这是因为不管是内存缓存还是存储设备缓存，它们的缓存大小都是有限制的，因为内存和诸如 SD 卡之类的存储设备都是有容量限制的，因此在使用缓存时总是要为缓存指定一个最大的容量。如果当缓存容量满了，但是程序还需要向其添加缓存，这个时候该怎么办呢？这就需要删除一些旧的缓存并添加新的缓存，如何定义缓存的新旧这就是一种策略，不同的策略就对应着不同的缓存算法，比如可以简单地根据文件的最后修改时间来定义缓存的新旧，当缓存满时就将最后修改时间较早的缓存移除，这就是一种缓存算法，但是这种算法并不算很完美。

目前常用的一种缓存算法是 LRU（Least Recently Used），LRU 是近期最少使用算法，它的核心思想是当缓存满时，会优先淘汰那些近期最少使用的缓存对象。采用 LRU 算法的缓存有两种：LruCache 和 DiskLruCache，LruCache 用于实现内存缓存，而 DiskLruCache 则充当了存储设备缓存，通过这二者的完美结合，就可以很方便地实现一个具有很高实用价值的 ImageLoader。本节首先会介绍 LruCache 和 DiskLruCache，然后利用 LruCache 和 DiskLruCache 来实现一个优秀的 ImageLoader，并且提供一个使用 ImageLoader 来从网络下载并展示图片的例子，在这个例子中体现了 ImageLoader 以及大批量网络图片加载所涉及的大量技术点。

12.2.1 LruCache

LruCache 是 Android 3.1 所提供的一个缓存类，通过 support-v4 兼容包可以兼容到早期的 Android 版本，目前 Android 2.2 以下的用户量已经很少了，因此我们开发的应用兼容到 Android 2.2 就已经足够了。为了能够兼容 Android 2.2 版本，在使用 LruCache 时建议采用 support-v4 兼容包中提供的 LruCache，而不要直接使用 Android 3.1 提供的 LruCache。

LruCache 是一个泛型类，它内部采用一个 LinkedHashMap 以强引用的方式存储外界的缓存对象，其提供了 get 和 put 方法来完成缓存的获取和添加操作，当缓存满时，LruCache 会移除较早使用的缓存对象，然后再添加新的缓存对象。这里读者要明白强引用、软引用和弱引用的区别，如下所示。

- 强引用：直接的对象引用；
- 软引用：当一个对象只有软引用存在时，系统内存不足时此对象会被 gc 回收；
- 弱引用：当一个对象只有弱引用存在时，此对象会随时被 gc 回收。

另外 LruCache 是线程安全的，下面是 LruCache 的定义：

```
public class LruCache<K, V> {
    private final LinkedHashMap<K, V> map;
    ...
}
```

LruCache 的实现比较简单，读者可以参考它的源码，这里仅介绍如何使用 LruCache 来实现内存缓存。仍然拿图片缓存来举例子，下面的代码展示了 LruCache 的典型的初始化过程：

```
int maxMemory = (int) (Runtime.getRuntime().maxMemory() / 1024);
int cacheSize = maxMemory / 8;
mMemoryCache = new LruCache<String, Bitmap>(cacheSize) {
    @Override
    protected int sizeOf(String key, Bitmap bitmap) {
        return bitmap.getRowBytes() * bitmap.getHeight() / 1024;
    }
};
```

在上面的代码中，只需要提供缓存的总容量大小并重写 sizeOf 方法即可。sizeOf 方法的作用是计算缓存对象的大小，这里大小的单位需要和总容量的单位一致。对于上面的示例代码来说，总容量的大小为当前进程的可用内存的 1/8，单位为 KB，而 sizeOf 方法则完成了 Bitmap 对象的大小计算。很明显，之所以除以 1024 也是为了将其单位转换为 KB。一些特殊情况下，还需要重写 LruCache 的 entryRemoved 方法，LruCache 移除旧缓存时会调用 entryRemoved 方法，因此可以在 entryRemoved 中完成一些资源回收工作（如果需要的话）。

除了 LruCache 的创建以外，还有缓存的获取和添加，这也很简单，从 LruCache 中获取一个缓存对象，如下所示。

```
mMemoryCache.get(key)
```

向 LruCache 中添加一个缓存对象，如下所示。

```
mMemoryCache.put(key, bitmap)
```

LruCache 还支持删除操作，通过 remove 方法即可删除一个指定的缓存对象。可以看到 LruCache 的实现以及使用都非常简单，虽然简单，但是仍然不影响它具有强大的功能，从 Android 3.1 开始，LruCache 就已经是 Android 源码的一部分了。

12.2.2 DiskLruCache

DiskLruCache 用于实现存储设备缓存，即磁盘缓存，它通过将缓存对象写入文件系统从而实现缓存的效果。DiskLruCache 得到了 Android 官方文档的推荐，但它不属于 Android SDK 的一部分，它的源码可以从如下网址得到：

```
https://android.googlesource.com/platform/libcore/+/android-4.1.1_r1/luni/src/main/java/libcore/io/DiskLruCache.java
```

需要注意的是，从上述网址获取的 DiskLruCache 的源码并不能直接在 Android 中使用，需要稍微修改编译错误。下面分别从 DiskLruCache 的创建、缓存查找和缓存添加这三个方面来介绍 DiskLruCache 的使用方式。

1. DiskLruCache 的创建

DiskLruCache 并不能通过构造方法来创建，它提供了 open 方法用于创建自身，如下所示。

```
public static DiskLruCache open(File directory, int appVersion, int valueCount, long maxSize)
```

open 方法有四个参数，其中第一个参数表示磁盘缓存在文件系统中的存储路径。缓存路径可以选择 SD 卡上的缓存目录，具体是指/sdcard/Android/data/package_name/cache 目录，其中 package_name 表示当前应用的包名，当应用被卸载后，此目录会一并被删除。当然也可以选择 SD 卡上的其他指定目录，还可以选择 data 下的当前应用的目录，具体可根据需要灵活设定。这里给出一个建议：如果应用卸载后就希望删除缓存文件，那么就选择 SD 卡上的缓存目录，如果希望保留缓存数据那就应该选择 SD 卡上的其他特定目录。

第二个参数表示应用的版本号，一般设为 1 即可。当版本号发生改变时 DiskLruCache 会清空之前所有的缓存文件，而这个特性在实际开发中作用并不大，很多情况下即使应用的版本号发生了改变缓存文件却仍然是有效的，因此这个参数设为 1 比较好。

第三个参数表示单个节点所对应的数据的个数，一般设为 1 即可。第四个参数表示缓存的总大小，比如 50MB，当缓存大小超出这个设定值后，DiskLruCache 会清除一些缓存从而保证总大小不大于这个设定值。下面是一个典型的 DiskLruCache 的创建过程：

```
private static final long DISK_CACHE_SIZE = 1024 * 1024 * 50; //50MB

File diskCacheDir = getDiskCacheDir(mContext, "bitmap");
if (!diskCacheDir.exists()) {
    diskCacheDir.mkdirs();
}
mDiskLruCache = DiskLruCache.open(diskCacheDir, 1, 1, DISK_CACHE_SIZE);
```

2. DiskLruCache 的缓存添加

DiskLruCache 的缓存添加的操作是通过 Editor 完成的，Editor 表示一个缓存对象的编辑对象。这里仍然以图片缓存举例，首先需要获取图片 url 所对应的 key，然后根据 key 就可以通过 edit() 来获取 Editor 对象，如果这个缓存正在被编辑，那么 edit() 会返回 null，即 DiskLruCache 不允许同时编辑一个缓存对象。之所以要把 url 转换成 key，是因为图片的 url 中很可能有特殊字符，这将影响 url 在 Android 中直接使用，一般采用 url 的 md5 值作为 key，如下所示。

```
private String hashKeyFormUrl(String url) {
    String cacheKey;
    try {
        final MessageDigest mDigest = MessageDigest.getInstance("MD5");
        mDigest.update(url.getBytes());
        cacheKey = bytesToHexString(mDigest.digest());
    } catch (NoSuchAlgorithmException e) {
        cacheKey = String.valueOf(url.hashCode());
    }
    return cacheKey;
}

private String bytesToHexString(byte[] bytes) {
    StringBuilder sb = new StringBuilder();
    for (int i = 0; i < bytes.length; i++) {
        String hex = Integer.toHexString(0xFF & bytes[i]);
        if (hex.length() == 1) {
            sb.append('0');
        }
        sb.append(hex);
    }
    return sb.toString();
}
```

将图片的 url 转成 key 以后，就可以获取 Editor 对象了。对于这个 key 来说，如果当前不存在其他 Editor 对象，那么 edit() 就会返回一个新的 Editor 对象，通过它就可以得到一个文件输出流。需要注意的是，由于前面在 DiskLruCache 的 open 方法中设置了一个节点只能有一个数据，因此下面的 DISK_CACHE_INDEX 常量直接设为 0 即可，如下所示。

```
String key = hashKeyFormUrl(url);
DiskLruCache.Editor editor = mDiskLruCache.edit(key);
if (editor != null) {
    OutputStream outputStream = editor.newOutputStream(DISK_CACHE_INDEX);
}
```

有了文件输出流,接下来要怎么做呢?其实是这样的,当从网络下载图片时,图片就可以通过这个文件输出流写入到文件系统上,这个过程的实现如下所示。

```java
public boolean downloadUrlToStream(String urlString,
        OutputStream outputStream) {
    HttpURLConnection urlConnection = null;
    BufferedOutputStream out = null;
    BufferedInputStream in = null;

    try {
        final URL url = new URL(urlString);
        urlConnection = (HttpURLConnection) url.openConnection();
        in = new BufferedInputStream(urlConnection.getInputStream(),
                IO_BUFFER_SIZE);
        out = new BufferedOutputStream(outputStream, IO_BUFFER_SIZE);

        int b;
        while ((b = in.read()) != -1) {
            out.write(b);
        }
        return true;
    } catch (IOException e) {
        Log.e(TAG, "downloadBitmap failed." + e);
    } finally {
        if (urlConnection != null) {
            urlConnection.disconnect();
        }
        MyUtils.close(out);
        MyUtils.close(in);
    }
    return false;
}
```

经过上面的步骤,其实并没有真正地将图片写入文件系统,还必须通过 Editor 的 commit() 来提交写入操作,如果图片下载过程发生了异常,那么还可以通过 Editor 的 abort() 来回退整个操作,这个过程如下所示。

```java
OutputStream outputStream = editor.newOutputStream(DISK_CACHE_INDEX);
if (downloadUrlToStream(url, outputStream)) {
    editor.commit();
```

```
} else {
    editor.abort();
}
mDiskLruCache.flush();
```

经过上面的几个步骤,图片已经被正确地写入到文件系统了,接下来图片获取的操作就不需要请求网络了。

3. DiskLruCache 的缓存查找

和缓存的添加过程类似,缓存查找过程也需要将 url 转换为 key,然后通过 DiskLruCache 的 get 方法得到一个 Snapshot 对象,接着再通过 Snapshot 对象即可得到缓存的文件输入流,有了文件输出流,自然就可以得到 Bitmap 对象了。为了避免加载图片过程中导致的 OOM 问题,一般不建议直接加载原始图片。在第 12.1 节中已经介绍了通过 BitmapFactory.Options 对象来加载一张缩放后的图片,但是那种方法对 FileInputStream 的缩放存在问题,原因是 FileInputStream 是一种有序的文件流,而两次 decodeStream 调用影响了文件流的位置属性,导致了第二次 decodeStream 时得到的是 null。为了解决这个问题,可以通过文件流来得到它所对应的文件描述符,然后再通过 BitmapFactory.decodeFileDescriptor 方法来加载一张缩放后的图片,这个过程的实现如下所示。

```
Bitmap bitmap = null;
String key = hashKeyFormUrl(url);
DiskLruCache.Snapshot snapShot = mDiskLruCache.get(key);
if (snapShot != null) {
    FileInputStream fileInputStream = (FileInputStream)snapShot.getInput-
    Stream(DISK_CACHE_INDEX);
    FileDescriptor fileDescriptor = fileInputStream.getFD();
    bitmap = mImageResizer.decodeSampledBitmapFromFileDescriptor
    (fileDescriptor,
        reqWidth, reqHeight);
    if (bitmap != null) {
        addBitmapToMemoryCache(key, bitmap);
    }
}
```

上面介绍了 DiskLruCache 的创建、缓存的添加和查找过程,读者应该对 DiskLruCache 的使用方式有了一个大致的了解,除此之外,DiskLruCache 还提供了 remove、delete 等方法用于磁盘缓存的删除操作。关于 DiskLruCache 的内部实现这里就不再介绍了,读者感兴趣的话可以查看它的源码实现。

12.2.3 ImageLoader 的实现

在本章的前面先后介绍了 Bitmap 的高效加载方式、LruCache 以及 DiskLruCache，现在我们来着手实现一个优秀的 ImageLoader。

一般来说，一个优秀的 ImageLoader 应该具备如下功能：

- 图片的同步加载；
- 图片的异步加载；
- 图片压缩；
- 内存缓存；
- 磁盘缓存；
- 网络拉取。

图片的同步加载是指能够以同步的方式向调用者提供所加载的图片，这个图片可能是从内存缓存中读取的，也可能是从磁盘缓存中读取的，还可能是从网络拉取的。图片的异步加载是一个很有用的功能，很多时候调用者不想在单独的线程中以同步的方式来获取图片，这个时候 ImageLoader 内部需要自己在线程中加载图片并将图片设置给所需的 ImageView。图片压缩的作用更毋庸置疑了，这是降低 OOM 概率的有效手段，ImageLoader 必须合适地处理图片的压缩问题。

内存缓存和磁盘缓存是 ImageLoader 的核心，也是 ImageLoader 的意义之所在，通过这两级缓存极大地提高了程序的效率并且有效地降低了对用户所造成的流量消耗，只有当这两级缓存都不可用时才需要从网络中拉取图片。

除此之外，ImageLoader 还需要处理一些特殊的情况，比如在 ListView 或者 GridView 中，View 复用既是它们的优点也是它们的缺点，优点想必读者都很清楚了，那缺点可能还不太清楚。考虑一种情况，在 ListView 或者 GridView 中，假设一个 item A 正在从网络加载图片，它对应的 ImageView 为 A，这个时候用户快速向下滑动列表，很可能 item B 复用了 ImageView A，然后等了一会之前的图片下载完毕了。如果直接给 ImageView A 设置图片，由于这个时候 ImageView A 被 item B 所复用，但是 item B 要显示的图片显然不是 item A 刚刚下载好的图片，这个时候就会出现 item B 中显示了 item A 的图片，这就是常见的列表的错位问题，ImageLoader 需要正确地处理这些特殊情况。

上面对 ImageLoader 的功能做了一个全面的分析，下面就可以一步步实现一个 ImageLoader 了，这里主要分为如下几步。

1. 图片压缩功能的实现

图片压缩在第 12.1 节中已经做了介绍，这里就不再多说了，为了有良好的设计风格，这里单独抽象了一个类用于完成图片的压缩功能，这个类叫 ImageResizer，它的实现如下所示。

```java
public class ImageResizer {
    private static final String TAG = "ImageResizer";

    public ImageResizer() {
    }

    public Bitmap decodeSampledBitmapFromResource(Resources res,
            int resId, int reqWidth, int reqHeight) {
        // First decode with inJustDecodeBounds=true to check dimensions
        final BitmapFactory.Options options = new BitmapFactory.Options();
        options.inJustDecodeBounds = true;
        BitmapFactory.decodeResource(res, resId, options);

        // Calculate inSampleSize
        options.inSampleSize = calculateInSampleSize(options, reqWidth,
                reqHeight);

        // Decode bitmap with inSampleSize set
        options.inJustDecodeBounds = false;
        return BitmapFactory.decodeResource(res, resId, options);
    }

    public Bitmap decodeSampledBitmapFromFileDescriptor(FileDescriptor fd,
    int reqWidth, int reqHeight) {
        // First decode with inJustDecodeBounds=true to check dimensions
        final BitmapFactory.Options options = new BitmapFactory.Options();
        options.inJustDecodeBounds = true;
        BitmapFactory.decodeFileDescriptor(fd, null, options);
```

```java
        // Calculate inSampleSize
        options.inSampleSize = calculateInSampleSize(options, reqWidth,
            reqHeight);

        // Decode bitmap with inSampleSize set
        options.inJustDecodeBounds = false;
        return BitmapFactory.decodeFileDescriptor(fd, null, options);
    }

    public int calculateInSampleSize(BitmapFactory.Options options,
        int reqWidth, int reqHeight) {
        if (reqWidth == 0 || reqHeight == 0) {
            return 1;
        }

        // Raw height and width of image
        final int height = options.outHeight;
        final int width = options.outWidth;
        Log.d(TAG, "origin, w=" + width + " h=" + height);
        int inSampleSize = 1;

        if (height > reqHeight || width > reqWidth) {
            final int halfHeight = height / 2;
            final int halfWidth = width / 2;

            // Calculate the largest inSampleSize value that is a power of 2
              and
            // keeps both
            // height and width larger than the requested height and width.
            while ((halfHeight / inSampleSize) >= reqHeight
                    && (halfWidth / inSampleSize) >= reqWidth) {
                inSampleSize *= 2;
            }
        }

        Log.d(TAG, "sampleSize:" + inSampleSize);
        return inSampleSize;
    }
}
```

2. 内存缓存和磁盘缓存的实现

这里选择 LruCache 和 DiskLruCache 来分别完成内存缓存和磁盘缓存的工作。在 ImageLoader 初始化时，会创建 LruCache 和 DiskLruCache，如下所示。

```
private LruCache<String, Bitmap> mMemoryCache;
private DiskLruCache mDiskLruCache;

private ImageLoader(Context context) {
    mContext = context.getApplicationContext();
    int maxMemory = (int) (Runtime.getRuntime().maxMemory() / 1024);
    int cacheSize = maxMemory / 8;
    mMemoryCache = new LruCache<String, Bitmap>(cacheSize) {
        @Override
        protected int sizeOf(String key, Bitmap bitmap) {
            return bitmap.getRowBytes() * bitmap.getHeight() / 1024;
        }
    };
    File diskCacheDir = getDiskCacheDir(mContext, "bitmap");
    if (!diskCacheDir.exists()) {
        diskCacheDir.mkdirs();
    }
    if (getUsableSpace(diskCacheDir) > DISK_CACHE_SIZE) {
        try {
            mDiskLruCache = DiskLruCache.open(diskCacheDir, 1, 1,
                DISK_CACHE_SIZE);
            mIsDiskLruCacheCreated = true;
        } catch (IOException e) {
            e.printStackTrace();
        }
    }
}
```

在创建磁盘缓存时，这里做了一个判断，即有可能磁盘剩余空间小于磁盘缓存所需的大小，一般是指用户的手机空间已经不足了，因此没有办法创建磁盘缓存，这个时候磁盘缓存就会失效。在上面的代码实现中，ImageLoader 的内存缓存的容量为当前进程可用内存的 1/8，磁盘缓存的容量为 50MB。

内存缓存和磁盘缓存创建完毕后，还需要提供方法来完成缓存的添加和获取功能。首先看内存缓存，它的添加和读取过程比较简单，如下所示。

```
private void addBitmapToMemoryCache(String key, Bitmap bitmap) {
    if (getBitmapFromMemCache(key) == null) {
        mMemoryCache.put(key, bitmap);
    }
}

private Bitmap getBitmapFromMemCache(String key) {
    return mMemoryCache.get(key);
}
```

而磁盘缓存的添加和读取功能稍微复杂一些,具体内容已经在 12.2.2 节中进行了详细的介绍,这里再简单说明一下。磁盘缓存的添加需要通过 Editor 来完成,Editor 提供了 commit 和 abort 方法来提交和撤销对文件系统的写操作,具体实现请参看下面的 loadBitmapFromHttp 方法。磁盘缓存的读取需要通过 Snapshot 来完成,通过 Snapshot 可以得到磁盘缓存对象对应的 FileInputStream,但是 FileInputStream 无法便捷地进行压缩,所以通过 FileDescriptor 来加载压缩后的图片,最后将加载后的 Bitmap 添加到内存缓存中,具体实现请参看下面的 loadBitmapFromDiskCache 方法。

```
private Bitmap loadBitmapFromHttp(String url, int reqWidth, int reqHeight)
        throws IOException {
    if (Looper.myLooper() == Looper.getMainLooper()) {
        throw new RuntimeException("can not visit network from UI Thread.");
    }
    if (mDiskLruCache == null) {
        return null;
    }

    String key = hashKeyFormUrl(url);
    DiskLruCache.Editor editor = mDiskLruCache.edit(key);
    if (editor != null) {
        OutputStream outputStream = editor.newOutputStream(DISK_CACHE_
        INDEX);
        if (downloadUrlToStream(url, outputStream)) {
            editor.commit();
        } else {
            editor.abort();
        }
        mDiskLruCache.flush();
    }
```

```
        return loadBitmapFromDiskCache(url, reqWidth, reqHeight);
}

private Bitmap loadBitmapFromDiskCache(String url, int reqWidth,
        int reqHeight) throws IOException {
    if (Looper.myLooper() == Looper.getMainLooper()) {
        Log.w(TAG, "load bitmap from UI Thread, it's not recommended!");
    }
    if (mDiskLruCache == null) {
        return null;
    }

    Bitmap bitmap = null;
    String key = hashKeyFormUrl(url);
    DiskLruCache.Snapshot snapShot = mDiskLruCache.get(key);
    if (snapShot != null) {
        FileInputStream fileInputStream = (FileInputStream)snapShot.
        getInputStream(DISK_CACHE_INDEX);
        FileDescriptor fileDescriptor = fileInputStream.getFD();
        bitmap = mImageResizer.decodeSampledBitmapFromFileDescriptor
        (fileDescriptor,
                reqWidth, reqHeight);
        if (bitmap != null) {
            addBitmapToMemoryCache(key, bitmap);
        }
    }

    return bitmap;
}
```

3. 同步加载和异步加载接口的设计

首先看同步加载，同步加载接口需要外部在线程中调用，这是因为同步加载很可能比较耗时，它的实现如下所示。

```
/**
 * load bitmap from memory cache or disk cache or network.
 * @param uri http url
 * @param reqWidth the width ImageView desired
 * @param reqHeight the height ImageView desired
```

```
 * @return bitmap, maybe null.
 */
public Bitmap loadBitmap(String uri, int reqWidth, int reqHeight) {
    Bitmap bitmap = loadBitmapFromMemCache(uri);
    if (bitmap != null) {
        Log.d(TAG, "loadBitmapFromMemCache,url:" + uri);
        return bitmap;
    }

    try {
        bitmap = loadBitmapFromDiskCache(uri, reqWidth, reqHeight);
        if (bitmap != null) {
            Log.d(TAG, "loadBitmapFromDisk,url:" + uri);
            return bitmap;
        }
        bitmap = loadBitmapFromHttp(uri, reqWidth, reqHeight);
        Log.d(TAG, "loadBitmapFromHttp,url:" + uri);
    } catch (IOException e) {
        e.printStackTrace();
    }

    if (bitmap == null && !mIsDiskLruCacheCreated) {
        Log.w(TAG, "encounter error, DiskLruCache is not created.");
        bitmap = downloadBitmapFromUrl(uri);
    }

    return bitmap;
}
```

从 loadBitmap 的实现可以看出，其工作过程遵循如下几步：首先尝试从内存缓存中读取图片，接着尝试从磁盘缓存中读取图片，最后才从网络中拉取图片。另外，这个方法不能在主线程中调用，否则就抛出异常。这个执行环境的检查是在 loadBitmapFromHttp 中实现的，通过检查当前线程的 Looper 是否为主线程的 Looper 来判断当前线程是否是主线程，如果是主线程就直接抛出异常中止程序，如下所示。

```
if (Looper.myLooper() == Looper.getMainLooper()) {
    throw new RuntimeException("can not visit network from UI Thread.");
}
```

接着看异步加载接口的设计，如下所示。

```java
public void bindBitmap(final String uri, final ImageView imageView,
        final int reqWidth, final int reqHeight) {
    imageView.setTag(TAG_KEY_URI, uri);
    Bitmap bitmap = loadBitmapFromMemCache(uri);
    if (bitmap != null) {
        imageView.setImageBitmap(bitmap);
        return;
    }

    Runnable loadBitmapTask = new Runnable() {

        @Override
        public void run() {
            Bitmap bitmap = loadBitmap(uri, reqWidth, reqHeight);
            if (bitmap != null) {
                LoaderResult result = new LoaderResult(imageView, uri, bitmap);
                mMainHandler.obtainMessage(MESSAGE_POST_RESULT, result).
                sendToTarget();
            }
        }
    };
    THREAD_POOL_EXECUTOR.execute(loadBitmapTask);
}
```

从 bindBitmap 的实现来看，bindBitmap 方法会尝试从内存缓存中读取图片，如果读取成功就直接返回结果，否则会在线程池中去调用 loadBitmap 方法，当图片加载成功后再将图片、图片的地址以及需要绑定的 imageView 封装成一个 LoaderResult 对象，然后再通过 mMainHandler 向主线程发送一个消息，这样就可以在主线程中给 imageView 设置图片了，之所以通过 Handler 来中转是因为子线程无法访问 UI。

bindBitmap 中用到了线程池和 Handler，这里看一下它们的实现，首先看线程池 THREAD_POOL_EXECUTOR 的实现，如下所示。可以看出它的核心线程数为当前设备的 CPU 核心数加 1，最大容量为 CPU 核心数的 2 倍加 1，线程闲置超时时长为 10 秒，关于线程池的详细介绍可以参看第 11 章的有关内容。

```java
private static final int CORE_POOL_SIZE = CPU_COUNT + 1;
private static final int MAXIMUM_POOL_SIZE = CPU_COUNT * 2 + 1;
private static final long KEEP_ALIVE = 10L;
```

```java
private static final ThreadFactory sThreadFactory = new ThreadFactory() {
    private final AtomicInteger mCount = new AtomicInteger(1);

    public Thread newThread(Runnable r) {
        return new Thread(r, "ImageLoader#" + mCount.getAndIncrement());
    }
};

public static final Executor THREAD_POOL_EXECUTOR = new ThreadPoolExecutor(
        CORE_POOL_SIZE, MAXIMUM_POOL_SIZE,
        KEEP_ALIVE, TimeUnit.SECONDS,
        new LinkedBlockingQueue<Runnable>(), sThreadFactory);
```

之所以采用线程池是有原因的，首先肯定不能采用普通的线程去做这个事，线程池的好处在第 11 章已经做了详细的说明。如果直接采用普通的线程去加载图片，随着列表的滑动这可能会产生大量的线程，这样并不利于整体效率的提升。另外一点，这里也没有选择采用 AsyncTask，AsyncTask 封装了线程池和 Handler，按道理它应该适合 ImageLoader 的场景。从第 11 章中对 AsyncTask 的分析可以知道，AsyncTask 在 3.0 的低版本和高版本上具有不同的表现，在 3.0 以上的版本 AsyncTask 无法实现并发的效果，这显然是不能接受的，因为 ImageLoader 就是需要并发特性，虽然可以通过改造 AsyncTask 或者使用 AsyncTask 的 executeOnExecutor 方法的形式来执行异步任务，但是这终归是不太自然的实现方式。鉴于以上两点原因，这里选择线程池和 Handler 来提供 ImageLoader 的并发能力和访问 UI 的能力。

分析完线程池的选择，下面看一下 Handler 的实现，如下所示。ImageLoader 直接采用主线程的 Looper 来构造 Handler 对象，这就使得 ImageLoader 可以在非主线程中构造了。另外为了解决由于 View 复用所导致的列表错位这一问题，在给 ImageView 设置图片之前都会检查它的 url 有没有发生改变，如果发生改变就不再给它设置图片，这样就解决了列表错位的问题。

```java
private Handler mMainHandler = new Handler(Looper.getMainLooper()) {
    @Override
    public void handleMessage(Message msg) {
        LoaderResult result = (LoaderResult) msg.obj;
        ImageView imageView = result.imageView;
        String uri = (String) imageView.getTag(TAG_KEY_URI);
        if (uri.equals(result.uri)) {
```

```
            imageView.setImageBitmap(result.bitmap);
        } else {
            Log.w(TAG, "set image bitmap,but url has changed, ignored!");
        }
    };
};
```

到此为止，ImageLoader 的细节都已经做了全面的分析，下面是 ImageLoader 的完整的代码。

```
public class ImageLoader {

    private static final String TAG = "ImageLoader";

    public static final int MESSAGE_POST_RESULT = 1;

    private static final int CPU_COUNT = Runtime.getRuntime().available-
Processors();
    private static final int CORE_POOL_SIZE = CPU_COUNT + 1;
    private static final int MAXIMUM_POOL_SIZE = CPU_COUNT * 2 + 1;
    private static final long KEEP_ALIVE = 10L;

    private static final int TAG_KEY_URI = R.id.imageloader_uri;
    private static final long DISK_CACHE_SIZE = 1024 * 1024 * 50;
    private static final int IO_BUFFER_SIZE = 8 * 1024;
    private static final int DISK_CACHE_INDEX = 0;
    private boolean mIsDiskLruCacheCreated = false;

    private static final ThreadFactory sThreadFactory = new ThreadFactory() {
        private final AtomicInteger mCount = new AtomicInteger(1);

        public Thread newThread(Runnable r) {
            return new Thread(r, "ImageLoader#" + mCount.getAndIncrement());
        }
    };

    public static final Executor THREAD_POOL_EXECUTOR = new ThreadPool-
Executor(
            CORE_POOL_SIZE, MAXIMUM_POOL_SIZE,
            KEEP_ALIVE, TimeUnit.SECONDS,
```

```java
            new LinkedBlockingQueue<Runnable>(), sThreadFactory);

private Handler mMainHandler = new Handler(Looper.getMainLooper()) {
    @Override
    public void handleMessage(Message msg) {
        LoaderResult result = (LoaderResult) msg.obj;
        ImageView imageView = result.imageView;
        String uri = (String) imageView.getTag(TAG_KEY_URI);
        if (uri.equals(result.uri)) {
            imageView.setImageBitmap(result.bitmap);
        } else {
            Log.w(TAG, "set image bitmap,but url has changed, ignored!");
        }
    };
};

private Context mContext;
private ImageResizer mImageResizer = new ImageResizer();
private LruCache<String, Bitmap> mMemoryCache;
private DiskLruCache mDiskLruCache;

private ImageLoader(Context context) {
    mContext = context.getApplicationContext();
    int maxMemory = (int) (Runtime.getRuntime().maxMemory() / 1024);
    int cacheSize = maxMemory / 8;
    mMemoryCache = new LruCache<String, Bitmap>(cacheSize) {
        @Override
        protected int sizeOf(String key, Bitmap bitmap) {
            return bitmap.getRowBytes() * bitmap.getHeight() / 1024;
        }
    };
    File diskCacheDir = getDiskCacheDir(mContext, "bitmap");
    if (!diskCacheDir.exists()) {
        diskCacheDir.mkdirs();
    }
    if (getUsableSpace(diskCacheDir) > DISK_CACHE_SIZE) {
        try {
            mDiskLruCache = DiskLruCache.open(diskCacheDir, 1, 1,
                DISK_CACHE_SIZE);
```

```java
            mIsDiskLruCacheCreated = true;
        } catch (IOException e) {
            e.printStackTrace();
        }
    }
}

/**
 * build a new instance of ImageLoader
 * @param context
 * @return a new instance of ImageLoader
 */
public static ImageLoader build(Context context) {
    return new ImageLoader(context);
}

private void addBitmapToMemoryCache(String key, Bitmap bitmap) {
    if (getBitmapFromMemCache(key) == null) {
        mMemoryCache.put(key, bitmap);
    }
}

private Bitmap getBitmapFromMemCache(String key) {
    return mMemoryCache.get(key);
}

/**
 * load bitmap from memory cache or disk cache or network async, then bind
   imageView and bitmap.
 * NOTE THAT: should run in UI Thread
 * @param uri http url
 * @param imageView bitmap's bind object
 */
public void bindBitmap(final String uri, final ImageView imageView) {
    bindBitmap(uri, imageView, 0, 0);
}

public void bindBitmap(final String uri, final ImageView imageView,
        final int reqWidth, final int reqHeight) {
    imageView.setTag(TAG_KEY_URI, uri);
```

```java
        Bitmap bitmap = loadBitmapFromMemCache(uri);
        if (bitmap != null) {
            imageView.setImageBitmap(bitmap);
            return;
        }

        Runnable loadBitmapTask = new Runnable() {

            @Override
            public void run() {
                Bitmap bitmap = loadBitmap(uri, reqWidth, reqHeight);
                if (bitmap != null) {
                    LoaderResult result = new LoaderResult(imageView, uri,
                        bitmap);
                    mMainHandler.obtainMessage(MESSAGE_POST_RESULT, result).
                    sendToTarget();
                }
            }
        };
        THREAD_POOL_EXECUTOR.execute(loadBitmapTask);
    }

    /**
     * load bitmap from memory cache or disk cache or network.
     * @param uri http url
     * @param reqWidth the width ImageView desired
     * @param reqHeight the height ImageView desired
     * @return bitmap, maybe null.
     */
    public Bitmap loadBitmap(String uri, int reqWidth, int reqHeight) {
        Bitmap bitmap = loadBitmapFromMemCache(uri);
        if (bitmap != null) {
            Log.d(TAG, "loadBitmapFromMemCache,url:" + uri);
            return bitmap;
        }

        try {
            bitmap = loadBitmapFromDiskCache(uri, reqWidth, reqHeight);
            if (bitmap != null) {
                Log.d(TAG, "loadBitmapFromDisk,url:" + uri);
```

```
            return bitmap;
        }
        bitmap = loadBitmapFromHttp(uri, reqWidth, reqHeight);
        Log.d(TAG, "loadBitmapFromHttp,url:" + uri);
    } catch (IOException e) {
        e.printStackTrace();
    }

    if (bitmap == null && !mIsDiskLruCacheCreated) {
        Log.w(TAG, "encounter error, DiskLruCache is not created.");
        bitmap = downloadBitmapFromUrl(uri);
    }

    return bitmap;
}

private Bitmap loadBitmapFromMemCache(String url) {
    final String key = hashKeyFormUrl(url);
    Bitmap bitmap = getBitmapFromMemCache(key);
    return bitmap;
}

private Bitmap loadBitmapFromHttp(String url, int reqWidth, int reqHeight)
        throws IOException {
    if (Looper.myLooper() == Looper.getMainLooper()) {
        throw new RuntimeException("can not visit network from UI
        Thread.");
    }
    if (mDiskLruCache == null) {
        return null;
    }

    String key = hashKeyFormUrl(url);
    DiskLruCache.Editor editor = mDiskLruCache.edit(key);
    if (editor != null) {
        OutputStream outputStream = editor.newOutputStream(DISK_CACHE_
        INDEX);
        if (downloadUrlToStream(url, outputStream)) {
            editor.commit();
```

```java
        } else {
            editor.abort();
        }
        mDiskLruCache.flush();
    }
    return loadBitmapFromDiskCache(url, reqWidth, reqHeight);
}

private Bitmap loadBitmapFromDiskCache(String url, int reqWidth,
        int reqHeight) throws IOException {
    if (Looper.myLooper() == Looper.getMainLooper()) {
        Log.w(TAG, "load bitmap from UI Thread, it's not recommended!");
    }
    if (mDiskLruCache == null) {
        return null;
    }

    Bitmap bitmap = null;
    String key = hashKeyFormUrl(url);
    DiskLruCache.Snapshot snapShot = mDiskLruCache.get(key);
    if (snapShot != null) {
        FileInputStream fileInputStream = (FileInputStream)snapShot.
        getInputStream(DISK_CACHE_INDEX);
        FileDescriptor fileDescriptor = fileInputStream.getFD();
        bitmap = mImageResizer.decodeSampledBitmapFromFileDescriptor
        (fileDescriptor, reqWidth, reqHeight);
        if (bitmap != null) {
            addBitmapToMemoryCache(key, bitmap);
        }
    }

    return bitmap;
}

public boolean downloadUrlToStream(String urlString,
        OutputStream outputStream) {
    HttpURLConnection urlConnection = null;
    BufferedOutputStream out = null;
    BufferedInputStream in = null;
```

```java
        try {
            final URL url = new URL(urlString);
            urlConnection = (HttpURLConnection) url.openConnection();
            in = new BufferedInputStream(urlConnection.getInputStream(),
                    IO_BUFFER_SIZE);
            out = new BufferedOutputStream(outputStream, IO_BUFFER_SIZE);

            int b;
            while ((b = in.read()) != -1) {
                out.write(b);
            }
            return true;
        } catch (IOException e) {
            Log.e(TAG, "downloadBitmap failed." + e);
        } finally {
            if (urlConnection != null) {
                urlConnection.disconnect();
            }
            MyUtils.close(out);
            MyUtils.close(in);
        }
        return false;
    }

    private Bitmap downloadBitmapFromUrl(String urlString) {
        Bitmap bitmap = null;
        HttpURLConnection urlConnection = null;
        BufferedInputStream in = null;

        try {
            final URL url = new URL(urlString);
            urlConnection = (HttpURLConnection) url.openConnection();
            in = new BufferedInputStream(urlConnection.getInputStream(),
                    IO_BUFFER_SIZE);
            bitmap = BitmapFactory.decodeStream(in);
        } catch (final IOException e) {
            Log.e(TAG, "Error in downloadBitmap: " + e);
        } finally {
            if (urlConnection != null) {
                urlConnection.disconnect();
```

```java
            }
            MyUtils.close(in);
        }
        return bitmap;
    }

    private String hashKeyFormUrl(String url) {
        String cacheKey;
        try {
            final MessageDigest mDigest = MessageDigest.getInstance("MD5");
            mDigest.update(url.getBytes());
            cacheKey = bytesToHexString(mDigest.digest());
        } catch (NoSuchAlgorithmException e) {
            cacheKey = String.valueOf(url.hashCode());
        }
        return cacheKey;
    }

    private String bytesToHexString(byte[] bytes) {
        StringBuilder sb = new StringBuilder();
        for (int i = 0; i < bytes.length; i++) {
            String hex = Integer.toHexString(0xFF & bytes[i]);
            if (hex.length() == 1) {
                sb.append('0');
            }
            sb.append(hex);
        }
        return sb.toString();
    }

    public File getDiskCacheDir(Context context, String uniqueName) {
        boolean externalStorageAvailable = Environment
                .getExternalStorageState().equals(Environment.MEDIA_MOUNTED);
        final String cachePath;
        if (externalStorageAvailable) {
            cachePath = context.getExternalCacheDir().getPath();
        } else {
            cachePath = context.getCacheDir().getPath();
        }
```

```java
            return new File(cachePath + File.separator + uniqueName);
        }

        @TargetApi(VERSION_CODES.GINGERBREAD)
        private long getUsableSpace(File path) {
            if (Build.VERSION.SDK_INT >= VERSION_CODES.GINGERBREAD) {
                return path.getUsableSpace();
            }
            final StatFs stats = new StatFs(path.getPath());
            return (long) stats.getBlockSize() * (long) stats.getAvailableBlocks();
        }

        private static class LoaderResult {
            public ImageView imageView;
            public String uri;
            public Bitmap bitmap;

            public LoaderResult(ImageView imageView, String uri, Bitmap bitmap) {
                this.imageView = imageView;
                this.uri = uri;
                this.bitmap = bitmap;
            }
        }
    }
```

12.3 ImageLoader 的使用

在 12.2.3 节中我们实现了一个功能完整的 ImageLoader，本节将演示如何通过 Image-Loader 来实现一个照片墙的效果，实际上我们会发现，通过 ImageLoader 打造一个照片墙是轻而易举的事情。最后针对如何提高列表的滑动流畅度这个问题，本节会给出一些针对性的建议供读者参考。

12.3.1 照片墙效果

实现照片墙效果需要用到 GridView，下面先准备好 GridView 所需的布局文件以及 item 的布局文件，如下所示。

```xml
// GridView的布局文件
<LinearLayout xmlns:android="http://schemas.android.com/apk/res/android"
    xmlns:tools="http://schemas.android.com/tools"
    android:layout_width="match_parent"
    android:layout_height="match_parent"
    android:orientation="vertical"
    android:padding="5dp" >

    <GridView
        android:id="@+id/gridView1"
        android:layout_width="match_parent"
        android:layout_height="match_parent"
        android:gravity="center"
        android:horizontalSpacing="5dp"
        android:verticalSpacing="5dp"
        android:listSelector="@android:color/transparent"
        android:numColumns="3"
        android:stretchMode="columnWidth" >
    </GridView>

</LinearLayout>

// GridView的item的布局文件
<?xml version="1.0" encoding="utf-8"?>
<LinearLayout xmlns:android="http://schemas.android.com/apk/res/android"
    android:layout_width="match_parent"
    android:layout_height="wrap_content"
    android:gravity="center"
    android:orientation="vertical" >

    <com.ryg.chapter_12.ui.SquareImageView
        android:id="@+id/image"
        android:layout_width="match_parent"
        android:layout_height="wrap_content"
        android:scaleType="centerCrop"
        android:src="@drawable/image_default" />

</LinearLayout>
```

也许读者已经注意到，GridView的item的布局文件中并没有采用ImageView，而是采用了一个叫 SquareImageView 的自定义控件。顾名思义，它的作用就是打造一个正方形的 ImageView，这样整个照片墙看起来会比较整齐美观。要实现一个宽、高相等的 ImageView 是非常简单的一件事，只需要在它的 onMeasure 方法中稍微做一下处理，如下所示。

```java
public class SquareImageView extends ImageView {

    public SquareImageView(Context context) {
        super(context);
    }

    public SquareImageView(Context context, AttributeSet attrs) {
        super(context, attrs);
    }

    public SquareImageView(Context context, AttributeSet attrs,int defStyle) {
        super(context, attrs, defStyle);
    }

    @Override
    protected void onMeasure(int widthMeasureSpec,int heightMeasureSpec) {
        super.onMeasure(widthMeasureSpec, widthMeasureSpec);
    }
}
```

可以看出，我们在 SquareImageView 的 onMeasure 方法中很巧妙地将 heightMeasureSpec 替换为 widthMeasureSpec，这样什么都不用做就可以一个宽、高相等的 ImageView 了。关于 View 的测量等过程的介绍，请读者参看第 4 章的有关内容，这里不再赘述了。

接着需要实现一个 BaseAdapter 给 GridView 使用，下面的代码展示了 ImageAdapter 的实现细节，其中 mUrList 中存储的是图片的 url：

```java
private class ImageAdapter extends BaseAdapter {
    ...
    @Override
    public int getCount() {
        return mUrList.size();
    }

    @Override
```

```java
    public String getItem(int position) {
        return mUrList.get(position);
    }

    @Override
    public long getItemId(int position) {
        return position;
    }

    @Override
    public View getView(int position, View convertView, ViewGroup parent){
        ViewHolder holder = null;
        if (convertView == null) {
            convertView = mInflater.inflate(R.layout.image_list_item,
            parent, false);
            holder = new ViewHolder();
            holder.imageView = (ImageView) convertView.findViewById(R.
            id.image);
            convertView.setTag(holder);
        } else {
            holder = (ViewHolder) convertView.getTag();
        }
        ImageView imageView = holder.imageView;
        final String tag = (String)imageView.getTag();
        final String uri = getItem(position);
        if (!uri.equals(tag)) {
            imageView.setImageDrawable(mDefaultBitmapDrawable);
        }
        if (mIsGridViewIdle && mCanGetBitmapFromNetWork) {
            imageView.setTag(uri);
            mImageLoader.bindBitmap(uri, imageView, mImageWidth, mImage-
            Width);
        }
        return convertView;
    }
}
```

从上述代码来看，ImageAdapter 的实现过程非常简捷，这几乎是最简洁的 BaseAdapter 的实现了。但是简洁并不等于简单，getView 方法中核心代码只有一句话，那就是：mImageLoader.bindBitmap(uri, imageView, mImageWidth, mImageWidth)。通过 bindBitmap 方

法很轻松地将复杂的图片加载过程交给了 ImageLoader，ImageLoader 加载图片以后会把图片自动设置给 imageView，而整个过程，包括内存缓存、磁盘缓存以及图片压缩等工作过程对 ImageAdapter 来说都是透明的。在这种设计思想下，ImageAdapter 什么也不需要知道，因此这是一个极其轻量级的 ImageAdapter。

接着将 ImageAdapter 设置给 GridView，如下所示。到此为止一个绚丽的图片墙就大功告成了，是不是惊叹于如此简捷而又优美的实现过程呢？

```
mImageGridView = (GridView) findViewById(R.id.gridView1);
mImageAdapter = new ImageAdapter(this);
mImageGridView.setAdapter(mImageAdapter);
```

最后，看一下我们亲手打造的图片墙的效果图，如图 12-1 所示。是不是看起来很优美呢？

图 12-1　采用 ImageLoader 实现的照片墙

另外，本节中的照片墙应用首次运行时会从网络中加载大量图片，这会消耗若干 MB 的流量，因此建议首次运行时选择 WiFi 环境，同时程序启动时也会有相应的提示，在非 WiFi 环境下，打开应用时会弹出如下提示，请读者运行时注意一下，避免消耗过多的流量。

```
if (!mIsWifi) {
    AlertDialog.Builder builder = new AlertDialog.Builder(this);
    builder.setMessage("初次使用会从网络中下载大概 5MB 的图片，确认要下载吗？");
    builder.setTitle("注意");
    builder.setPositiveButton("是", new OnClickListener() {
        @Override
        public void onClick(DialogInterface dialog, int which) {
            mCanGetBitmapFromNetWork = true;
            mImageAdapter.notifyDataSetChanged();
        }
    });
    builder.setNegativeButton("否", null);
    builder.show();
}
```

12.3.2　优化列表的卡顿现象

这个问题困扰了很多开发者，其实答案很简单，不要在主线程中做太耗时的操作即可提高滑动的流畅度，可以从三个方面来说明这个问题。

首先，不要在 getView 中执行耗时操作。对于上面的例子来说，如果直接在 getView 方法中加载图片，肯定会导致卡顿，因为加载图片是一个耗时的操作，这种操作必须通过异步的方式来处理，就像 ImageLoader 实现的那样。

其次，控制异步任务的执行频率。这一点也很重要，对于列表来说，仅仅在 getView 中采用异步操作是不够的。考虑一种情况，以照片墙来说，在 getView 方法中会通过 ImageLoader 的 bindBitmap 方法来异步加载图片，但是如果用户刻意地频繁上下滑动，这就会在一瞬间产生上百个异步任务，这些异步任务会造成线程池的拥堵并随即带来大量的 UI 更新操作，这是没有意义的。由于一瞬间存在大量的 UI 更新操作，这些 UI 操作是运行在主线程的，这就会造成一定程度的卡顿。如何解决这个问题呢？可以考虑在列表滑动的时候停止加载图片，尽管这个过程是异步的，等列表停下来以后再加载图片仍然可以获得良好的用户体验。具体实现时，可以给 ListView 或者 GridView 设置 setOnScrollListener，并在 OnScrollListener 的 onScrollStateChanged 方法中判断列表是否处于滑动状态，如果是的话就停止加载图片，如下所示。

```
public void onScrollStateChanged(AbsListView view, int scrollState) {
    if (scrollState == OnScrollListener.SCROLL_STATE_IDLE) {
        mIsGridViewIdle = true;
        mImageAdapter.notifyDataSetChanged();
    } else {
        mIsGridViewIdle = false;
    }
}
```

然后在 getView 方法中，仅当列表静止时才能加载图片，如下所示。

```
if (mIsGridViewIdle && mCanGetBitmapFromNetWork) {
    imageView.setTag(uri);
    mImageLoader.bindBitmap(uri, imageView, mImageWidth, mImageWidth);
}
```

一般来说，经过上面两个步骤，列表都不会有卡顿现象，但是在某些特殊情况下，列表还是会有偶尔的卡顿现象，这个时候还可以开启硬件加速。绝大多数情况下，硬件加速都可以解决莫名的卡顿问题，通过设置 android:hardwareAccelerated="true" 即可为 Activity 开启硬件加速。

第 13 章 综 合 技 术

本章介绍的主题在日常开发中使用频率略低，但是对它们有一定的了解仍然是很有必要的，下面分别介绍它们的使用场景。

我们知道，不管程序怎么写都很难避免不 crash，当程序 crash 后虽然无法让其再继续运行，但是如果能够知道程序 crash 的原因，那么就可以修复错误。但是很多时候产品发布后，如果用户在使用时发生了 crash，这个 crash 信息是很难获取到的，这非常不利于一个产品的持续发展。其实可以通过 CrashHandler 来监视应用的 crash 信息，给程序设置一个 CrashHandler，这样当程序 crash 时就会调用 CrashHandler 的 uncaughtException 方法。在这个方法中我们可以获取 crash 信息并上传到服务器，通过这种方式服务端就能监控程序的运行状况了，在后续的版本开发中，开发人员就可以对一些错误进行修复了。

在 Android 中，有一个限制，那就是整个应用的方法数不能超过 65536，否则就会出现编译错误，并且程序也无法成功地安装到手机上。当项目日益庞大后这个问题就比较容易遇到，Google 提供了 multidex 方案专门用于解决这个问题，通过将一个 dex 文件拆分为多个 dex 文件来避免单个 dex 文件方法数越界的问题。

方法数越界的另一种解决方案是动态加载。动态加载可以直接加载一个 dex 形式的文件，将部分代码打包到一个单独的 dex 文件中（也可以是 dex 格式的 jar 或者 apk），并在程序运行时根据需要去动态加载 dex 中的类，这种方式既可以解决缓解方法数越界的问题，也可以为程序提供按需加载的特性，同时这还为应用按模块更新提供了可能性。

反编译在应用开发中用得不是很多，但是很多时候我们需要研究其他产品的实现思路，这个时候就需要反编译了。在 Android 中反编译主要通过 dex2jar 以及 apktool 来完成。dex2jar

可以将一个 apk 转成一个 jar 包,这个 jar 包再通过反编译工具 jd-gui 来打开就可以查看到反编译后的 Java 代码了。Apktool 主要用于应用的解包和二次打包,实际上通过 Apktool 的二次打包可以做很多事情,甚至是一些违法的事情。目前不少公司都有专门的反编译团队,也称逆向团队,他们做的事情会更加深入,但是对于应用开发者来说并不需要了解那么多深入的逆向知识,因此本章仅仅介绍一些简单常用的反编译方法。

13.1 使用 CrashHandler 来获取应用的 crash 信息

Android 应用不可避免地会发生 crash,也称之为崩溃,无论你的程序写得多么完美,总是无法完全避免 crash 的发生,可能是由于 Android 系统底层的 bug,也可能是由于不充分的机型适配或者是糟糕的网络状况。当 crash 发生时,系统会 kill 掉正在执行的程序,现象就是闪退或者提示用户程序已停止运行,这对用户来说是很不友好的,也是开发者所不愿意看到的。更糟糕的是,当用户发生了 crash,开发者却无法得知程序为何 crash,即便开发人员想去解决这个 crash,但是由于无法知道用户当时的 crash 信息,所以往往也无能为力。幸运的是,Android 提供了处理这类问题的方法,请看下面 Thread 类中的一个方法 setDefaultUncaughtExceptionHandler:

```
/**
 * Sets the default uncaught exception handler. This handler is invoked in
 * case any Thread dies due to an unhandled exception.
 *
 * @param handler
 *          The handler to set or null.
 */
public static void setDefaultUncaughtExceptionHandler(UncaughtException-
Handler handler) {
    Thread.defaultUncaughtHandler = handler;
}
```

从方法的字面意义来看,这个方法好像可以设置系统的默认异常处理器,其实这个方法就可以解决上面所提到的 crash 问题。当 crash 发生的时候,系统就会回调 UncaughtExceptionHandler 的 uncaughtException 方法,在 uncaughtException 方法中就可以获取到异常信息,可以选择把异常信息存储到 SD 卡中,然后在合适的时机通过网络将 crash 信息上传到服务器上,这样开发人员就可以分析用户 crash 的场景从而在后面的版本中修复

此类 crash。我们还可以在 crash 发生时，弹出一个对话框告诉用户程序 crash 了，然后再退出，这样做比闪退要温和一点。

有了上面的分析，现在读者肯定知道获取应用 crash 信息的方式了。首先需要实现一个 UncaughtExceptionHandler 对象，在它的 uncaughtException 方法中获取异常信息并将其存储在 SD 卡中或者上传到服务器供开发人员分析，然后调用 Thread 的 setDefaultUncaughtExceptionHandler 方法将它设置为线程默认的异常处理器，由于默认异常处理器是 Thread 类的静态成员，因此它的作用对象是当前进程的所有线程。这么来看监听应用的 crash 信息实际上是很简单的一件事，下面是一个典型的异常处理器的实现：

```java
public class CrashHandler implements UncaughtExceptionHandler {
    private static final String TAG = "CrashHandler";
    private static final boolean DEBUG = true;

    private static final String PATH = Environment.getExternal-
StorageDirectory().getPath() + "/CrashTest/log/";
    private static final String FILE_NAME = "crash";
    private static final String FILE_NAME_SUFFIX = ".trace";

    private static CrashHandler sInstance = new CrashHandler();
    private UncaughtExceptionHandler mDefaultCrashHandler;
    private Context mContext;

    private CrashHandler() {
    }

    public static CrashHandler getInstance() {
        return sInstance;
    }

    public void init(Context context) {
        mDefaultCrashHandler = Thread.getDefaultUncaughtExceptionHandler();
        Thread.setDefaultUncaughtExceptionHandler(this);
        mContext = context.getApplicationContext();
    }

    /**
     * 这个是最关键的函数，当程序中有未被捕获的异常，系统将会自动调用#uncaught-
     Exception 方法
```

```java
 * thread 为出现未捕获异常的线程，ex 为未捕获的异常，有了这个 ex，我们就可以得到异
 * 常信息
 */
@Override
public void uncaughtException(Thread thread, Throwable ex) {
    try {
        //导出异常信息到 SD 卡中
        dumpExceptionToSDCard(ex);
        //这里可以上传异常信息到服务器，便于开发人员分析日志从而解决 bug
        uploadExceptionToServer();
    } catch (IOException e) {
        e.printStackTrace();
    }

    ex.printStackTrace();
    //如果系统提供了默认的异常处理器，则交给系统去结束程序，否则就由自己结束自己
    if (mDefaultCrashHandler != null) {
        mDefaultCrashHandler.uncaughtException(thread, ex);
    } else {
        Process.killProcess(Process.myPid());
    }
}

private void dumpExceptionToSDCard(Throwable ex) throws IOException {
    //如果 SD 卡不存在或无法使用，则无法把异常信息写入 SD 卡
    if (!Environment.getExternalStorageState().equals(Environment.
    MEDIA_MOUNTED))     {
        if (DEBUG) {
            Log.w(TAG, "sdcard unmounted,skip dump exception");
            return;
        }
    }

    File dir = new File(PATH);
    if (!dir.exists()) {
        dir.mkdirs();
    }
    long current = System.currentTimeMillis();
    String time = new SimpleDateFormat("yyyy-MM-dd HH:mm:ss").format(new
    Date(current));
```

```java
            File file = new File(PATH + FILE_NAME + time + FILE_NAME_SUFFIX);

            try {
                PrintWriter pw = new PrintWriter(new BufferedWriter(new
                FileWriter(file)));
                pw.println(time);
                dumpPhoneInfo(pw);
                pw.println();
                ex.printStackTrace(pw);
                pw.close();
            } catch (Exception e) {
                Log.e(TAG, "dump crash info failed");
            }
        }

        private void dumpPhoneInfo(PrintWriter pw) throws NameNotFoundException{
            PackageManager pm = mContext.getPackageManager();
            PackageInfo pi = pm.getPackageInfo(mContext.getPackageName(),
            PackageManager.GET_ACTIVITIES);
            pw.print("App Version: ");
            pw.print(pi.versionName);
            pw.print('_');
            pw.println(pi.versionCode);

            //Android 版本号
            pw.print("OS Version: ");
            pw.print(Build.VERSION.RELEASE);
            pw.print("_");
            pw.println(Build.VERSION.SDK_INT);

            //手机制造商
            pw.print("Vendor: ");
            pw.println(Build.MANUFACTURER);

            //手机型号
            pw.print("Model: ");
            pw.println(Build.MODEL);

            //CPU 架构
```

```
        pw.print("CPU ABI: ");
        pw.println(Build.CPU_ABI);
    }

    private void uploadExceptionToServer() {
        //TODO Upload Exception Message To Your Web Server
    }

}
```

从上面的代码可以看出，当应用崩溃时，CrashHandler 会将异常信息以及设备信息写入 SD 卡，接着将异常交给系统处理，系统会帮我们中止程序，如果系统没有默认的异常处理机制，那么就自行中止。当然也可以选择将异常信息上传到服务器，本节中的 CrashHandler 并没有实现这个逻辑，但是在实际开发中一般都需要将异常信息上传到服务器。

如何使用上面的 CrashHandler 呢？也很简单，可以选择在 Application 初始化的时候为线程设置 CrashHandler，如下所示。

```
public class TestApp extends Application {

    private static TestApp sInstance;

    @Override
    public void onCreate() {
        super.onCreate();
        sInstance = this;
        //在这里为应用设置异常处理，然后程序才能获取未处理的异常
        CrashHandler crashHandler = CrashHandler.getInstance();
        crashHandler.init(this);
    }

    public static TestApp getInstance() {
        return sInstance;
    }

}
```

经过上面两个步骤，程序就可以处理未处理的异常了，就再也不怕程序 crash 了，同时

还可以很方便地从服务器上查看用户的 crash 信息。需要注意的是，代码中被 catch 的异常不会交给 CrashHandler 处理，CrashHandler 只能收到那些未被捕获的异常。下面我们就模拟一下发生 crash 的情形，看程序是如何处理的，如下所示。

```java
public class CrashActivity extends Activity implements OnClickListener {

    private Button mButton;

    @Override
    protected void onCreate(Bundle savedInstanceState) {
        super.onCreate(savedInstanceState);
        setContentView(R.layout.activity_crash);
        initView();
    }

    private void initView() {
        mButton = (Button) findViewById(R.id.button1);
        mButton.setOnClickListener(this);
    }

    @Override
    public void onClick(View v) {
        if (v == mButton) {
            // 在这里模拟异常抛出情况，人为地抛出一个运行时异常
            throw new RuntimeException("自定义异常：这是自己抛出的异常");
        }
    }

}
```

在上面的测试代码中，给按钮加一个单击事件，在 onClick 中人为抛出一个运行时异常，这个时候程序就 crash 了，看看异常处理器为我们做了什么。从图 13-1 中可以看出，异常处理器为我们创建了一个日志文件，打开日志文件，可以看到手机的信息以及异常发生时的调用栈，有了这些内容，开发人员就很容易定位问题了。从图 13-1 中的函数调用栈可以看出，CrashActivity 的 28 行发生了 RuntimeException，再看一下 CrashActivity 的代码，发现 28 行的确抛出了一个 RuntimeException，这说明 CrashHandler 已经成功地获取了未被捕获的异常信息，从现在开始，为应用加上默认异常事件处理器吧。

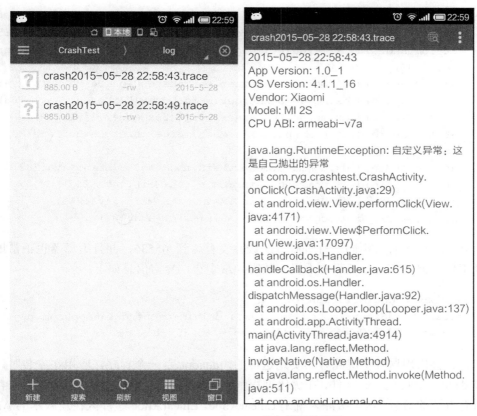

图 13-1　CrashHandler 获取的异常信息

13.2　使用 multidex 来解决方法数越界

在 Android 中单个 dex 文件所能够包含的最大方法数为 65536，这包含 Android FrameWork、依赖的 jar 包以及应用本身的代码中的所有方法。65536 是一个很大的数，一般来说一个简单应用的方法数的确很难达到 65536，但是对于一些比较大型的应用来说，65536 就很容易达到了。当应用的方法数达到 65536 后，编译器就无法完成编译工作并抛出类似下面的异常：

```
UNEXPECTED TOP-LEVEL EXCEPTION:
com.android.dex.DexIndexOverflowException: method ID not in [0, 0xffff]:
```

```
65536
    at com.android.dx.merge.DexMerger$6.updateIndex(DexMerger.java:502)
    at com.android.dx.merge.DexMerger$IdMerger.mergeSorted(DexMerger.
java:283)
    at com.android.dx.merge.DexMerger.mergeMethodIds(DexMerger.java:491)
    at com.android.dx.merge.DexMerger.mergeDexes(DexMerger.java:168)
    at com.android.dx.merge.DexMerger.merge(DexMerger.java:189)
    at com.android.dx.command.dexer.Main.mergeLibraryDexBuffers(Main.
java:454)
    at com.android.dx.command.dexer.Main.runMonoDex(Main.java:303)
    at com.android.dx.command.dexer.Main.run(Main.java:246)
    at com.android.dx.command.dexer.Main.main(Main.java:215)
    at com.android.dx.command.Main.main(Main.java:106)
```

另外一种情况有所不同，有时候方法数并没有达到 65536，并且编译器也正常地完成了编译工作，但是应用在低版本手机安装时异常中止，异常信息如下：

```
E/dalvikvm: Optimization failed
E/installd: dexopt failed on '/data/dalvik-cache/data@app@com.ryg.
multidextest-2.apk@classes.dex' res = 65280
```

为什么会出现这种情况呢？其实是这样的，dexopt 是一个程序，应用在安装时，系统会通过 dexopt 来优化 dex 文件，在优化过程中 dexopt 采用一个固定大小的缓冲区来存储应用中所有方法的信息，这个缓冲区就是 LinearAlloc。LinearAlloc 缓冲区在新版本的 Android 系统中其大小是 8MB 或者 16MB，但是在 Android 2.2 和 2.3 中却只有 5MB，当待安装的 apk 中的方法数比较多时，尽管它还没有达到 65536 这个上限，但是它的存储空间仍然有可能超出 5MB，这种情况下 dexopt 程序就会报错，从而导致安装失败，这种情况主要在 2.x 系列的手机上出现。

可以看到，不管是编译时方法数越界还是安装时的 dexopt 错误，它们都给开发过程带来了很大的困扰。从目前的 Android 版本的市场占有率来说，Android 3.0 以下的手机仍然占据着不到 10%的比率，目前主流的应用都不可能放弃 Android 3.0 以下的用户，对于这些应用来说，方法数越界就是一个必须要解决的问题了。

如何解决方法数越界的问题呢？我们首先想到的肯定是删除无用的代码和第三方库。没错，这的确是必须要做的工作，但是很多情况下即使删除了无用的代码，方法数仍然越界，这个时候该怎么办呢？针对这个问题，之前很多应用都会考虑采用插件化的机制来动态加载部分 dex，通过将一个 dex 拆分成两个或多个 dex，这就在一定程度上解决了方法数越界的问题。但是插件化是一套重量级的技术方案，并且其兼容性问题往往较多，从单纯解决方法数越界的角度来说，插件化并不是一个非常适合的方案，关于插件化的意义将在第 13.3 节中进行介绍。为了解决这个问题，Google 在 2014 年提出了 multidex 的解决方案，通过 multidex 可以很好地解决方法数越界的问题，并且使用起来非常简单。

在 Android 5.0 以前使用 multidex 需要引入 Google 提供的 android-support-multidex.jar 这个 jar 包，这个 jar 包可以在 Android SDK 目录下的 extras/android/support/multidex/library/libs 下面找到。从 Android 5.0 开始，Android 默认支持了 multidex，它可以从 apk 中加载多个 dex 文件。Multidex 方案主要是针对 AndroidStudio 和 Gradle 编译环境的，如果是 Eclipse 和 ant 那就复杂一些，而且由于 AndroidStudio 作为官方 IDE 其最终会完全替代 Eclipse ADT，因此本节中也不再介绍 Eclipse 中配置 multidex 的细节了。

在 AndroidStudio 和 Gradle 编译环境中，如果要使用 multidex，首先要使用 Android SDK Build Tools 21.1 及以上版本，接着修改工程中 app 目录下的 build.gradle 文件，在 defaultConfig 中添加 multiDexEnabled true 这个配置项，如下所示。关于如何使用 AndroidStudio 和 Gradle 请读者自行查看相关资料，这里不再介绍。

```
android {
    compileSdkVersion 22
    buildToolsVersion "22.0.1"

    defaultConfig {
        applicationId "com.ryg.multidextest"
        minSdkVersion 8
        targetSdkVersion 22
        versionCode 1
        versionName "1.0"

        // enable multidex support
        multiDexEnabled true
    }
    ...
}
```

接着还需要在 dependencies 中添加 multidex 的依赖：compile 'com.android.support:multidex:1.0.0'，如下所示。

```
dependencies {
```

```
    compile fileTree(dir: 'libs', include: ['*.jar'])
    compile 'com.android.support:appcompat-v7:22.1.1'
    compile 'com.android.support:multidex:1.0.0'
}
```

最终配置完成的 build.gradle 文件如下所示，其中加粗的部分是专门为 multidex 所添加的配置项：

```
apply plugin: 'com.android.application'

android {
    compileSdkVersion 22
    buildToolsVersion "22.0.1"

    defaultConfig {
        applicationId "com.ryg.multidextest"
        minSdkVersion 8
        targetSdkVersion 22
        versionCode 1
        versionName "1.0"

        // enable multidex support
        multiDexEnabled true
    }
    buildTypes {
        release {
            minifyEnabled false
            proguardFiles getDefaultProguardFile('proguard-android.txt'),
            'proguard-rules.pro'
        }
    }
}

dependencies {
    compile fileTree(dir: 'libs', include: ['*.jar'])
    compile 'com.android.support:appcompat-v7:22.1.1'
    compile 'com.android.support:multidex:1.0.0'
}
```

经过了上面的过程，还需要做另一项工作，那就是在代码中加入支持 multidex 的功能，

这个过程是比较简单的，有三种方案可以选。

第一种方案，在 manifest 文件中指定 Application 为 MultiDexApplication，如下所示。

```xml
<application
    android:name="android.support.multidex.MultiDexApplication"
    android:allowBackup="true"
    android:icon="@mipmap/ic_launcher"
    android:label="@string/app_name"
    android:theme="@style/AppTheme" >
    ...
</application>
```

第二种方案，让应用的 Application 继承 MultiDexApplication，比如：

```java
public class TestApplication extends MultiDexApplication {
    …
}
```

第三种方案，如果不想让应用的 Application 继承 MultiDexApplication，还可以选择重写 Application 的 attachBaseContext 方法，这个方法比 Application 的 onCreate 要先执行，如下所示。

```java
public class TestApplication extends Application {

    @Override
    protected void attachBaseContext(Context base) {
        super.attachBaseContext(base);
        MultiDex.install(this);
    }

}
```

现在所有的工作都已经完成了，可以发现应用不但可以编译通过了并且还可以在 Android 2.x 手机上面正常安装了。可以发现，multidex 使用起来还是很简单的，对于一个使用 multidex 方案的应用，采用了上面的配置项，如果这个应用的方法数没有越界，那么 Gradle 并不会生成多个 dex 文件，如果方法数越界后，Gradle 就会在 apk 中打包 2 个或多个 dex 文件，具体会打包多少个 dex 文件要看当前项目的代码规模。图 13-2 展示了采用 multidex 方案的 apk 中多个 dex 的分布情形。

459

 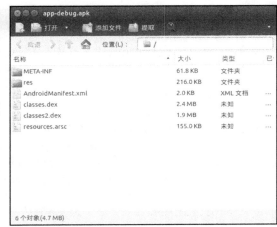

图 13-2 普通 apk 和采用 multidex 方案的 apk

上面介绍的是 multidex 默认的配置，还可以通过 build.gradle 文件中一些其他配置项来定制 dex 文件的生成过程。在有些情况下，可能需要指定主 dex 文件中所要包含的类，这个时候就可以通过--main-dex-list 选项来实现这个功能。下面是修改后的 build.gradle 文件，在里面添加了 afterEvaluate 区域，在 afterEvaluate 区域内部采用了--main-dex-list 选项来指定主 dex 中要包含的类，如下所示。

```
apply plugin: 'com.android.application'

android {
    compileSdkVersion 22
    buildToolsVersion "22.0.1"

    defaultConfig {
        applicationId "com.ryg.multidextest"
        minSdkVersion 8
        targetSdkVersion 22
        versionCode 1
        versionName "1.0"

        // enable multidex support
        multiDexEnabled true
    }
    buildTypes {
```

```
        release {
            minifyEnabled false
            proguardFiles getDefaultProguardFile('proguard-android.txt'),
            'proguard-rules.pro'
        }
    }
}

afterEvaluate {
    println "afterEvaluate"
    tasks.matching {
        it.name.startsWith('dex')
    }.each { dx ->
        def listFile = project.rootDir.absolutePath + '/app/maindexlist.txt'
        println "root dir:" + project.rootDir.absolutePath
        println "dex task found: " + dx.name
        if (dx.additionalParameters == null) {
            dx.additionalParameters = []
        }
        dx.additionalParameters += '--multi-dex'
        dx.additionalParameters += '--main-dex-list=' + listFile
        dx.additionalParameters += '--minimal-main-dex'
    }
}

dependencies {
    compile fileTree(dir: 'libs', include: ['*.jar'])
    compile 'com.android.support:appcompat-v7:22.1.1'
    compile 'com.android.support:multidex:1.0.0'
}
```

在上面的配置文件中，--multi-dex 表示当方法数越界时则生成多个 dex 文件，--main-dex-list 指定了要在主 dex 中打包的类的列表，--minimal-main-dex 表明只有 --main-dex-list 所指定的类才能打包到主 dex 中。它的输入是一个文件，在上面的配置中，它的输入是工程中 app 目录下的 maindexlist.txt 这个文件，在 maindexlist.txt 中则指定了一系列的类，所有在 maindexlist.txt 中的类都会被打包到主 dex 中。注意 maindexlist.txt 这个文件名是可以修改的，但是它的内容必须要遵守一定的格式，下面是一个示例，这种格式是固定的。

```
com/ryg/multidextest/TestApplication.class
com/ryg/multidextest/MainActivity.class

// multidex
android/support/multidex/MultiDex.class
android/support/multidex/MultiDexApplication.class
android/support/multidex/MultiDexExtractor.class
android/support/multidex/MultiDexExtractor$1.class
android/support/multidex/MultiDex$V4.class
android/support/multidex/MultiDex$V14.class
android/support/multidex/MultiDex$V19.class
android/support/multidex/ZipUtil.class
android/support/multidex/ZipUtil$CentralDirectory.class
```

程序编译后可以反编译 apk 中生成的主 dex 文件，可以发现主 dex 文件的确只有 maindexlist.txt 文件中所声明的类，读者可以自行尝试。maindexlist.txt 这个文件很多时候都是可以通过脚本来自动生成内容的，这个脚本需要根据当前的项目自行实现，如果不采用脚本，人工编辑 maindexlist.txt 也是可以的。

需要注意的是，multidex 的 jar 包中的 9 个类必须也要打包到主 dex 中，否则程序运行时会抛出异常，告知无法找到 multidex 相关的类。这是因为 Application 对象被创建以后会在 attachBaseContext 方法中通过 MultiDex.install(this)来加载其他 dex 文件，这个时候如果 MultiDex 相关的类不在主 dex 中，很显然这些类是无法被加载的，那么程序执行就会出错。同时由于 Application 的成员和代码块会先于 attachBaseContext 方法而初始化，而这个时候其他 dex 文件还没有被加载，因此不能在 Application 的成员以及代码块中访问其他 dex 中的类，否则程序也会因为无法加载对应的类而中止执行。在下面的代码中，模拟了这种场景，在 Application 的成员中使用了其他 dex 文件中的类 View1。

```java
public class TestApplication extends Application {
    private View1 view1 = new View1();

    @Override
    protected void attachBaseContext(Context base) {
        super.attachBaseContext(base);
        MultiDex.install(this);
    }
}
```

上面的代码会导致如下运行错误，因此在实际开发中要避免这个错误。

```
E/AndroidRuntime: FATAL EXCEPTION: main
Process: com.ryg.multidextest, PID: 12709
java.lang.NoClassDefFoundError: com.ryg.multidextest.ui.View1
        at com.ryg.multidextest.TestApplication.<init>(TestApplication.
java:14)
        at java.lang.Class.newInstanceImpl(Native Method)
        at java.lang.Class.newInstance(Class.java:1208)
        at android.app.Instrumentation.newApplication(Instrumentation.
java:990)
        at android.app.Instrumentation.newApplication(Instrumentation.
java:975)
        at android.app.LoadedApk.makeApplication(LoadedApk.java:504)
        ...
```

Multidex 方法虽然很好地解决了方法数越界这个问题，但它也是有一些局限性的，下面是采用 multidex 可能带来的问题：

（1）应用启动速度会降低。由于应用启动时会加载额外的 dex 文件，这将导致应用的启动速度降低，甚至可能出现 ANR 现象，尤其是其他 dex 文件较大的时候，因此要避免生成较大的 dex 文件。

（2）由于 Dalvik linearAlloc 的 bug，这可能导致使用 multidex 的应用无法在 Android 4.0 以前的手机上运行，因此需要做大量的兼容性测试。同时由于 Dalvik linearAlloc 的 bug，有可能出现应用在运行中由于采用了 multidex 方案从而产生大量的内存消耗的情况，这会导致应用崩溃。

在实际的项目中，（1）中的现象是客观存在的，但是（2）中的现象目前极少遇到，综合来说，multidex 还是一个解决方法数越界非常好的方案，可以在实际项目中使用。

13.3　Android 的动态加载技术

动态加载技术（也叫插件化技术）在技术驱动型的公司中扮演着相当重要的角色，当项目越来越庞大的时候，需要通过插件化来减轻应用的内存和 CPU 占用，还可以实现热插拔，即在不发布新版本的情况下更新某些模块。动态加载是一项很复杂的技术，这里主要介绍动态加载技术中的三个基础性问题，至于完整的动态加载技术的实现请参考笔者发起的开源插件化框架 DL：https://github.com/singwhatiwanna/dynamic-load-apk。项目期间有多位开发人员一起贡献代码。

不同的插件化方案各有各的特色，但是它们都必须要解决三个基础性问题：资源访问、Activity 生命周期的管理和 ClassLoader 的管理。在介绍它们之前，首先要明白宿主和插件的概念，宿主是指普通的 apk，而插件一般是指经过处理的 dex 或者 apk，在主流的插件化框架中多采用经过特殊处理的 apk 来作为插件，处理方式往往和编译以及打包环节有关，另外很多插件化框架都需要用到代理 Activity 的概念，插件 Activity 的启动大多数是借助一个代理 Activity 来实现的。

1. 资源访问

我们知道，宿主程序调起未安装的插件 apk，一个很大的问题就是资源如何访问，具体来说就是插件中凡是以 R 开头的资源都不能访问了。这是因为宿主程序中并没有插件的资源，所以通过 R 来加载插件的资源是行不通的，程序会抛出异常：无法找到某某 id 所对应的资源。针对这个问题，有人提出了将插件中的资源在宿主程序中也预置一份，这虽然能解决问题，但是这样就会产生一些弊端。首先，这样就需要宿主和插件同时持有一份相同的资源，增加了宿主 apk 的大小；其次，在这种模式下，每次发布一个插件都需要将资源复制到宿主程序中，这意味着每发布一个插件都要更新一下宿主程序，这就和插件化的思想相违背了。因为插件化的目的就是要减小宿主程序 apk 包的大小，同时降低宿主程序的更新频率并做到自由装载模块，所以这种方法不可取，它限制了插件的线上更新这一重要特性。还有人提供了另一种方式，首先将插件中的资源解压出来，然后通过文件流去读取资源，这样做理论上是可行的，但是实际操作起来还是有很大难度的。首先不同资源有不同的文件流格式，比如图片、XML 等，其次针对不同设备加载的资源可能是不一样的，如何选择合适的资源也是一个需要解决的问题，基于这两点，这种方法也不建议使用，因为它实现起来有较大难度。为了方便地对插件进行资源管理，下面给出一种合理的方式。

我们知道，Activity 的工作主要是通过 ContextImpl 来完成的，Activity 中有一个叫 mBase 的成员变量，它的类型就是 ContextImpl。注意到 Context 中有如下两个抽象方法，看起来是和资源有关的，实际上 Context 就是通过它们来获取资源的。这两个抽象方法的真正实现在 ContextImpl 中，也就是说，只要实现这两个方法，就可以解决资源问题了。

```
/** Return an AssetManager instance for your application's package. */
public abstract AssetManager getAssets();
```

```
/** Return a Resources instance for your application's package. */
public abstract Resources getResources();
```
下面给出具体的实现方式，首先要加载 apk 中的资源，如下所示。

```
protected void loadResources() {
    try {
        AssetManager assetManager = AssetManager.class.newInstance();
        Method addAssetPath = assetManager.getClass().getMethod
        ("addAssetPath", String.class);
        addAssetPath.invoke(assetManager, mDexPath);
        mAssetManager = assetManager;
    } catch (Exception e) {
        e.printStackTrace();
    }
    Resources superRes = super.getResources();
    mResources = new Resources(mAssetManager, superRes.getDisplay-
    Metrics(),
            superRes.getConfiguration());
    mTheme = mResources.newTheme();
    mTheme.setTo(super.getTheme());
}
```

从 loadResources() 的实现可以看出，加载资源的方法是通过反射，通过调用 AssetManager 中的 addAssetPath 方法，我们可以将一个 apk 中的资源加载到 Resources 对象中，由于 addAssetPath 是隐藏 API 我们无法直接调用，所以只能通过反射。下面是它的声明，通过注释我们可以看出，传递的路径可以是 zip 文件也可以是一个资源目录，而 apk 就是一个 zip，所以直接将 apk 的路径传给它，资源就加载到 AssetManager 中了。然后再通过 AssetManager 来创建一个新的 Resources 对象，通过这个对象我们就可以访问插件 apk 中的资源了，这样一来问题就解决了。

```
/**
 * Add an additional set of assets to the asset manager. This can be
 * either a directory or ZIP file. Not for use by applications. Returns
 * the cookie of the added asset, or 0 on failure.
 * {@hide}
 */
public final int addAssetPath(String path) {
    synchronized (this) {
```

```
        int res = addAssetPathNative(path);
        makeStringBlocks(mStringBlocks);
        return res;
    }
}
```

接着在代理 Activity 中实现 getAssets() 和 getResources()，如下所示。关于代理 Activity 的含义请参看 DL 开源插件化框架的实现细节，这里不再详细描述了。

```
@Override
public AssetManager getAssets() {
    return mAssetManager == null ? super.getAssets() : mAssetManager;
}

@Override
public Resources getResources() {
    return mResources == null ? super.getResources() : mResources;
}
```

通过上述这两个步骤，就可以通过 R 来访问插件中的资源了。

2．Activity 生命周期的管理

管理 Activity 生命周期的方式各种各样，这里只介绍两种：反射方式和接口方式。反射的方式很好理解，首先通过 Java 的反射去获取 Activity 的各种生命周期方法，比如 onCreate、onStart、onResume 等，然后在代理 Activity 中去调用插件 Activity 对应的生命周期方法即可，如下所示。

```
@Override
protected void onResume() {
    super.onResume();
    Method onResume = mActivityLifecircleMethods.get("onResume");
    if (onResume != null) {
        try {
            onResume.invoke(mRemoteActivity, new Object[] { });
        } catch (Exception e) {
            e.printStackTrace();
        }
    }
}
```

```java
@Override
protected void onPause() {
    Method onPause = mActivityLifecircleMethods.get("onPause");
    if (onPause != null) {
        try {
            onPause.invoke(mRemoteActivity, new Object[] { });
        } catch (Exception e) {
            e.printStackTrace();
        }
    }
    super.onPause();
}
```

使用反射来管理插件 Activity 的生命周期是有缺点的，一方面是反射代码写起来比较复杂，另一方面是过多使用反射会有一定的性能开销。下面介绍接口方式，接口方式很好地解决了反射方式的不足之处，这种方式将 Activity 的生命周期方法提取出来作为一个接口（比如叫 DLPlugin），然后通过代理 Activity 去调用插件 Activity 的生命周期方法，这样就完成了插件 Activity 的生命周期管理，并且没有采用反射，这就解决了性能问题。同时接口的声明也比较简单，下面是 DLPlugin 的声明：

```java
public interface DLPlugin {
    public void onStart();
    public void onRestart();
    public void onActivityResult(int requestCode, int resultCode, Intent data);
    public void onResume();
    public void onPause();
    public void onStop();
    public void onDestroy();
    public void onCreate(Bundle savedInstanceState);
    public void setProxy(Activity proxyActivity, String dexPath);
    public void onSaveInstanceState(Bundle outState);
    public void onNewIntent(Intent intent);
    public void onRestoreInstanceState(Bundle savedInstanceState);
    public boolean onTouchEvent(MotionEvent event);
    public boolean onKeyUp(int keyCode, KeyEvent event);
    public void onWindowAttributesChanged(LayoutParams params);
    public void onWindowFocusChanged(boolean hasFocus);
```

```
public void onBackPressed();
…
}
```

在代理 Activity 中只需要按如下方式即可调用插件 Activity 的生命周期方法，这就完成了插件 Activity 的生命周期的管理。

```
...
@Override
protected void onStart() {
    mRemoteActivity.onStart();
    super.onStart();
}

@Override
protected void onRestart() {
    mRemoteActivity.onRestart();
    super.onRestart();
}

@Override
protected void onResume() {
    mRemoteActivity.onResume();
    super.onResume();
}
...
```

通过上述代码应该不难理解接口方式对插件 Activity 生命周期的管理思想，其中 mRemoteActivity 就是 DLPlugin 的实现。

3. 插件 ClassLoader 的管理

为了更好地对多插件进行支持，需要合理地去管理各个插件的 DexClassLoader，这样同一个插件就可以采用同一个 ClassLoader 去加载类，从而避免了多个 ClassLoader 加载同一个类时所引发的类型转换错误。在下面的代码中，通过将不同插件的 ClassLoader 存储在一个 HashMap 中，这样就可以保证不同插件中的类彼此互不干扰。

```
public class DLClassLoader extends DexClassLoader {
    private static final String TAG = "DLClassLoader";

    private static final HashMap<String, DLClassLoader> mPluginClassLoaders
```

```
        = new HashMap<String, DLClassLoader>();

    protected DLClassLoader(String dexPath, String optimizedDirectory,
String libraryPath, ClassLoader parent) {
        super(dexPath, optimizedDirectory, libraryPath, parent);
    }

    /**
     * return a available classloader which belongs to different apk
     */
    public static DLClassLoader getClassLoader(String dexPath, Context
context, ClassLoader parentLoader) {
        DLClassLoader dLClassLoader = mPluginClassLoaders.get(dexPath);
        if (dLClassLoader != null)
            return dLClassLoader;

        File dexOutputDir = context.getDir("dex", Context.MODE_PRIVATE);
        final String dexOutputPath = dexOutputDir.getAbsolutePath();
        dLClassLoader = new DLClassLoader(dexPath, dexOutputPath, null,
parentLoader);
        mPluginClassLoaders.put(dexPath, dLClassLoader);

        return dLClassLoader;
    }
}
```

事实上插件化的技术细节非常多，这绝非一个章节的内容所能描述清楚的，另外插件化作为一种核心技术，需要开发者有较深的开发功底才能够很好地理解，因此本节的内容更多是让读者对插件化开发有一个感性的了解，细节上还需要读者自己去钻研，也可以通过 DL 插件化框架去深入地学习。

13.4 反编译初步

反编译属于逆向工程中的一种，反编译有很多高级的手段和工具，本节只是为了让读者掌握初级的反编译手段，毕竟对于一个不是专业做逆向的开发人员来说，的确没有必要花大量时间去研究反编译的一些高级技巧。本节主要介绍两方面的内容，一方面是介绍使

用 dex2jar 和 jd-gui 来反编译 apk 的方式，另一方面是介绍使用 apktool 来对 apk 进行二次打包的方式。下面是这三个反编译工具的下载地址。

apktool：http://ibotpeaches.github.io/Apktool/

dex2jar：https://github.com/pxb1988/dex2jar

jd-gui：http://jd.benow.ca/

13.4.1　使用 dex2jar 和 jd-gui 反编译 apk

Dex2jar 和 jd-gui 在很多操作系统上都可以使用，本节只介绍它们在 Windows 和 Linux 上的使用方式。Dex2jar 是一个将 dex 文件转换为 jar 包的工具，它在 Windows 和 Linux 上都有对应的版本，dex 文件来源于 apk，将 apk 通过 zip 包的方式解压缩即可提取出里面的 dex 文件。有了 jar 包还不行，因为 jar 包中都是 class 文件，这个时候还需要 jd-gui 将 jar 包进一步转换为 Java 代码，jd-gui 仍然支持 Windows 和 Linux，不管是 dex2jar 还是 jd-gui，它们在不同的操作系统中的使用方式都是一致的。

Dex2jar 是命令行工具，它的使用方式如下：

```
Linux (Ubuntu)：./dex2jar.sh classes.dex
Windows：dex2jar.bat classes.dex
```

Jd-gui 是图形化工具，直接双击打开后通过菜单打开 jar 包即可查看 jar 包的源码。下面做一个示例，通过 dex2jar 和 jd-gui 来反编译 13.1 节中的示例程序的 apk。首先将 apk 解压后提取出 classes.dex 文件，接着通过 dex2jar 反编译 classes.dex，然后通过 jd-gui 来打开反编译后的 jar 包，如图 13-3 所示。可以发现反编译后的结果和第 13.1 节中 CrashActivity 的源代码已经比较接近了，通过左边的菜单可以查看其他类的反编译结果。

13.4.2　使用 apktool 对 apk 进行二次打包

在 13.4.1 节中介绍了 dex2jar 和 jd-gui 的使用方式，通过它们可以将一个 dex 文件反编译为 Java 代码，但是它们无法反编译出 apk 中的二进制数据资源，但是采用 apktool 就可以做到这一点。apktool 另外一个常见的用途是二次打包，也就是常见的山寨版应用。将官方应用二次打包为山寨应用，这是一种不被提倡的行为，甚至是违法的，建议开发者不要去这么做，但是掌握以下二次打包的技术对于个人技术的提高还是很有意义的。apktool 同样有多个版本，这里同样只介绍 Windows 版和 Linux 版，apktool 是一个命令行工具，它的使用方式如下所示。这里仍然拿 13.1 节中的示例程序为例，假定 apk 的文件名为 CrashTest.apk。

Linux（Ubuntu）

解包：./apktool d -f CrashTest.apk CrashTest。

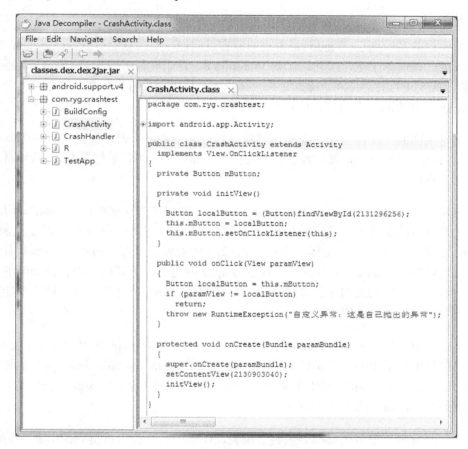

图 13-3　使用 jd-gui 反编译 jar 包

二次打包：./apktool b CrashTest CrashTest-fake.apk。

签名：java -jar signapk.jar testkey.x509.pem testkey.pk8 CrashTest-fake.apk CrashTest-fake-signed.apk。

Windows

解包：apktool.bat d -f CrashTest.apk CrashTest。

二次打包：apktool.bat b CrashTest CrashTest-fake.apk。

签名：java -jar signapk.jar testkey.x509.pem testkey.pk8 CrashTest-fake.apk CrashTest-fake-signed.apk。

需要注意的是，由于 Windows 系统的兼容性问题，有时候会导致 apktool.bat 无法在 Windows 的一些版本上正常工作，比如 Windows 8，这个时候可以安装 Cygwin，然后采用 Linux 的方式来进行打包即可。除此之外，部分 apk 也可能会打包失败，以笔者的个人经验来说，apktool 在 Linux 上的打包成功率要比 Windows 高。

这里对上面的二次打包的命令稍作解释，解包命令中，d 表示解包，CrashTest.apk 表示待解包的 apk，CrashTest 表示解包后的文件的存储路径，-f 表示如果 CrashTest 目录已经存在，那么直接覆盖它。

打包命令中，b 表示打包，CrashTest 表示解包后的文件的存储路径，CrashTest-fake.apk 表示二次打包后的文件名。通过 apktool 解包以后，可以查看到 apk 中的资源以及 smali 文件，smali 文件是 dex 文件反编译（不同于 dex2jar 的反编译过程）的结果。smali 有自己的语法并且可以修改，修改后可以被二次打包为 apk，通过这种方式就可以修改 apk 的执行逻辑，显然这让山寨应用变得十分危险。需要注意的是，apk 经过二次打包后并不能直接安装，必须要经过签名后才能安装。

签名命令中，采用 signapk.jar 来完成签名，签名后生成的 apk 就是一个山寨版的 apk，因为签名过程中所采用的签名文件不是官方的，最终 CrashTest-fake-signed.apk 就是二次打包形成的一个山寨版的 apk。对于本文中的例子来说，CrashTest-fake-signed.apk 安装后其功能和正版的 CrashTest.apk 是没有区别的。

在实际开发中，很多产品都会做签名校验，简单的二次打包所得到的山寨版 apk 安装后无法运行。尽管如此，还是可以通过修改 smali 的方式来绕过签名校验，这就是为什么市面上仍然有那么多的山寨版应用的原因。一般来说山寨版应用都具有一定的危害性，我们要抵制山寨版应用以防止自己的财产遭受到损失。关于 smali 的语法以及如何修改 smali，这就属于比较深入的话题了，本节不再对它们进行深入的介绍，感兴趣的话可以阅读逆向方面的专业书籍，也可以阅读笔者之前写的一篇介绍应用破解的文章：http://blog.csdn.net/singwhatiwanna/article/details/18797493。

第 14 章 JNI 和 NDK 编程

Java JNI 的本意是 Java Native Interface（Java 本地接口），它是为了方便 Java 调用 C、C++等本地代码所封装的一层接口。我们都知道，Java 的优点是跨平台，但是作为优点的同时，其在和本地交互的时候就出现了短板。Java 的跨平台特性导致其本地交互的能力不够强大，一些和操作系统相关的特性 Java 无法完成，于是 Java 提供了 JNI 专门用于和本地代码交互，这样就增强了 Java 语言的本地交互能力。通过 Java JNI，用户可以调用用 C、C++所编写的本地代码。

NDK 是 Android 所提供的一个工具集合，通过 NDK 可以在 Android 中更加方便地通过 JNI 来访问本地代码，比如 C 或者 C++。NDK 还提供了交叉编译器，开发人员只需要简单地修改 mk 文件就可以生成特定 CPU 平台的动态库。使用 NDK 有如下好处：

（1）提高代码的安全性。由于 so 库反编译比较困难，因此 NDK 提高了 Android 程序的安全性。

（2）可以很方便地使用目前已有的 C/C++开源库。

（3）便于平台间的移植。通过 C/C++实现的动态库可以很方便地在其他平台上使用。

（4）提高程序在某些特定情形下的执行效率，但是并不能明显提升 Android 程序的性能。

由于 JNI 和 NDK 比较适合在 Linux 环境下开发，因此本文选择 Ubuntu 14.10（64 位操作系统）作为开发环境，同时选择 AndroidStuio 作为 IDE。至于 Windows 环境下的 NDK 开发，整体流程是类似的，有差别的只是和操作系统相关的特性，这里就不再单独介绍了。在 Linux

环境中,JNI 和 NDK 开发所用到的动态库的格式是以.so 为后缀的文件,下面统一简称为 so 库。另外,由于 JNI 和 NDK 主要用于底层和嵌入式开发,在 Android 的应用层开发中使用较少,加上它们本身更加侧重于 C 和 C++方面的编程,因此本章只介绍 JNI 和 NDK 的基础知识,其他更加深入的知识点如果读者感兴趣的话可以查看专门介绍 JNI 和 NDK 的书籍。

14.1 JNI 的开发流程

JNI 的开发流程有如下几步,首先需要在 Java 中声明 native 方法,接着用 C 或者 C++ 实现 native 方法,然后就可以编译运行了。

1. 在 Java 中声明 native 方法

创建一个类,这里叫做 JniTest.java,代码如下所示。

```java
package com.ryg;

import java.lang.System;

public class JniTest {

    static {
        System.loadLibrary("jni-test");
    }

    public static void main(String args[]) {
        JniTest jniTest = new JniTest();
        System.out.println(jniTest.get());
        jniTest.set("hello world");
    }

    public native String get();
    public native void set(String str);
}
```

可以看到上面的代码中,声明了两个 native 方法:get 和 set(String),这两个就是需要在 JNI 中实现的方法。在 JniTest 的头部有一个加载动态库的过程,其中 jni-test 是 so 库的标识，so 库完整的名称为 libjni-test.so，这是加载 so 库的规范。

2. 编译 Java 源文件得到 class 文件，然后通过 javah 命令导出 JNI 的头文件

具体的命令如下：

```
javac com/ryg/JniTest.java
javah com.ryg.JniTest
```

在当前目录下，会产生一个 com_ryg_JniTest.h 的头文件，它是 javah 命令自动生成的，内容如下所示。

```
/* DO NOT EDIT THIS FILE - it is machine generated */
#include <jni.h>
/* Header for class com_ryg_JniTest */

#ifndef _Included_com_ryg_JniTest
#define _Included_com_ryg_JniTest
#ifdef __cplusplus
extern "C" {
#endif
/*
 * Class:     com_ryg_JniTest
 * Method:    get
 * Signature: ()Ljava/lang/String;
 */
JNIEXPORT jstring JNICALL Java_com_ryg_JniTest_get
  (JNIEnv *, jobject);

/*
 * Class:     com_ryg_JniTest
 * Method:    set
 * Signature: (Ljava/lang/String;)V
 */
JNIEXPORT void JNICALL Java_com_ryg_JniTest_set
  (JNIEnv *, jobject, jstring);

#ifdef __cplusplus
}
#endif
#endif
```

上面的代码需要做一下说明，首先函数名的格式遵循如下规则：Java_包名_类名_方法

名。比如 JniTest 中的 set 方法，到这里就变成了 JNIEXPORT void JNICALL Java_com_ryg_JniTest_set(JNIEnv *, jobject, jstring)，其中 com_ryg 是包名，JniTest 是类名，jstring 是代表的是 set 方法的 String 类型的参数。关于 Java 和 JNI 的数据类型之间的对应关系会在 14.3 节中进行介绍，这里只需要知道 Java 的 String 对应于 JNI 的 jstring 即可。JNIEXPORT、JNICALL、JNIEnv 和 jobject 都是 JNI 标准中所定义的类型或者宏，它们的含义如下：

- JNIEnv*：表示一个指向 JNI 环境的指针，可以通过它来访问 JNI 提供的接口方法；

- jobject：表示 Java 对象中的 this；

- JNIEXPORT 和 JNICALL：它们是 JNI 中所定义的宏，可以在 jni.h 这个头文件中查找到。

下面的宏定义是必需的，它指定 extern "C" 内部的函数采用 C 语言的命名风格来编译。否则当 JNI 采用 C++ 来实现时，由于 C 和 C++ 编译过程中对函数的命名风格不同，这将导致 JNI 在链接时无法根据函数名查找到具体的函数，那么 JNI 调用就无法完成。更多的细节实际上是有关 C 和 C++ 编译时的一些问题，这里就不再展开了。

```
#ifdef __cplusplus
extern "C" {
#endif
```

3. 实现 JNI 方法

JNI 方法是指 Java 中声明的 native 方法，这里可以选择用 C++ 或者 C 来实现，它们的实现过程是类似的，只有少量的区别，下面分别用 C++ 和 C 来实现 JNI 方法。首先，在工程的主目录下创建一个子目录，名称随意，这里选择 jni 作为子目录的名称，然后将之前通过 javah 生成的头文件 com_ryg_JniTest.h 复制到 jni 目录下，接着创建 test.cpp 和 test.c 两个文件，它们的实现如下所示。

```
// test.cpp
#include "com_ryg_JniTest.h"
#include <stdio.h>

JNIEXPORT jstring JNICALL Java_com_ryg_JniTest_get(JNIEnv *env, jobject thiz) {
    printf("invoke get in c++\n");
    return env->NewStringUTF("Hello from JNI !");
```

```
}

JNIEXPORT void JNICALL Java_com_ryg_JniTest_set(JNIEnv *env, jobject thiz,
jstring string) {
    printf("invoke set from C++\n");
    char* str = (char*)env->GetStringUTFChars(string,NULL);
    printf("%s\n", str);
    env->ReleaseStringUTFChars(string, str);
}

// test.c
#include "com_ryg_JniTest.h"
#include <stdio.h>

JNIEXPORT jstring JNICALL Java_com_ryg_JniTest_get(JNIEnv *env, jobject
thiz) {
    printf("invoke get from C\n");
    return (*env)->NewStringUTF(env, "Hello from JNI !");
}

JNIEXPORT void JNICALL Java_com_ryg_JniTest_set(JNIEnv *env, jobject thiz,
jstring string) {
    printf("invoke set from C\n");
    char* str = (char*)(*env)->GetStringUTFChars(env,string,NULL);
    printf("%s\n", str);
    (*env)->ReleaseStringUTFChars(env, string, str);
}
```

可以发现，test.cpp 和 test.c 的实现很类似，但是它们对 env 的操作方式有所不同，因此用 C++和 C 来实现同一个 JNI 方法，它们的区别主要集中在对 env 的操作上，其他都是类似的，如下所示。

```
C++: env->NewStringUTF("Hello from JNI !");
C:   (*env)->NewStringUTF(env, "Hello from JNI !")
```

4．编译 so 库并在 Java 中调用

so 库的编译这里采用 gcc，切换到 jni 目录中，对于 test.cpp 和 test.c 来说，它们的编译指令如下所示。

```
C++: gcc -shared -I /usr/lib/jvm/java-7-openjdk-amd64/include -fPIC test.cpp
-o libjni-test.so
```

```
C: gcc -shared -I /usr/lib/jvm/java-7-openjdk-amd64/include -fPIC test.c -o libjni-test.so
```

上面的编译命令中，/usr/lib/jvm/java-7-openjdk-amd64 是本地的 jdk 的安装路径，在其他环境编译时将其指向本机的 jdk 路径即可。而 libjni-test.so 则是生成的 so 库的名字，在 Java 中可以通过如下方式加载：System.loadLibrary("jni-test")，其中 so 库名字中的"lib"和".so"是不需要明确指出的。so 库编译完成后，就可以在 Java 程序中调用 so 库了，这里通过 Java 指令来执行 Java 程序，切换到主目录，执行如下指令：java -Djava.library.path=jni com.ryg.JniTest，其中-Djava.library.path=jni 指明了 so 库的路径。

首先，采用 C++产生 so 库，程序运行后产生的日志如下所示。

```
invoke get in c++
Hello from JNI !
invoke set from C++
hello world
```

然后，采用 C 产生 so 库，程序运行后产生的日志如下所示。

```
invoke get from C
Hello from JNI !
invoke set from C
hello world
```

通过上面的日志可以发现，在 Java 中成功地调用了 C/C++的代码，这就是 JNI 典型的工作流程。

14.2 NDK 的开发流程

NDK 的开发是基于 JNI 的，其主要由如下几个步骤。

1. 下载并配置 NDK

首先要从 Android 官网上下载 NDK，下载地址为 https://developer.android.com/ndk/downloads/index.html，本章中采用的 NDK 的版本是 android-ndk-r10d。下载完成以后，将 NDK 解压到一个目录，然后为 NDK 配置环境变量，步骤如下所示。

首先打开当前用户的环境变量配置文件：

```
vim ~/.bashrc
```

然后在文件后面添加如下信息：export PATH=~/Android/android-ndk-r10d:$PATH，其中 ~/Android/android-ndk-r10d 是本地的 NDK 的存放路径。

添加完毕后，执行 source ~/.bashrc 来立刻刷新刚刚设置的环境变量。设置完环境变量后，ndk-build 命令就可以使用了，通过 ndk-build 命令就可以编译产生 so 库。

2. 创建一个 Android 项目，并声明所需的 native 方法

```java
package com.ryg.JniTestApp;

import android.support.v7.app.ActionBarActivity;
import android.os.Bundle;
import android.util.Log;
import android.view.Menu;
import android.view.MenuItem;
import android.widget.TextView;

public class MainActivity extends ActionBarActivity {

    static {
        System.loadLibrary("jni-test");
    }

    @Override
    protected void onCreate(Bundle savedInstanceState) {
        super.onCreate(savedInstanceState);
        setContentView(R.layout.activity_main);
        TextView textView = (TextView)findViewById(R.id.msg);
        textView.setText(get());
        set("hello world from JniTestApp");
    }

    public native String get();
    public native void set(String str);
}
```

3. 实现 Android 项目中所声明的 native 方法

在外部创建一个名为 jni 的目录，然后在 jni 目录下创建 3 个文件：test.cpp、Android.mk

和 Application.mk，它们的实现如下所示。

```cpp
// test.cpp
#include <jni.h>
#include <stdio.h>

#ifdef __cplusplus
extern "C" {
#endif

jstring Java_com_ryg_JniTestApp_MainActivity_get(JNIEnv *env,jobject thiz){
    printf("invoke get in c++\n");
    return env->NewStringUTF("Hello from JNI in libjni-test.so !");
}

void Java_com_ryg_JniTestApp_MainActivity_set(JNIEnv *env, jobject thiz, jstring string) {
    printf("invoke set from C++\n");
    char* str = (char*)env->GetStringUTFChars(string,NULL);
    printf("%s\n", str);
    env->ReleaseStringUTFChars(string, str);
}

#ifdef __cplusplus
}
#endif

// Android.mk
# Copyright (C) 2009 The Android Open Source Project
#
# Licensed under the Apache License, Version 2.0 (the "License");
# you may not use this file except in compliance with the License.
# You may obtain a copy of the License at
#
#      http://www.apache.org/licenses/LICENSE-2.0
#
# Unless required by applicable law or agreed to in writing, software
# distributed under the License is distributed on an "AS IS" BASIS,
# WITHOUT WARRANTIES OR CONDITIONS OF ANY KIND, either express or implied.
# See the License for the specific language governing permissions and
# limitations under the License.
#
```

```
LOCAL_PATH := $(call my-dir)

include $(CLEAR_VARS)

LOCAL_MODULE    := jni-test
LOCAL_SRC_FILES := test.cpp

include $(BUILD_SHARED_LIBRARY)

// Application.mk
APP_ABI := armeabi
```

这里对 Android.mk 和 Application.mk 做一下简单的介绍。在 Android.mk 中，LOCAL_MODULE 表示模块的名称，LOCAL_SRC_FILES 表示需要参与编译的源文件。Application.mk 中常用的配置项是 APP_ABI，它表示 CPU 的架构平台的类型，目前市面上常见的架构平台有 armeabi、x86 和 mips，其中在移动设备中占据主要地位的是 armeabi，这也是大部分 apk 中只包含 armeabi 类型的 so 库的原因。默认情况下 NDK 会编译产生各个 CPU 平台的 so 库，通过 APP_ABI 选项即可指定 so 库的 CPU 平台的类型，比如 armeabi，这样 NDK 就只会编译 armeabi 平台下的 so 库了，而 all 则表示编译所有 CPU 平台的 so 库。

4．切换到 jni 目录的父目录，然后通过 ndk-build 命令编译产生 so 库

这个时候 NDK 会创建一个和 jni 目录平级的目录 libs，libs 下面存放的就是 so 库的目录，如图 14-1 所示。需要注意的是，ndk-build 命令会默认指定 jni 目录为本地源码的目录，如果源码存放的目录名不是 jni，那么 ndk-build 则无法成功编译。

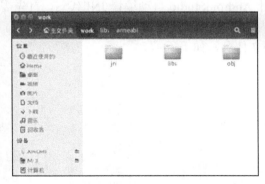 

图 14-1　通过 NDK 编译产生的 so 库

然后在 app/src/main 中创建一个名为 jniLibs 的目录，将生成的 so 库复制到 jniLibs 目录中，然后通过 AndroidStudio 编译运行即可，运行效果如图 14-2 所示。这说明从 Android 中调用 so 库中的方法已经成功了。

图 14-2　Android 中调用 so 库中的方法示例

在上面的步骤中，需要将 NDK 编译的 so 库放置到 jniLibs 目录下，这个是 AndroidStudio 所识别的默认目录，如果想使用其他目录，可以按照如下方式修改 App 的 build.gradle 文件，其中 jniLibs.srcDir 选项指定了新的存放 so 库的目录。

```
android {
    ...
    sourceSets.main {
        jniLibs.srcDir 'src/main/jni_libs'
    }
}
```

除了手动使用 ndk-build 命令创建 so 库，还可以通过 AndroidStudio 来自动编译产生 so 库，这个操作过程要稍微复杂一些。为了能够让 AndroidStudio 自动编译 JNI 代码，首先需要在 App 的 build.gradle 的 defaultConfig 区域内添加 NDK 选项，其中 moduleName 指定了模块的名称，这个名称指定了打包后的 so 库的文件名，如下所示。

```
android {
    ...
    defaultConfig {
        applicationId "com.ryg.JniTestApp"
        minSdkVersion 8
        targetSdkVersion 22
        versionCode 1
        versionName "1.0"
```

```
    ndk {
        moduleName "jni-test"
    }
  }
}
```

接着需要将 JNI 的代码放在 app/src/main/jni 目录下，注意存放 JNI 代码的目录名必须为 jni，如果不想采用 jni 这个名称，可以通过如下方式来指定 JNI 的代码路径，其中 jni.srcDirs 指定了 JNI 代码的路径：

```
android {
    ...
    sourceSets.main {
        jni.srcDirs 'src/main/jni_src'
    }
}
```

经过了上面的步骤，AndroidStudio 就可以自动编译 JNI 代码了，但是这个时候 AndroidStudio 会把所有 CPU 平台的 so 库都打包到 apk 中，一般来说实际开发中只需要打包 armeabi 平台的 so 库即可。要解决这个问题也很简单，按照如下方式修改 build.gradle 的配置，然后在 Build Variants 面板中选择 armDebug 选项进行编辑就可以了。

```
android {
    ...
    productFlavors {
        arm {
            ndk {
                abiFilter "armeabi"
            }
        }
        x86 {
            ndk {
                abiFilter "x86"
            }
        }
    }
}
```

如图 14-3 所示，可以看到 apk 中就只有 armeabi 平台的 so 库了。

图 14-3　AndroidStudio 打包后的 apk

14.3　JNI 的数据类型和类型签名

JNI 的数据类型包含两种：基本类型和引用类型。基本类型主要有 jboolean、jchar、jint 等，它们和 Java 中的数据类型的对应关系如表 14-1 所示。

表 14-1　JNI 基本数据类型的对应关系

JNI 类型	Java 类型	描　　述
jboolean	boolean	无符号 8 位整型
jbyte	byte	有符号 8 位整型
jchar	char	无符号 16 位整型
jshort	short	有符号 16 位整型
jint	int	32 位整型
jlong	long	64 位整型
jfloat	float	32 位浮点型
jdouble	double	64 位浮点型
void	void	无类型

JNI 中的引用类型主要有类、对象和数组，它们和 Java 中的引用类型的对应关系如表 14-2 所示。

表 14-2　JNI 引用类型的对应关系

JNI 类型	Java 类型	描　　述
jobject	Object	Object 类型
jclass	Class	Class 类型
jstring	String	字符串
jobjectArray	Object[]	对象数组
jbooleanArray	boolean[]	boolean 数组
jbyteArray	byte[]	byte 数组
jcharArray	char[]	char 数组
jshortArray	short[]	short 数组
jintArray	int[]	int 数组
jlongArray	long[]	long 数组
jfloatArray	float[]	float 数组
jdoubleArray	double[]	double 数组
jthrowable	Throwable	Throwable

JNI 的类型签名标识了一个特定的 Java 类型，这个类型既可以是类和方法，也可以是数据类型。

类的签名比较简单，它采用 "L+包名+类名+;" 的形式，只需要将其中的.替换为/即可。比如 java.lang.String，它的签名为 Ljava/lang/String;，注意末尾的;也是签名的一部分。

基本数据类型的签名采用一系列大写字母来表示，如表 14-3 所示。

表 14-3　基本数据类型的签名

Java 类型	签　　名	Java 类型	签　　名
boolean	Z	long	J
byte	B	float	F
char	C	double	D
short	S	void	V
int	I		

从表 14-3 可以看出，基本数据类型的签名是有规律的，一般为首字母的大写，但是 boolean 除外，因为 B 已经被 byte 占用了，而 long 的签名之所以不是 L，那是因为 L 表示的是类的签名。

对象和数组的签名稍微复杂一些。对于对象来说，它的签名就是对象所属的类的签名，比如 String 对象，它的签名为 Ljava/lang/String;。对于数组来说，它的签名为[+类型签名，比如 int 数组，其类型为 int，而 int 的签名为 I，所以 int 数组的签名就是[I，同理就可以得出如下的签名对应关系：

```
char[]        [C
float[]       [F
double[]      [D
long[]        [J
String[]      [Ljava/lang/String;
Object[]      [Ljava/lang/Object;
```

对于多维数组来说，它的签名为 n 个[+类型签名，其中 n 表示数组的维度，比如，int[][] 的签名为[[I，其他情况可以依此类推。

方法的签名为（参数类型签名）+返回值类型签名，这有点不好理解。举个例子，如下方法：boolean fun1(int a, double b, int[] c)，根据签名的规则可以知道，它的参数类型的签名连在一起是 ID[I，返回值类型的签名为 Z，所以整个方法的签名就是(ID[I)Z。再举个例子，下面的方法：boolean fun1(int a, String b, int[] c)，它的签名是(ILjava/lang/String;[I)Z。为了能够更好地理解方法的签名格式，下面再给出两个示例：

```
int fun1()           签名为 ()I
void fun1(int i)     签名为 (I)V
```

14.4　JNI 调用 Java 方法的流程

JNI 调用 Java 方法的流程是先通过类名找到类，然后再根据方法名找到方法的 id，最后就可以调用这个方法了。如果是调用 Java 中的非静态方法，那么需要构造出类的对象后才能调用它。下面的例子演示了如何在 JNI 中调用 Java 的静态方法，至于调用非静态方法只是多了一步构造对象的过程，这里就不再介绍了。

首先需要在 Java 中定义一个静态方法供 JNI 调用，如下所示。

```java
public static void methodCalledByJni(String msgFromJni) {
    Log.d(TAG, "methodCalledByJni,msg: " + msgFromJni);
}
```

然后在 JNI 中调用上面定义的静态方法：

```cpp
void callJavaMethod(JNIEnv *env, jobject thiz) {
    jclass clazz = env->FindClass("com/ryg/JniTestApp/MainActivity");
    if (clazz == NULL) {
        printf("find class MainActivity error!");
        return;
    }
    jmethodID id = env->GetStaticMethodID(clazz, "methodCalledByJni",
    "(Ljava/lang/String;)V");
    if (id == NULL) {
        printf("find method methodCalledByJni error!");
    }
    jstring msg = env->NewStringUTF("msg send by callJavaMethod in test.cpp.");
    env->CallStaticVoidMethod(clazz, id, msg);
}
```

从 callJavaMethod 的实现可以看出，程序首先根据类名 com/ryg/JniTestApp/MainActivity 找到类，然后再根据方法名 methodCalledByJni 找到方法，其中(Ljava/lang/String;)V 是 methodCalledByJni 方法的签名，接着再通过 JNIEnv 对象的 CallStaticVoidMethod 方法来完成最终的调用过程。

最后在 Java_com_ryg_JniTestApp_MainActivity_get 方法中调用 callJavaMethod 方法，如下所示。

```cpp
jstring Java_com_ryg_JniTestApp_MainActivity_get(JNIEnv *env,jobject thiz){
    printf("invoke get in c++\n");
    callJavaMethod(env, thiz);
    return env->NewStringUTF("Hello from JNI in libjni-test.so !");
}
```

由于 MainActivity 会调用 JNI 中的 Java_com_ryg_JniTestApp_MainActivity_get 方法，Java_com_ryg_JniTestApp_MainActivity_get 方法又会调用 callJavaMethod 方法，而

callJavaMethod 方法又会反过来调用 MainActivity 的 methodCalledByJni 方法，这样一来就完成了一次从 Java 调用 JNI 然后再从 JNI 中调用 Java 方法的过程。安装运行程序，可以看到如下日志，这说明程序已经成功地从 JNI 中调用了 Java 中的 methodCalledByJni 方法。

```
D/MainActivity: methodCalledByJni, msg: msg send by callJavaMethod in test.cpp.
```

我们可以发现，JNI 调用 Java 的过程和 Java 中方法的定义有很大关联，针对不同类型的 Java 方法，JNIEnv 提供了不同的接口去调用，本章作为一个 JNI 的入门章节就不再对它们一一进行介绍了，毕竟大部分应用层的开发人员并不需要那么深入地了解 JNI，如果读者感兴趣可以自行阅读相关的 JNI 专业书籍。

第 15 章 Android 性能优化

本章是本书的最后一章，所介绍的主题是 Android 的性能优化方法和程序设计的一些思想。通过本章的内容，读者可以掌握常见的性能优化方法，这将有助于提高 Android 程序的性能；另一方面，本章还讲解了 Android 程序设计的一些思想，这将有助于提高程序的可维护性和可扩展性。另外，2015 年 Google 在 YouTube 上发布了关于 Android 性能优化典范的专题，通过一系列短视频来帮助开发者创建更快更优秀的 Android 应用，课程专题不仅仅介绍了 Android 系统中有关性能问题的底层工作原理，同时也介绍了如何通过工具来找出性能问题以及提升性能的建议，地址是：

https://www.youtube.com/playlist?list=PLWz5rJ2EKKc9CBxr3BVjPTPoDPLdPIFCE。

Android 设备作为一种移动设备，不管是内存还是 CPU 的性能都受到了一定的限制，无法做到像 PC 设备那样具有超大的内存和高性能的 CPU。鉴于这一点，这也意味着 Android 程序不可能无限制地使用内存和 CPU 资源，过多地使用内存会导致程序内存溢出，即 OOM。而过多地使用 CPU 资源，一般是指做大量的耗时任务，会导致手机变得卡顿甚至出现程序无法响应的情况，即 ANR。由此来看，Android 程序的性能问题就变得异常突出了，这对开发人员也提出了更高的要求。为了提高应用程序的性能，本章第一节介绍了一些有效的性能优化方法，主要内容包括布局优化、绘制优化、内存泄露优化、响应速度优化、ListView 优化、Bitmap 优化、线程优化以及一些性能优化建议，同时在介绍响应速度优化的同时还介绍了 ANR 日志的分析方法。

性能优化中一个很重要的问题就是内存泄露，内存泄露并不会导致程序功能异常，但是它会导致 Android 程序的内存占用过大，这将提高内存溢出的发生几率。如何避免写出内存泄露的代码，这和开发人员的水平和意识有很大关系，甚至很多情况下内存泄露的原因是很难直接发现的，这个时候就需要借助一些内存泄露分析工具，在本章的第二节将介绍内存泄露分析工具 MAT 的使用，通过 MAT 就可以发现一些开发过程中难以发现的内存泄露问题。

在做程序设计时，除了要完成功能开发、提高程序的性能以外，还有一个问题也是不容忽视的，那就是代码的可维护性和可扩展性。如果一个程序的可维护性和可扩展性很差，那就意味着后续的代码维护代价是相当高的，比如需要对一个功能做一些调整，这可能会出现牵一发而动全身的局面。另外添加新功能时也觉得无从下手，整个代码看起来可读性很差，这的确是一份很糟糕的代码。关于代码的可维护性和可扩展性，看起来是一个很抽象的问题，其实它并不抽象，它是可以通过一些合理的设计原则去完成的，比如良好的代码风格、清晰的代码层级、代码的可扩展性和合理的设计模式，在本章的第三节对这些设计原则做了介绍，这将在一定程度上提高程序的可维护性和可扩展性。

15.1 Android 的性能优化方法

本节介绍了一些有效的性能优化方法，主要内容包括布局优化、绘制优化、内存泄露优化、响应速度优化、ListView 优化、Bitmap 优化、线程优化以及一些性能优化建议，在介绍响应速度优化的同时还介绍了 ANR 日志的分析方法。

15.1.1 布局优化

布局优化的思想很简单，就是尽量减少布局文件的层级，这个道理是很浅显的，布局中的层级少了，这就意味着 Android 绘制时的工作量少了，那么程序的性能自然就高了。

如何进行布局优化呢？首先删除布局中无用的控件和层级，其次有选择地使用性能较低的 ViewGroup，比如 RelativeLayout。如果布局中既可以使用 LinearLayout 也可以使用 RelativeLayout，那么就采用 LinearLayout，这是因为 RelativeLayout 的功能比较复杂，它的布局过程需要花费更多的 CPU 时间。FrameLayout 和 LinearLayout 一样都是一种简单高效的 ViewGroup，因此可以考虑使用它们，但是很多时候单纯通过一个 LinearLayout 或者 FrameLayout 无法实现产品效果，需要通过嵌套的方式来完成。这种情况下还是建议采用 RelativeLayout，因为 ViewGroup 的嵌套就相当于增加了布局的层级，同样会降低程序的性能。

布局优化的另外一种手段是采用<include>标签、<merge>标签和 ViewStub。<include>标签主要用于布局重用，<merge>标签一般和<include>配合使用，它可以降低减少布局的层级，而 ViewStub 则提供了按需加载的功能，当需要时才会将 ViewStub 中的布局加载到内存，这提高了程序的初始化效率，下面分别介绍它们的使用方法。

<include>标签

<include>标签可以将一个指定的布局文件加载到当前的布局文件中，如下所示。

```xml
<LinearLayout xmlns:android="http://schemas.android.com/apk/res/android"
    android:orientation="vertical"
    android:layout_width="match_parent"
    android:layout_height="match_parent"
    android:background="@color/app_bg"
    android:gravity="center_horizontal">

    <include layout="@layout/titlebar"/>

    <TextView android:layout_width="match_parent"
        android:layout_height="wrap_content"
        android:text="@string/text"
        android:padding="5dp" />
    ...
</LinearLayout>
```

上面的代码中，@layout/titlebar 指定了另外一个布局文件，通过这种方式就不用把 titlebar 这个布局文件的内容再重复写一遍了，这就是<include>的好处。<include>标签只支持以 android:layout_ 开头的属性，比如 android:layout_width、android:layout_height，其他属性是不支持的，比如 android:background。当然, android:id 这个属性是个特例，如果<include>指定了这个 id 属性，同时被包含的布局文件的根元素也指定了 id 属性，那么以<include>指定的 id 属性为准。需要注意的是，如果<include>标签指定了 android:layout_*这种属性，那么要求 android:layout_width 和 android:layout_height 必须存在，否则其他 android:layout_*形式的属性无法生效，下面是一个指定了 android:layout_*属性的示例。

```xml
<include android:id="@+id/new_title"
    android:layout_width="match_parent"
    android:layout_height="match_parent"
    layout="@layout/title"/>
```

<merge>标签

<merge>标签一般和<include>标签一起使用从而减少布局的层级。在上面的示例中，由于当前布局是一个竖直方向的 LinearLayout，这个时候如果被包含的布局文件中也采用了竖直方向的 LinearLayout，那么显然被包含的布局文件中的 LinearLayout 是多余的，通过 <merge>标签就可以去掉多余的那一层 LinearLayout，如下所示。

```
<merge xmlns:android="http://schemas.android.com/apk/res/android">

    <Button
        android:layout_width="wrap_content"
        android:layout_height="wrap_content"
        android:text="@string/one"/>

    <Button
        android:layout_width="wrap_content"
        android:layout_height="wrap_content"
        android:text="@string/two"/>

</merge>
```

ViewStub

ViewStub 继承了 View，它非常轻量级且宽/高都是 0，因此它本身不参与任何的布局和绘制过程。ViewStub 的意义在于按需加载所需的布局文件，在实际开发中，有很多布局文件在正常情况下不会显示，比如网络异常时的界面，这个时候就没有必要在整个界面初始化的时候将其加载进来，通过 ViewStub 就可以做到在使用的时候再加载，提高了程序初始化时的性能。下面是一个 ViewStub 的示例：

```
<ViewStub
    android:id="@+id/stub_import"
    android:inflatedId="@+id/panel_import"
    android:layout="@layout/layout_network_error"
    android:layout_width="match_parent"
    android:layout_height="wrap_content"
    android:layout_gravity="bottom" />
```

其中 stub_import 是 ViewStub 的 id，而 panel_import 是 layout/layout_network_error 这个布局的根元素的 id。如何做到按需加载呢？在需要加载 ViewStub 中的布局时，可以按照如

下两种方式进行：

```
((ViewStub) findViewById(R.id.stub_import)).setVisibility(View.VISIBLE);
```

或者

```
View importPanel = ((ViewStub) findViewById(R.id.stub_import)).inflate();
```

当 ViewStub 通过 setVisibility 或者 inflate 方法加载后，ViewStub 就会被它内部的布局替换掉，这个时候 ViewStub 就不再是整个布局结构中的一部分了。另外，目前 ViewStub 还不支持<merge>标签。

15.1.2　绘制优化

绘制优化是指 View 的 onDraw 方法要避免执行大量的操作，这主要体现在两个方面。

首先，onDraw 中不要创建新的局部对象，这是因为 onDraw 方法可能会被频繁调用，这样就会在一瞬间产生大量的临时对象，这不仅占用了过多的内存而且还会导致系统更加频繁 gc，降低了程序的执行效率。

另外一方面，onDraw 方法中不要做耗时的任务，也不能执行成千上万次的循环操作，尽管每次循环都很轻量级，但是大量的循环仍然十分抢占 CPU 的时间片，这会造成 View 的绘制过程不流畅。按照 Google 官方给出的性能优化典范中的标准，View 的绘制帧率保证 60fps 是最佳的，这就要求每帧的绘制时间不超过 16ms（16ms = 1000 / 60），虽然程序很难保证 16ms 这个时间，但是尽量降低 onDraw 方法的复杂度总是切实有效的。

15.1.3　内存泄露优化

内存泄露在开发过程中是一个需要重视的问题，但是由于内存泄露问题对开发人员的经验和开发意识有较高的要求，因此这也是开发人员最容易犯的错误之一。内存泄露的优化分为两个方面，一方面是在开发过程中避免写出有内存泄露的代码，另一方面是通过一些分析工具比如 MAT 来找出潜在的内存泄露继而解决。本节主要介绍一些常见的内存泄露的例子，通过这些例子读者可以很好地理解内存泄露的发生场景并积累规避内存泄露的经验。关于如何通过工具分析内存泄露将在 15.2 节中专门介绍。

场景 1：静态变量导致的内存泄露

下面这种情形是一种最简单的内存泄露，相信读者都不会这么干，下面的代码将导致

Activity 无法正常销毁，因此静态变量 sContext 引用了它。

```
public class MainActivity extends Activity {
    private static final String TAG = "MainActivity";

    private static Context sContext;

    @Override
    protected void onCreate(Bundle savedInstanceState) {
        super.onCreate(savedInstanceState);
        setContentView(R.layout.activity_main);
        sContext = this;
    }
}
```

上面的代码也可以改造一下，如下所示。sView 是一个静态变量，它内部持有了当前 Activity，所以 Activity 仍然无法释放，估计读者也都明白。

```
public class MainActivity extends Activity {
    private static final String TAG = "MainActivity";

    private static View sView;

    @Override
    protected void onCreate(Bundle savedInstanceState) {
        super.onCreate(savedInstanceState);
        setContentView(R.layout.activity_main);
        sView = new View(this);
    }
}
```

场景 2：单例模式导致的内存泄露

静态变量导致的内存泄露都太过于明显，相信读者都不会犯这种错误，而单例模式所带来的内存泄露是我们容易忽视的，如下所示。首先提供一个单例模式的 TestManager，TestManager 可以接收外部的注册并将外部的监听器存储起来。

```
public class TestManager {

    private List<OnDataArrivedListener> mOnDataArrivedListeners = new
```

```
    ArrayList<OnDataArrivedListener>();

    private static class SingletonHolder {
        public static final TestManager INSTANCE = new TestManager();
    }

    private TestManager() {
    }

    public static TestManager getInstance() {
        return SingletonHolder.INSTANCE;
    }

    public synchronized void registerListener(OnDataArrivedListener
    listener) {
        if (!mOnDataArrivedListeners.contains(listener)) {
            mOnDataArrivedListeners.add(listener);
        }
    }

    public synchronized void unregisterListener(OnDataArrivedListener
    listener) {
        mOnDataArrivedListeners.remove(listener);
    }

    public interface OnDataArrivedListener {
        public void onDataArrived(Object data);
    }
}
```

接着再让 Activity 实现 OnDataArrivedListener 接口并向 TestManager 注册监听，如下所示。下面的代码由于缺少解注册的操作所以会引起内存泄露，泄露的原因是 Activity 的对象被单例模式的 TestManager 所持有，而单例模式的特点是其生命周期和 Application 保持一致，因此 Activity 对象无法被及时释放。

```
    protected void onCreate(Bundle savedInstanceState) {
        super.onCreate(savedInstanceState);
        setContentView(R.layout.activity_main);
```

```
    TestManager.getInstance().registerListener(this);
}
```

场景 3：属性动画导致的内存泄露

从 Android 3.0 开始，Google 提供了属性动画，属性动画中有一类无限循环的动画，如果在 Activity 中播放此类动画且没有在 onDestroy 中去停止动画，那么动画会一直播放下去，尽管已经无法在界面上看到动画效果了，并且这个时候 Activity 的 View 会被动画持有，而 View 又持有了 Activity，最终 Activity 无法释放。下面的动画是无限动画，会泄露当前 Activity，解决方法是在 Activity 的 onDestroy 中调用 animator.cancel() 来停止动画。

```
protected void onCreate(Bundle savedInstanceState) {
    super.onCreate(savedInstanceState);
    setContentView(R.layout.activity_main);
    mButton = (Button) findViewById(R.id.button1);
    ObjectAnimator animator = ObjectAnimator.ofFloat(mButton, "rotation",
            0, 360).setDuration(2000);
    animator.setRepeatCount(ValueAnimator.INFINITE);
    animator.start();
    //animator.cancel();
}
```

15.1.4 响应速度优化和 ANR 日志分析

响应速度优化的核心思想是避免在主线程中做耗时操作，但是有时候的确有很多耗时操作，怎么办呢？可以将这些耗时操作放在线程中去执行，即采用异步的方式执行耗时操作。响应速度过慢更多地体现在 Activity 的启动速度上面，如果在主线程中做太多事情，会导致 Activity 启动时出现黑屏现象，甚至出现 ANR。Android 规定，Activity 如果 5 秒钟之内无法响应屏幕触摸事件或者键盘输入事件就会出现 ANR，而 BroadcastReceiver 如果 10 秒钟之内还未执行完操作也会出现 ANR。在实际开发中，ANR 是很难从代码上发现的，如果在开发过程中遇到了 ANR，那么怎么定位问题呢？其实当一个进程发生 ANR 了以后，系统会在/data/anr 目录下创建一个文件 traces.txt，通过分析这个文件就能定位出 ANR 的原因，下面模拟一个 ANR 的场景。下面的代码在 Activity 的 onCreate 中休眠 30s，程序运行后持续点击屏幕，应用一定会出现 ANR：

```
protected void onCreate(Bundle savedInstanceState) {
    super.onCreate(savedInstanceState);
    setContentView(R.layout.activity_main);
```

```
        SystemClock.sleep(30 * 1000);
}
```

这里先假定我们无法从代码中看出 ANR，为了分析 ANR 的原因，可以到处 traces 文件，如下所示，其中.表示当前目录：

```
adb pull /data/anr/traces.txt .
```

traces 文件一般是非常长的，下面是 traces 文件的部分内容：

```
----- pid 29395 at 2015-05-31 16:14:36 -----
Cmd line: com.ryg.chapter_15

DALVIK THREADS:
(mutexes: tll=0 tsl=0 tscl=0 ghl=0)

"main" prio=5 tid=1 TIMED_WAIT
  | group="main" sCount=1 dsCount=0 obj=0x4185b700 self=0x4012d0b0
  | sysTid=29395 nice=0 sched=0/0 cgrp=apps handle=1073954608
  | schedstat=( 0 0 0 ) utm=3 stm=2 core=2
  at java.lang.VMThread.sleep(Native Method)
  at java.lang.Thread.sleep(Thread.java:1031)
  at java.lang.Thread.sleep(Thread.java:1013)
  at android.os.SystemClock.sleep(SystemClock.java:114)
  at com.ryg.chapter_15.MainActivity.onCreate(MainActivity.java:42)
  at android.app.Activity.performCreate(Activity.java:5086)
  at android.app.Instrumentation.callActivityOnCreate(Instrumentation.java:1079)
  at android.app.ActivityThread.performLaunchActivity(ActivityThread.java:2056)
  at android.app.ActivityThread.handleLaunchActivity(ActivityThread.java:2117)
  at android.app.ActivityThread.access$600(ActivityThread.java:140)
  at android.app.ActivityThread$H.handleMessage(ActivityThread.java:1213)
  at android.os.Handler.dispatchMessage(Handler.java:99)
  at android.os.Looper.loop(Looper.java:137)
  at android.app.ActivityThread.main(ActivityThread.java:4914)
  at java.lang.reflect.Method.invokeNative(Native Method)
  at java.lang.reflect.Method.invoke(Method.java:511)
  at com.android.internal.os.ZygoteInit$MethodAndArgsCaller.run
```

```
         (ZygoteInit.java:808)
       at com.android.internal.os.ZygoteInit.main(ZygoteInit.java:575)
       at dalvik.system.NativeStart.main(Native Method)

  "Binder_2" prio=5 tid=10 NATIVE
    | group="main" sCount=1 dsCount=0 obj=0x42296d80 self=0x69068848
    | sysTid=29407 nice=0 sched=0/0 cgrp=apps handle=1750664088
    | schedstat=( 0 0 0 ) utm=0 stm=0 core=1
    #00  pc 0000cc50  /system/lib/libc.so (__ioctl+8)
    #01  pc 0002816d  /system/lib/libc.so (ioctl+16)
    #02  pc 00016f9d  /system/lib/libbinder.so (android::IPCThreadState::
    talkWithDriver(bool)+124)
    #03  pc 0001768f  /system/lib/libbinder.so (android::IPCThreadState::
    joinThreadPool(bool)+154)
    #04  pc 0001b4e9  /system/lib/libbinder.so
    #05  pc 00010f7f  /system/lib/libutils.so (android::Thread::_threadLoop
    (void*)+114)
    #06  pc 00048ba5  /system/lib/libandroid_runtime.so (android::Android-
    Runtime::javaThreadShell(void*)+44)
    #07  pc 00010ae5  /system/lib/libutils.so
    #08  pc 00012ff0  /system/lib/libc.so (__thread_entry+48)
    #09  pc 00012748  /system/lib/libc.so (pthread_create+172)
       at dalvik.system.NativeStart.run(Native Method)
```

从 traces 的内容可以看出，主线程直接 sleep 了，而原因就是 MainActivity 的 42 行。第 42 行刚好就是 SystemClock.sleep(30 * 1000)，这样一来就可以定位问题了。当然这个例子太直接了，下面再模拟一个稍微复杂点的 ANR 的例子。

下面的代码也会导致 ANR，原因是这样的，在 Activity 的 onCreate 中开启了一个线程，在线程中执行 testANR()，而 testANR()和 initView()都被加了同一个锁，为了百分之百让 testANR()先获得锁，特意在执行 initView()之前让主线程休眠了 10ms，这样一来 initView() 肯定会因为等待 testANR()所持有的锁而被同步住，这样就产生了一个稍微复杂些的 ANR。这个 ANR 是很参考意义的，这样的代码很容易在实际开发中出现，尤其是当调用关系比较复杂时，这个时候分析 ANR 日志就显得异常重要了。下面的代码中虽然已经将耗时操作放在线程中了，按道理就不会出现 ANR 了，但是仍然要注意子线程和主线程抢占同步锁的情况。

```
protected void onCreate(Bundle savedInstanceState) {
```

```java
    super.onCreate(savedInstanceState);
    setContentView(R.layout.activity_main);

    new Thread(new Runnable() {
        @Override
        public void run() {
            testANR();
        }
    }).start();

    SystemClock.sleep(10);
    initView();
}

private synchronized void testANR() {
    SystemClock.sleep(30 * 1000);
}

private synchronized void initView() {

}
```

为了分析问题，需要从 traces 文件着手，如下所示。

```
----- pid 32662 at 2015-05-31 16:40:21 -----
Cmd line: com.ryg.chapter_15

DALVIK THREADS:
(mutexes: tll=0 tsl=0 tscl=0 ghl=0)

"main" prio=5 tid=1 MONITOR
  | group="main" sCount=1 dsCount=0 obj=0x4185b700 self=0x4012d0b0
  | sysTid=32662 nice=0 sched=0/0 cgrp=apps handle=1073954608
  | schedstat=( 0 0 0 ) utm=0 stm=4 core=0
  at com.ryg.chapter_15.MainActivity.initView(MainActivity.java:~62)
  - waiting to lock <0x422a0120> (a com.ryg.chapter_15.MainActivity) held
  by tid=11 (Thread-13248)
  at com.ryg.chapter_15.MainActivity.onCreate(MainActivity.java:53)
  at android.app.Activity.performCreate(Activity.java:5086)
  at android.app.Instrumentation.callActivityOnCreate(Instrumentation.
```

```
    java:1079)
    at android.app.ActivityThread.performLaunchActivity(ActivityThread.
java:2056)
    at android.app.ActivityThread.handleLaunchActivity(ActivityThread.
java:2117)
    at android.app.ActivityThread.access$600(ActivityThread.java:140)
    at android.app.ActivityThread$H.handleMessage(ActivityThread.java:1213)
    at android.os.Handler.dispatchMessage(Handler.java:99)
    at android.os.Looper.loop(Looper.java:137)
    at android.app.ActivityThread.main(ActivityThread.java:4914)
    at java.lang.reflect.Method.invokeNative(Native Method)
    at java.lang.reflect.Method.invoke(Method.java:511)
    at com.android.internal.os.ZygoteInit$MethodAndArgsCaller.run
(ZygoteInit.java:808)
    at com.android.internal.os.ZygoteInit.main(ZygoteInit.java:575)
    at dalvik.system.NativeStart.main(Native Method)

"Thread-13248" prio=5 tid=11 TIMED_WAIT
    | group="main" sCount=1 dsCount=0 obj=0x422b0ed8 self=0x683d20c0
    | sysTid=32687 nice=0 sched=0/0 cgrp=apps handle=1751804288
    | schedstat=( 0 0 0 ) utm=0 stm=0 core=0
    at java.lang.VMThread.sleep(Native Method)
    at java.lang.Thread.sleep(Thread.java:1031)
    at java.lang.Thread.sleep(Thread.java:1013)
    at android.os.SystemClock.sleep(SystemClock.java:114)
    at com.ryg.chapter_15.MainActivity.testANR(MainActivity.java:57)
    at com.ryg.chapter_15.MainActivity.access$0(MainActivity.java:56)
    at com.ryg.chapter_15.MainActivity$1.run(MainActivity.java:49)
    at java.lang.Thread.run(Thread.java:856)
```

上面的情况稍微复杂一些，需要逐步分析。首先看主线程，如下所示。可以看得出主线程在 initView 方法中正在等待一个锁<0x422a0120>，这个锁的类型是一个 MainActivity 对象，并且这个锁已经被线程 id 为 11（即 tid=11）的线程持有了，因此需要再看一下线程 11 的情况。

```
    at com.ryg.chapter_15.MainActivity.initView(MainActivity.java:~62)
    - waiting to lock <0x422a0120> (a com.ryg.chapter_15.MainActivity) held
    by tid=11 (Thread-13248)
```

tid 是 11 的线程就是 "Thread-13248"，就是它持有了主线程所需的锁，可以看出

"Thread-13248"正在 sleep，sleep 的原因是 MainActivity 的 57 行，即 testANR 方法。这个时候可以发现 testANR 方法和主线程的 initView 方法都加了 synchronized 关键字，表明它们在竞争同一个锁，即当前 Activity 的对象锁，这样一来 ANR 的原因就明确了，接着就可以修改代码了。

上面分析了两个 ANR 的实例，尤其是第二个 ANR 在实际开发中很容易出现，我们首先要有意识地避免出现 ANR，其次出现 ANR 了也不要着急，通过分析 traces 文件即可定位问题。

15.1.5　ListView 和 Bitmap 优化

ListView 的优化在第 12 章已经做了介绍，这里再简单回顾一下。主要分为三个方面：首先要采用 ViewHolder 并避免在 getView 中执行耗时操作；其次要根据列表的滑动状态来控制任务的执行频率，比如当列表快速滑动时显然是不太适合开启大量的异步任务的；最后可以尝试开启硬件加速来使 Listview 的滑动更加流畅。注意 Listview 的优化策略完全适用于 GridView。

Bitmap 的优化同样在第 12 章已经做了详细的介绍，主要是通过 BitmapFactory.Options 来根据需要对图片进行采样，采样过程中主要用到了 BitmapFactory.Options 的 inSampleSize 参数，详情这里就不再重复了，请参考第 12 章的有关内容。

15.1.6　线程优化

线程优化的思想是采用线程池，避免程序中存在大量的 Thread。线程池可以重用内部的线程，从而避免了线程的创建和销毁所带来的性能开销，同时线程池还能有效地控制线程池的最大并发数，避免大量的线程因互相抢占系统资源从而导致阻塞现象的发生。因此在实际开发中，我们要尽量采用线程池，而不是每次都要创建一个 Thread 对象，关于线程池的详细介绍请参考第 11 章的内容。

15.1.7　一些性能优化建议

本节介绍的是一些性能优化的小建议，通过它们可以在一定程度上提高性能。

- 避免创建过多的对象；
- 不要过多使用枚举，枚举占用的内存空间要比整型大；
- 常量请使用 static final 来修饰；

- 使用一些 Android 特有的数据结构，比如 SparseArray 和 Pair 等，它们都具有更好的性能；
- 适当使用软引用和弱引用；
- 采用内存缓存和磁盘缓存；
- 尽量采用静态内部类，这样可以避免潜在的由于内部类而导致的内存泄露。

15.2 内存泄露分析之 MAT 工具

MAT 的全称是 Eclipse Memory Analyzer，它是一款强大的内存泄露分析工具，MAT 不需要安装，下载后解压即可使用，下载地址为 http://www.eclipse.org/mat/downloads.php。对于 Eclipse 来说，MAT 也有插件版，但是不建议使用插件版，因为独立版使用起来更加方便，即使不安装 Eclipse 也可以正常使用，当然前提是有内存分析后的 hprof 文件。

为了采用 MAT 来分析内存泄露，下面模拟一种简单的内存泄露情况，下面的代码肯定会造成内存泄露：

```java
public class MainActivity extends Activity {
    private static final String TAG = "MainActivity";

    private static Context sContext;

    @Override
    protected void onCreate(Bundle savedInstanceState) {
        super.onCreate(savedInstanceState);
        setContentView(R.layout.activity_main);
        sContext = this;
    }
}
```

编译安装，然后打开 DDMS 界面，其中 AndroidStudio 的 DDMS 位于 Monitor 中。接着用鼠标选中要分析的进程，然后使用待分析应用的一些功能，这样做是为了将尽量多的内存泄露暴露出来，然后单击 Dump HPROF file 这个按钮（对应图 15-1 中底部有黑线的按钮），等待一小段时间即可导出一个 hprof 后缀的文件，如图 15-1 所示。

第 15 章 Android 性能优化

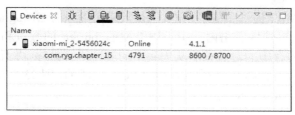

图 15-1　DDMS 视图

导出 hprof 文件后并不能使用它来进行分析，因为它不能被 MAT 直接识别，需要通过 hprof-conv 命令转换一下。hprof-conv 命令是 Android SDK 提供的工具，它位于 Android SDK 的 platform-tools 目录下：

```
hprof-conv com.ryg.chapter_15.hprof com.ryg.chapter_15-conv.hprof
```

当然如果使用的是 Eclipse 插件版的 MAT 的话，就可以不进行格式转换了，可以直接用 MAT 插件打开。

经过了上面的步骤，接下来就可以直接通过 MAT 来进行内存分析了。打开 MAT，通过菜单打开刚才转换后的 com.ryg.chapter_15-conv.hprof 这个文件，打开后的界面如图 15-2 所示。

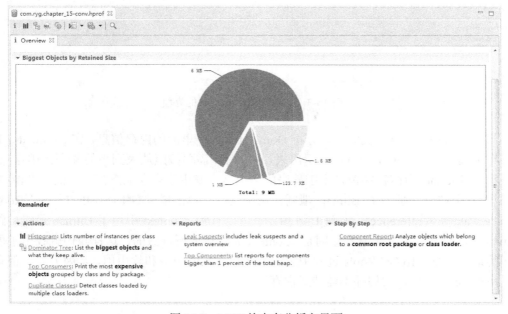

图 15-2　MAT 的内存分析主界面

如图 15-2 所示，MAT 提供了很多功能，但是最常用的只有 Histogram 和 Dominator Tree，通过 Histogram 可以直观地看出内存中不同类型的 buffer 的数量和占用的内存大小，而 Dominator Tree 则把内存中的对象按照从大到小的顺序进行排序，并且可以分析对象之间的引用关系，内存泄露分析就是通过 Dominator Tree 来完成的。图 15-3 和图 15-4 分别是 MAT 中 Histogram 和 Dominator Tree 的界面。

图 15-3　MAT 中 Histogram 的界面

为了分析内存泄露，我们需要分析 Dominator Tree 里面的内存信息，在 Dominator Tree 中内存泄露的原因一般不会直接显示出来，这个时候需要按照从大到小的顺序去排查一遍。一般来说 Bitmap 泄露往往都是由于程序的某个地方发生了内存泄露都引起的，在图 15-4 中的第 2 个结果就是一个 Bitmap 泄露，选中它然后单击鼠标右键->Path To GC Roots->exclude wake/soft references，如图 15-5 所示。可以看到 sContext 引用了 Bitmap 最终导致了 Bitmap 无法释放，但其实根本原因是 sContext 无法释放所导致的，这样我们就找出了内存泄露的地方。Path To GC Roots 过程中之所以选择排除弱引用和软引用，是因为二者都有较大几率被 gc 回收掉，它们并不能造成内存泄露。

第15章 Android 性能优化

Class Name	Shallow Heap	Retained Heap	Percentage
class android.content.res.Resources @ 0x4188a3c0 System Class	40	6,301,512	67.01%
android.graphics.Bitmap @ 0x419b9c00	48	1,048,656	11.15%
class android.text.Html$HtmlParser @ 0x41a460d8 System Class	8	126,632	1.35%
class libcore.icu.TimeZones @ 0x41856380 System Class	40	108,888	1.16%
class org.apache.harmony.security.fortress.Services @ 0x4198ba50 System Cla	32	70,280	0.75%
class com.android.internal.R$styleable @ 0x4187f848 System Class	6,656	55,208	0.59%
class android.R$styleable @ 0x418daf30 System Class	6,064	51,240	0.54%
class android.view.View @ 0x4187dc58 System Class	1,056	46,664	0.50%
com.android.org.bouncycastle.jce.provider.BouncyCastleProvider @ 0x4199db	112	45,064	0.48%
miui.content.res.ThemeZipFile @ 0x41916038	40	41,824	0.44%
class android.content.res.MiuiResources @ 0x418e7e30 System Class	40	35,848	0.38%
class android.text.AutoText @ 0x41a389c0 System Class	56	29,088	0.31%
miui.content.res.ThemeZipFile @ 0x41928050	40	20,864	0.22%
class libcore.net.MimeUtils @ 0x41b66b78 System Class	16	17,576	0.19%
class java.io.Console @ 0x418548b0 System Class	16	17,520	0.19%
char[8194] @ 0x41aac420 Africa/Abidjan Africa/Accra Africa/Addis_Ababa Africa	16,400	16,400	0.17%
class libcore.icu.LocaleData @ 0x418539c8 System Class, Native Stack	8	15,856	0.17%
class java.lang.ref.FinalizerReference @ 0x41850f88 System Class	16	13,312	0.14%
android.graphics.NinePatch @ 0x4198e250	32	13,152	0.14%
class com.android.org.bouncycastle.asn1.pkcs.PKCSObjectIdentifiers @ 0x419a	504	12,816	0.14%
class android.media.MediaFile @ 0x4195ca88 System Class	336	11,472	0.12%
class libcore.io.OsConstants @ 0x41856650 System Class	1,600	11,192	0.12%
class org.apache.harmony.security.utils.AlgNameMapper @ 0x41b892f0 Syster	24	11,152	0.12%
class android.view.KeyEvent @ 0x4186aae8 System Class, Native Stack	1,112	10,008	0.11%
class com.android.i18n.phonenumbers.PhoneNumberUtil @ 0x41a364b8 Syste	184	9,640	0.10%
class android.opengl.GLES20 @ 0x418a81e0 System Class	1,216	8,480	0.09%

图 15-4 MAT 中 Ddominator Tree 的界面

Class Name	Shallow Heap	Retained Heap
android.graphics.Bitmap @ 0x419b9c00	48	1,048,656
mBitmap android.graphics.drawable.BitmapDrawable$BitmapState @ 0x418ac188	40	120
[98] java.lang.Object[254] @ 0x41ab9d90	1,032	6,298,592
mValues android.util.LongSparseArray @ 0x4190bed8	24	6,300,664
sPreloadedDrawables class android.content.res.Resources @ 0x4188a3c	40	6,301,512
mBitmap android.graphics.drawable.BitmapDrawable @ 0x422c9ec8	64	136
mCurrDrawable android.graphics.drawable.StateListDrawable @ 0x422c9dc0	96	536
mBackground com.android.internal.policy.impl.PhoneWindow$DecorView @	544	1,472
mDecor com.android.internal.policy.impl.PhoneWindow @ 0x422c6d90	216	2,824
mWindow com.ryg.chapter_15.MainActivity @ 0x422c6a38	184	2,200
sContext class com.ryg.chapter_15.MainActivity @ 0x422a1530 Sys	16	104
mOuterContext android.app.ContextImpl @ 0x422c6bf8	96	480
Σ Total: 2 entries		

图 15-5 Path To GC Roots 后的结果

在 Dominator Tree 界面中是可以使用搜索功能的，比如我们尝试搜索 MainActivity，因为这里我们已经知道 MainActivity 存在内存泄露了，搜索后的结果如图 15-6 所示。我们发现里面有 6 个 MainActivity 的对象，这是因为每次按 back 键退出再重新进入 MainActivity，系统都会重新创建一个新的 MainActivity，但是由于老的 MainActivity 无法被回收，所以就

505

出现了多个 MainActivity 对象的情形。另外 MAT 还有很多其他功能，这里就不再一一介绍了，请读者自己体验吧。

图 15-6　Dominator Tree 的搜索功能

15.3　提高程序的可维护性

　　本节所讲述的内容是 Android 的程序设计思想，主旨是如何提高代码的可维护性和可扩展性，而程序的可维护性本质上也包含可扩展性。本节的切入点为：代码风格、代码的层次性和单一职责原则、面向扩展编程以及设计模式，下面围绕着它们分别展开。

　　可读性是代码可维护性的前提，一段别人很难读懂的代码的可维护性显然是极差的。而良好的代码风格在一定程度上可以提高程序的可读性。代码风格包含很多方面，比如命名规范、代码的排版以及是否写注释等。到底什么样的代码风格是良好的？这是个仁者见仁的问题，下面是笔者的一些看法。

　　（1）命名要规范，要能正确地传达出变量或者方法的含义，少用缩写，关于变量的前缀可以参考 Android 源码的命名方式，比如私有成员以 m 开头，静态成员以 s 开头，常量则全部用大写字母表示，等等。

　　（2）代码的排版上需要留出合理的空白来区分不同的代码块，其中同类变量的声明要放在一组，两类变量之间要留出一行空白作为区分。

（3）仅为非常关键的代码添加注释，其他地方不写注释，这就对变量和方法的命名风格提出了很高的要求，一个合理的命名风格可以让读者阅读源码就像在阅读注释一样，因此根本不需要为代码额外写注释。

代码的层次性是指代码要有分层的概念，对于一段业务逻辑，不要试图在一个方法或者一个类中去全部实现，而要将它分成几个子逻辑，然后每个子逻辑做自己的事情，这样既显得代码层次分明，又可以分解任务从而实现简化逻辑的效果。单一职责是和层次性相关联的，代码分层以后，每一层仅仅关注少量的逻辑，这样就做到了单一职责。代码的层次性和单一职责原则可以以公司的组织结构为例来说明，比如现在有一个复杂的需求来到了部门经理面前，如果部门经理需要给每个员工来安排具体的任务，那显然他会显得很累，因为他必须要了解每个员工的工作并最终收集每个员工的完成情况，这个时候整个工作过程就缺少了层次性，并且也违背了单一职责的原则，毕竟经理的主要工作是管理团队而不是给员工安排任务。如果采用分层的思想要怎么做呢？首先经理可以将复杂的任务分成若干份，每一份交给一个主管处理，然后剩下的事情经理就不用管了，他只需要管理主管即可。对于主管来说，分配给他的任务相对于整个任务就简单了不少，这个时候他再拆解任务给组员，这个时候真正到达组员手里的任务其实就没有那么复杂了，这其实类似于分治策略。这样一来整个工作过程就具有了三层的结构，并且每一层有不同的职责，一旦出现了错误也可以很方便地定位到具体的地方。

程序的扩展性标志着开发人员是否有足够的经验，很多时候在开发过程中我们无法保证已经做好的需求不在后面的版本发生变更，因此在写程序的过程中要时刻考虑到扩展，考虑着如果这个逻辑后面发生了改变那么需要做哪些修改，以及怎么样才能降低修改的工作量，面向扩展编程会使程序具有很好的扩展性。

恰当地使用设计模式可以提高代码的可维护性和可扩展性，但是 Android 程序容易有性能瓶颈，因此要控制设计的度，设计不能太牵强，否则就是过度设计了。常见的设计模式有很多，比如单例模式、工厂模式以及观察者模式等，由于本书不是专门介绍设计模式的书，因此这里就不对设计模式进行详细的介绍了，读者可以参看《大话设计模式》和《Android 源码设计模式解析与实战》这两本书，另外设计模式需要理解后灵活运用才能发挥更好的效果。

反侵权盗版声明

电子工业出版社依法对本作品享有专有出版权。任何未经权利人书面许可，复制、销售或通过信息网络传播本作品的行为；歪曲、篡改、剽窃本作品的行为，均违反《中华人民共和国著作权法》，其行为人应承担相应的民事责任和行政责任，构成犯罪的，将被依法追究刑事责任。

为了维护市场秩序，保护权利人的合法权益，我社将依法查处和打击侵权盗版的单位和个人。欢迎社会各界人士积极举报侵权盗版行为，本社将奖励举报有功人员，并保证举报人的信息不被泄露。

举报电话：（010）88254396；（010）88258888

传　　真：（010）88254397

E-mail：　dbqq@phei.com.cn

通信地址：北京市海淀区万寿路 173 信箱
　　　　　电子工业出版社总编办公室

邮　　编：100036

精品畅销书系列

Cocos2d-x 3.x 游戏开发之旅

畅销书《Cocos2d-x 游戏开发之旅》的升级版，被誉为最适合初学者的入门图书。基于全新的3.0版本而作，新增了对 CocoStudio、UI 编辑器、Cocos2d-x 3.0 新特性以及网络方面的知识点。泰然网创始人杨雍和蜉蝣极客组织者章旭东力荐！

Cocos2d-x 3.X 手游开发实例详解

最新最简 Cocos2d-x 手机游戏开发学习方法，以热门游戏 2048、卡牌为例，完整再现手游的开发过程，实例丰富，代码完备，Cocos2d-x 作者之一林顺和泰然网创始人杨雍力荐。

Cocos2d-x 3.x 游戏开发详解

本书是基于全新的 Cocos2d-x 3.x API 来编写的游戏引擎开发书籍，从最基本的环境搭建到最后的项目实战，内容全面并配有大量案例，无论是游戏初学者还是多年的游戏开发高手，都能从中受益。全书最后以《疯狂地鼠》案例，全面系统地梳理 Cocos2d-x 游戏开发的完整过程，Cocos2d-x 的大部分内容进行综合应用。

我所理解的 Cocos2d-x

本书针对最新的 Cocos2d-x 3.x 版本，介绍了 Cocos2d-x 游戏引擎的基本架构、渲染机制，以及各个子模块的功能和原理，并结合 OpenGL ES 图形渲染管线，深入探讨了游戏开发中涉及的图形学知识。本书偏重讲解每个知识模块的概念及原理，使读者能够透过现象看到其背后的工作机制，本书并不是围绕 Cocos2d-x 接口的使用组织内容，而是按照通用游戏引擎架构及图形学的知识组织内容。

Cocos2d-x 3.x 游戏开发实战

本书是一本介绍 Cocos2d-x 游戏引擎的实用图书，全面介绍了最新的 Cocos2d-x 3.2 游戏引擎各方面的知识。本书针对每一个知识点都通过相应的程序给出了示范，并结合 Cocos2d-x 自带的 cpp-tests 实例进行讲解，在总结一些知识点之后还开发了一些小游戏，目的是让读者全面掌握 Cocos2d-x 的基础理论和基本使用。

Cocos2d-x 3.X 游戏开发入门精解

畅销书《Cocos2d-x 游戏开发之旅》的升级版，被誉为最适合初学者的入门图书。基于全新的3.0 版本而作，新增了对 CocoStudio、UI 编辑器、Cocos2d-x 3.0 新特性以及网络方面的知识点。泰然网创始人杨雍和蜉蝣极客组织者章旭东力荐！

欢迎投稿：

投稿信箱： jsj@phei.com.cn
　　　　　 editor@broadview.com.cn

读者信箱： market@broadview.com.cn

电　　话： 010-51260888

更多信息请关注：

博文视点官方网站：
http://www.broadview.com.cn

博文视点官方微博：
http://t.sina.com.cn/broadviewbj